energy science

principles, technologies, and impacts

energy science

principles, technologies, and impacts

John Andrews and Nick Jelley

OXFORD
UNIVERSITY PRESS

Great Clarendon Street, Oxford OX2 6DP

Oxford University Press is a department of the University of Oxford.
It furthers the University's objective of excellence in research, scholarship,
and education by publishing worldwide in

Oxford New York

Auckland Cape Town Dar es Salaam Hong Kong Karachi
Kuala Lumpur Madrid Melbourne Mexico City Nairobi
New Delhi Shanghai Taipei Toronto

With offices in

Argentina Austria Brazil Chile Czech Republic France Greece
Guatemala Hungary Italy Japan Poland Portugal Singapore
South Korea Switzerland Thailand Turkey Ukraine Vietnam

Published in the United States
by Oxford University Press Inc., New York

British Library Cataloguing in Publication Data
Data available

Library of Congress Cataloging in Publication Data
Data available

Typeset by Laserwords Private Limited, Chennai, India
Printed in Great Britain
on acid-free paper by
CPI Bath Ltd, Bath

ISBN 978–0–19–928112–1

1 3 5 7 9 10 8 6 4 2

Preface

Harnessing the Earth's energy resources has been a source of inspiration since ancient times. Energy devices have transformed civilization beyond the wildest imagination of our predecessors. These days, energy is always in the news. Why is this so? How is the global demand for energy going to be satisfied in the future? Can we avoid global warming becoming an insurmountable crisis? What are the options?

These questions have to be answered. What the present civilization does with the remaining energy resources will have a profound effect on the lives of future generations and the state of the planet. One approach is to do nothing and to assume that market forces and governments will sort it all out. The energy field is not short of uninformed or politically motivated opinions and commercial interests. This book is for those who prefer to make up their own minds through an understanding of the science involved and the impact of the various technologies on society.

Using this book

The idea for writing this book originated from undergraduate lecture courses given by the authors in Bristol and Oxford. The main focus of the book is to explain the physical principles underlying each technology and to discuss its environmental, economic, and social impact. It describes all the key areas of energy science, covering fossil fuels, nuclear energy, alternative energy, and the emerging energy technologies. Energy science is a broad subject that crosses the boundaries between the traditional scientific and engineering disciplines. It is not essential to have a background in any particular discipline in order to use this book, apart from a general knowledge of science and mathematics to about high school standard. The aim is to enable students, professionals, and lay-readers to make quantitative estimates and form sound judgements.

The content of this book is presented in such a way that it can be understood on different levels. All the important results are described qualitatively using straightforward mathematical methods, with numerical **Examples**. More difficult mathematical derivations are set apart from the text as **Derivations**, while items of supplementary information are contained in **Boxes**. These boxes can be bypassed by those who do not wish to consider such detail. Readers are encouraged to work through the **Exercises** at the end of each chapter. These exercises are designed to be informative and thought-provoking, but not intimidating! For those who want to stretch their cerebral muscles and gain a deeper understanding of some of the more difficult points, some starred exercises have been added in each chapter. Numerical answers to exercises are provided at the end of the book.

Getting started

Chapter 1 presents a brief history of energy technology, together with a review of the long-term energy trends and the evidence for global warming. Chapters 2 and 3 provide the essentials of thermal physics and fluid mechanics required for some of the later chapters in the book. Many readers will already be aware of the basic concepts contained in Chapters 2 and 3 but may not be familiar with the details of the thermodynamic cycles used in thermal plants, the greenhouse effect, the nature of fossil fuels, geothermal power, and energy-related applications of fluid mechanics. Chapters 4–10 cover a wide range of areas of energy technology; these chapters are largely self-contained and can be read in virtually any order. Finally, Chapter 11 appraises the central issues surrounding the impact of energy on society, paying particular attention to the dangers of global warming, how it may be combated, and the energy scene of the future.

Online resource centre

The book is accompanied by an Online Resource Centre at www.oxfordtextbooks.co.uk/orc/andrews_jelley/, which features additional materials for both students and lecturers.

For students:

- A library of all weblinks cited in the book, grouped by chapter.

For lecturers:

- Figures from the book in electronic format.
- Full solutions to end-of-chapter exercises.

Acknowledgements

The authors would like to express their appreciation to their families, their editor Jonathan Crowe, and to their friends and colleagues for numerous ideas, criticisms, and helpful suggestions, in particular, to David Andrews, Katherine Blundell, Peter Cook, George Doucas, Nigel Dowrick, Mahieddine Emziane, Kieran Finan, Godfrey Gardner, John Hannay, James Loach, Helen O'Keefe, Robert Paynter, Bruce Pilsworth, John Pye, Rachel Quarrel, and Justin Wark.

John Andrews
Department of Physics
University of Bristol

Nick Jelley
Department of Physics and Lincoln College
University of Oxford

Contents

Acknowledgement of sources

Tables

Table 1.2 is adapted from Boyle, G. (ed.) (1996). *Renewable energy*. Oxford University Press, Oxford.

Table 1.3 uses data from the US Department of Energy, Energy Information Administration, *International energy outlook* (2006), www.eia.doe.gov/iea

Table 1.4 uses data from Bourdier, J.-P. (2000). Bois, charbon, pétrole, gaz, nucléaire et autres. Mêmes problémes. *La Jaune et la Rouge*, May 2000.

In Table 1.6, data on the concentration of CO_2 are from the Hadley Centre for Climate Prediction and Research (crown copyright), *www.metoffice.com/research/hadleycentre*; data on global temperature difference are from the US National Oceanic and Atmospheric Administration (NOAA) Mauna Loa Observatory, *www.mlo.noaa.gov*.

Table 2.1 uses data from the World Energy Council, *www.worldenergy.org*.

Tables 4.1 and 4.2 use data from the World Energy Council, *www.worldenergy.org/wec-geis/publications/reports/ser/hydro/hydro.asp* and *reports/ser/tide/tide.asp*

Table 5.3 uses data from the European Wind Energy Association (EWEA) (2004). *Wind energy—the facts*. EWEA. *www.ewea.org* with acknowledgements to Airtricity, *www.airtricity.com*.

Table 5.4 uses data from R.H. Williams (2001), *www.princeton.edu/~energy/* and from the US Energy Information Administration, *www.eia.doe.gov*.

Table 5.5 uses data from the UK Department of the Environment, 1993.

Table 5.6 uses data from *www.strategy.gov.uk/downloads/files/PIUh.pdf*.

Table 6.3 uses data from Dresdner Kleinwort Wasserstein Research (DKWR) estimates.

Table 7.2 uses data from the US National Renewable Energy Laboratory (NREL), *www.nrel.gov*, the US Department of the Environment (DOE), and the US Department of Agriculture, USDA.

Table 8.1 uses data from Lilley, J. (2001). *Nuclear physics*. Wiley, New York.

Table 8.3 uses data from Lamarsh, J.R. and Baratta, A.J. (2001). *Introduction to nuclear engineering*. Prentice-Hall Englewood Cliffs, NJ.

Table 8.4 uses data from Rasmussen, N. *et al.* (1975). *Reactor safety study*, WASH-1400. US Nuclear Regulatory Commission (NRC), Washington, DC.

Table 8.5 uses data from the International Commission on Radiological Protection (ICRP) (1996). Conversion coefficients for use in radiological protection against external radiation. *Annals of the ICRP* **26** (3), 1–205.

Table 8.6 uses data from the UK National Radiological Protection Board (1994).

Table 8.7 uses data from the MIT report (2003) on the 'Future of nuclear power'. *web.mit.edu/nuclearpower*.

Table 8.8 uses data from the World Nuclear Association, *www.world-nuclear.org/info/reactors.htm*.

Table 10.1 uses data from *hydrogen.pnl.gov/filedownloads/hydrogen/datasheets/lower_and_higher_heating_values.xls* and Burke, A. and Gardiner, M. (2005). *Hydrogen storage options: technologies and comparisons for light duty vehicle applications*, University of California at Davis Institute of Transportation Studies Report, UCD-ITS-RR-05–01.

Figures

Figure 1.5 uses data adapted from the International Monetary Fund (2006), *www.imf.org/external/pubs/ft/weo/2006/01/data/index.htm*, and the International Energy Agency (2004), *earthtrends.wri.org/searchable_db/results*.
Figures 1.6 and 1.7 use data adapted from the US Department of Energy, Energy Information Administration, *International energy outlook* (2004), *www.eia.doe.gov/iea/*.
Figure 1.8 uses data adapted from the US Department of Energy, Energy Information Administration, *www.eia.doe.gov/oiaf/ieo/emissions*.
Figure 1.9 is by Gerald Meehl, NCAR, www.ucar.edu/research/climate/warming.shtml from Meehl, G.A., Washington, W.M., Ammann, C., Arblaster, J.M., Wigley, T.M.L., and Tebaldi, C. (2004). Combinations of natural and anthropogenic forcings and 20th century climate. *J. Climate* **17**, 3721–27.
Figure 4.8 is adapted from Boyle, G. (ed.) (2004). *Renewable energy*, 2nd edn. Oxford University Press, Oxford.
Figures 5.9 and 5.10 use data from Sandia National Laboratories report, SAND99–0089.
Figure 5.14 uses data from Grubb, M.J. and Meyer, N.I. (1993). *Renewable energy*. Island Press, Washington, DC.
Figure 5.15 uses data from the European Wind Energy Association (2004). *Wind energy—the facts*. EWEA, Brussels, *www.ewea.org*, and from the American Wind Energy Association, *www.awea.org*.
Figure 6.13 is adapted from the US National Renewable Energy Laboratory (NREL) report, SR-520–3624.
Figure 6.19 uses data from a report by Strategies Unlimited (2003).
Figure 7.5 is adapted from Guidotti, G. (2002). *Biogas from excreta: treatment of faecal sludge in developing countries*.
Figure 7.7 uses data from Sarlos, G. (2005). The transportation sector—evolution of significant indicators. APECATC-Workshop on Biomass Utilization. Tokyo, Japan January 19–21, 2005.
Figure 7.10 uses data from Smith, S. (2004). The future role of biomass. GCEP Energy Workshop, Stanford, April 2004.
Figure 8.8(a) is adapted from a US DOE drawing.
Figure 8.11 is adapted from Generation IV International Forum (GIF) (2002). Generation IV roadmap final system screening evaluation methodology R & D report, GIF-012–00.
Figure 8.13 uses data from the Energy Information Agency, *www.eia.doe.gov*.
Figure 9.3 uses data from Lister, J. and Weisen, H. (2005). What will we learn from ITER? *Europhysics News* March/April 2005, 47–51.
Figures 9.4 and 9.18 are adapted from Unterberg, B. and Samm, U. (2004). *Overview of tokamak results* available at *www.jet.efda.org*.
Figure 9.11 is adapted from material available at *www.jet.efda.org*.
Figure 9.16 is adapted from Wesson, J. *The science of JET*, a book available at *www.jet.efda.org/documents/wesson/wesson.html*.
Figure 9.17 is adapted from Samm, U. *Controlled thermonuclear fusion enters with ITER into a new era* available from *www.jet.efda.org*.
Figure 10.9 uses data from 'Methanol fuel cell miniaturization' under the heading 'Chemical engineering projects 2005', available at *chemelab.ucsd.edu/fuel05*.
Figure 11.1(a) is by Lisa Buckley, Earth Policy Institute, www.earth-policy.org/Indicators/CO2/CO2_data.htm # fig6
Figure 11.1(b) is by Phillippe Rekacewicz, UNEP/GRID-Arendal. Source: Petit, J.R., Jonzel, J., *et al.* (1999). Climate and atmospheric history of the past 400 000 years from the Vostok ice core in Antarctica, *Nature* **399**, 429–36. www.grida.no/climate/vital/02.htm

Figure 11.2 is adapted from 'Technical and economic potential of renewable energy generating technologies: potentials and cost reductions to 2020', available at *www.strategy.gov.uk*.
Figure 11.3 is adapted from Lichtenstein, S., *et al.* (1978). Judged frequency of lethal events. *J. Exp. Psychol.* **4**, 551–78.
Figure 11.4 is adapted from Pacala, S. and Socolow, R. (2004). Stabilization wedges. *Science* **305**, 968.

Sections

5.15 WEC (1994): World Energy Council (1994). *New renewable energy resources: a guide to the future*, Kogan Page, London.
5.15 Grubb and Meyer (1993): Grubb, M.J., and Meyer, N.I. (1993). *Renewable energy*, Island Press, Washington, DC.
5.15 Archer and Jacobson (2005): Evaluation of global wind power, J. Geophys. Res.—Atm. **110** D12110.
7.3.1 FAO: Food and Agriculture Organisation of the United Nations. *http://www.fao.org.docrep/004/T0423E/T0423E07.htm*.
8.7 Chernobyl Forum (2006): *http://about.greenfacts.org*
Constants, symbols, and conversion factors: *http://bioenergy.ornl.gov/papers/misc/energy_conv.html*

1 Introduction

List of Topics

- ☐ Early energy devices
- ☐ Steam engines
- ☐ Electromagnetic induction
- ☐ Heat and energy
- ☐ Development of power stations
- ☐ Energy trends
- ☐ Greenhouse effect
- ☐ Units and dimensional analysis

Introduction

For millions of years the impact of life on Earth was energy-neutral: animals and plants coexisted in a continuous cycle. When primitive human beings discovered how to make fire, the balance of nature began to shift irreversibly. With fire, it was possible to cook meat, deter predators, and fashion metals into tools and deadly weapons. In the last two centuries, mankind has discovered how to convert heat into electricity, the most versatile and convenient form of energy. Electricity has enabled astonishing advances to be made in science and engineering, transforming civilization and making life far more comfortable than that of our predecessors. However, in the process it has created a consumer society that treats electricity and other forms of energy as commodities that should be available on demand.

In the relatively short period since the start of the Industrial Revolution (in the second half of the eighteenth century), a significant fraction of the fossil reserves of the planet, which took hundreds of millions of years to evolve, has been significantly depleted. The emission of carbon dioxide and other products of combustion is now having a noticeable impact on the global climate. The threat to life in the relatively near future could be dire unless mankind can rise to the greatest challenge it has faced since its emergence as the dominant species on Earth.

Energy conversion is a disparate subject, but it is possible to obtain a good understanding of the essentials by applying basic physical principles. Energy issues tend to be open-ended and controversial. Addressing them with an independent mind is a rewarding and intellectually stimulating exercise. It is important to be objective, to pay attention to fact rather than opinion, to challenge assumptions, and to always look for constructive solutions.

1.1 A brief history of energy technology

Throughout history the harnessing of energy in its various forms has presented great intellectual challenges and stimulated scientific discovery. The technologies of today are the result of advances in scientific understanding, inspiration, and gradual improvements in engineering design over many centuries. By the time of the Roman Empire, water engineering was already a well-established technology. Thousands of years earlier, irrigation systems had greatly enlarged the area of farming land around the River Nile and increased the prosperity of ancient Egypt. An illustration of the ingenuity of the early engineers is demonstrated by **Archimedes' screw** (Fig. 1.1), a device used to extract water from rivers, empty grain from the holds of ships, and clear water from flooded mines (e.g. Rio Tinto in Spain). There are records showing that the device was used to irrigate the Hanging Gardens of Babylon (before the birth of Archimedes!), and it is still widely used today by the water and chemical industries.

In order to raise water, Archimedes' screw is encased in a hollow cylinder. One end is immersed in water, and the water trapped in the hollows of the screw is transported upwards by rotating the device about its axis (see Exercises 1.6 and 1.14).

Waterwheels (Chapter 4) existed in the ancient world and by 1000 AD were common throughout Western Europe. The early waterwheels were very inefficient but designs gradually improved over the centuries (Fig. 4.1). A technological breakthrough was made in 1832 with the invention of the **Fourneyron turbine** (Section 4.1), which used fixed guide vanes to direct water between the blades of a rotating runner. The design of the vanes and the blades enabled most of the kinetic energy of the incident flow to be captured. The Fourneyron turbine pioneered the development of modern turbines, leading to the emergence of hydropower as one of the major providers of electricity today.

The earliest recorded steam engine was a toy device, invented by Hero of Alexandria in the first century AD (Fig. 1.2). It is essentially a hollow metal sphere filled with steam, supported by two pivots. The steam is allowed to escape through two bent spouts and the momentum of the steam jets produces a reaction in the opposite direction on the spouts which causes the sphere to rotate. The Industrial Revolution was made possible by the emergence of steam power. However, before the first commercial steam engines appeared on the scene there were serious misunderstandings about the nature of vacuum and air pressure that needed to be resolved. In 1644 Torricelli (a follower of Galileo) invented the mercury barometer. He proved that the rise of the column of mercury was due to the difference in pressure between the atmosphere and the vacuum above the mercury inside the column. The next major breakthrough came about through the invention of the piston air pump by von Guericke in 1650. In one spectacular demonstration von Guericke took two identical metal hemispheres (known as the 'Magdeburg

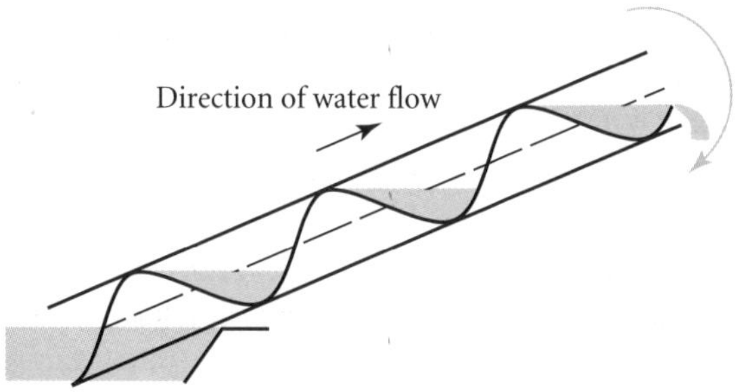

Direction of water flow

Fig. 1.1 Archimedes screw.

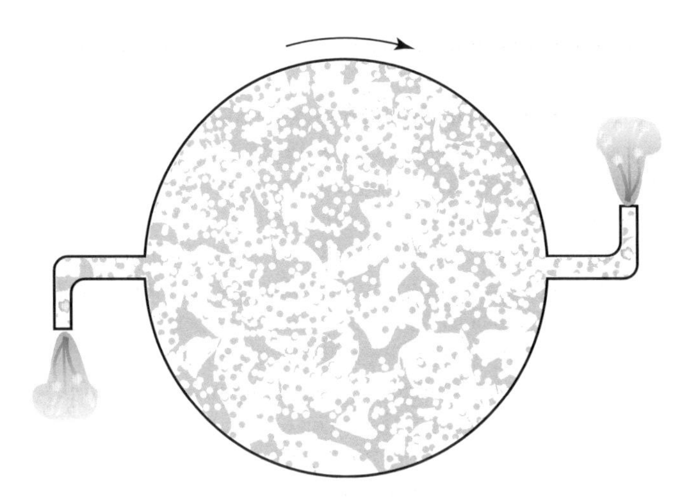

Fig. 1.2 Hero's steam engine.

hemispheres', after the town where von Guericke was burgomaster and a military engineer during the Thirty Years' War) and placed them together to form a complete sphere. The hemispheres were in touching contact along a flange, but there was no method of fixing. Von Guericke pumped out the air inside the sphere and invited two teams of eight horses to pull the hemispheres apart; they were unsuccessful, because the force exerted by the air was too great (see Exercise 1.4).

In 1666, the Academy des Sciences was established in Paris and the Dutch scientist, Christiaan Huygens, was made its first president. Huygens appointed two young assistants, Gottfried Leibniz and Denis Papin, who modified a von Guericke air pump to conduct experiments with gunpowder. A small charge of gunpowder was exploded inside a cylindrical chamber containing a tightly fitting piston. The air was expelled through two leather valves, thereby creating a partial vacuum. The piston then collapsed due to the imbalance in air pressure across it, and useful work was done in the process. Huygens realized the enormous potential of this discovery, from new forms of transport to powerful engines that could revolutionize industry.

Not surprisingly, gunpowder explosions proved to be too difficult to control! Papin, with help from Leibniz, tried using water instead of gunpowder, exploiting the fact that water expands to 160 times its original volume when converted to steam. Papin placed a quantity of water inside a piston chamber and heated it from the outside by a flame. The pressure of the steam raised the piston against the pressure of the air. In order to return the piston to its original position, Papin poured cold water over the outside of the cylinder so that the steam condensed back to water and the space inside the cylinder dropped to below atmospheric pressure.

Papin was forced to flee France in 1685, due to religious persecution. He settled in London and worked at the Royal Society, where he continued to develop his steam engine. In 1690 he applied for a patent to build commercial steam engines, but the patent was awarded instead to Thomas Savery, a military engineer, who had been strongly influenced by Papin's ideas. Papin's friendship with Leibniz meant that he did not get the support of Isaac Newton, who was locked in a bitter dispute with Leibniz as to which of them had been the inventor of calculus. Papin died a pauper in 1712.

Savery's steam engines consumed huge amounts of coal and proved to be uneconomic. The first successful steam engine was built in 1712 by Thomas Newcomen, a blacksmith. Newcomen had discovered by accident that the steam inside a Savery steam engine condensed suddenly after some cold water had leaked into the steam chamber. Newcomen exploited this effect by installing a pipe to squirt a jet of cold water directly into the steam chamber. The Newcomen steam engine was a large structure with a long horizontal beam that rocked to and fro, with the rise and fall of the piston. Though it could only perform about five or six strokes a minute, it was capable of lifting large volumes of water from flooded mines. In the early Newcomen engines, a boy attendant was employed to open and shut two taps, one that allowed steam into the cylinder and the other that turned on a jet of cold water to condense the steam. This was a very monotonous job but, one day, a young lad called Humphrey Potter had a bright idea. He wanted to play with his friends and decided to connect a cord from each of the taps to the beam, so that the taps opened and closed at just the right moments in the cycle. It worked, and his invention was incorporated into all Newcomen steam engines.

A much more efficient steam engine was patented by James Watt in 1769. Watt was an instrument-maker and was working as an assistant to Professor Black, who had discovered the latent heat of steam (see Section 2.1). Watt took a keen interest in steam engines and the properties of steam. When asked to repair a malfunctioning model of a Newcomen steam engine, he calculated that about 80% of the heat was lost in heating the walls of the steam cylinder. Watt deduced that it would be much better to condense the steam in a separate chamber (known as the **condenser**) so that the temperature of the walls in the piston chamber could be maintained and thereby conserve heat (Fig. 1.3(c)).

Watt needed money to exploit his idea and formed a partnership with a wealthy iron foundry owner, Matthew Boulton. This gave Watt the finance to develop a commercial steam engine but he soon ran into a major technical hitch: the cylinder castings were distorted and allowed too much steam to escape. Fortunately for Watt, a breakthrough in the manufacture of cannons provided him with the solution he needed. Cannons were constructed as thick-walled cylindrical tubes, but irregularities in the casting process meant that cannonballs often missed their target. In 1775 John 'Iron Mad' Wilkinson produced a solid cast iron block, from which he bored a smooth cylindrical hole of exactly the right shape and size. This improved the

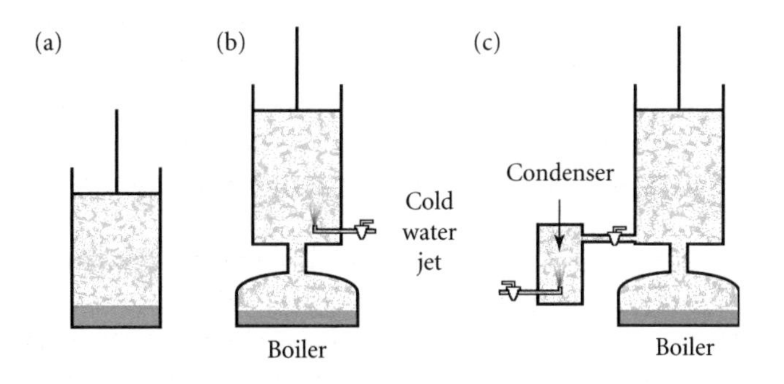

Fig. 1.3 Evolution of design of early steam engines: (a) Papin; (b) Newcomen; (c) Watt.

ballistics of cannon balls and enabled Watt to build leak-tight steam engines. The first Boulton and Watt steam engines were sold in 1776 and by 1824 they had produced 1164 machines.

Surprisingly, perhaps, it was not until the middle of the nineteenth century that the concept that heat is a form of energy was finally accepted. Heat was originally thought to be a fluid, known as caloric, that was deemed to flow from hot bodies to cold bodies and could not be created or destroyed. The caloric theory was a remnant of the science of the ancient Greeks, who believed that all matter consisted of four basic substances: air, fire, earth, and water. The caloric theory was shown to be erroneous by Benjamin Rumford, an American scientist who had worked as a spy for the British in the American War of Independence. Rumford fled to Europe, where he married the widow of Antoine Lavoisier, one of the joint discovers of oxygen (guillotined during the French Revolution in 1794). Rumford found occupation in Bavaria, where he improved the manufacture of cannons and was made a Count of the Holy Roman Empire. Cannons were bored under water and the boring process made the water boil. Rumford observed that the water boiled for as long as the boring process was continued. He deduced that caloric was apparently being produced by friction, in contradiction to the belief that it was uncreatable. Later, in the 1840s, an amateur scientist, James Joule, proved that heat and mechanical energy are equivalent, and that energy is conserved.

In 1824, a young French scientist, Nicholas Carnot, proved that the maximum possible efficiency of an ideal heat engine depends only on the values of the hot and cold temperatures between which it operates. Carnot proved that the result was independent of the nature of the working fluid, but his explanation assumed the validity of the caloric theory, 'The motive power of heat is independent of the agents employed to realise it; its quantity is fixed solely by the temperatures of the bodies between which is effected, finally, the transfer of caloric'.

Steam power continued to advance through the nineteenth century with the development of steam trains and ships, powered by reciprocating steam engines. In 1884, Charles Parsons invented the rotary steam turbine. In order to demonstrate the superiority of his invention to a sceptical British Admiralty, he fitted his vessel, Turbinia, with three rotary steam turbines. He had the audacity to appear with Turbinia, uninvited, at the display of Her Majesty's fleet for Queen Victoria's Diamond Jubilee at Spithead. He was chased by the Navy's fastest patrol boats but they were no match for Turbinia. Parsons had made his point and the rotary steam turbine was accepted.

Earlier in the nineteenth century Michael Faraday, a self-taught scientist, was making exciting discoveries in electromagnetism. Faraday had begun his career at the age of 12 as a bookbinder. This gave him the opportunity to read learned books on science and he was eventually appointed by Sir Humphry Davy as an Assistant at the Royal Institution, in London. During the 1820s, Faraday was intrigued by two recent discoveries: one by Oersted, that the needle of a compass is deflected at right angles to the direction of flow of an electric current in a wire, and the other by Ampere, that two current-carrying wires exert a force on each another. In 1831, Faraday published his laws of electromagnetic induction, based on his own discoveries that a current is set up in a closed circuit by a changing magnetic field and also in a loop of wire when moved through a stationary magnetic field.

Faraday also showed that a steady current, I, is induced across a rotating copper disc between the poles of a strong magnet (Fig. 1.4). This result led to the invention of the dynamo, which helped the introduction of electric lighting. The early dynamos produced very spiky

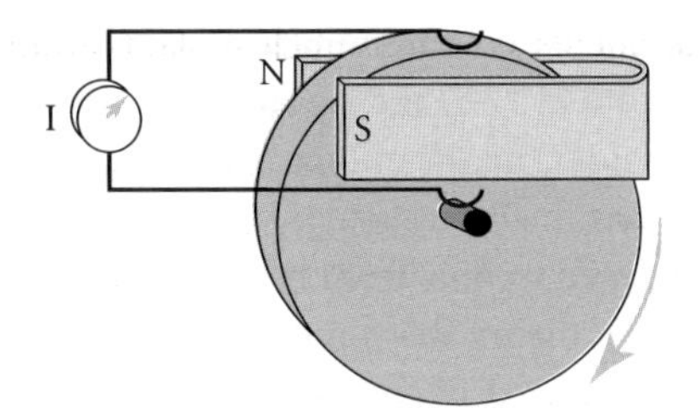

Fig. 1.4 Faraday's rotating disc.

outputs; the first device to produce a smooth current was the Gramme dynamo, using a continuous loop of wire wrapped around a rotating iron ring. Electric telegraphy and electroplating were two of the first useful applications of electricity, followed by arc-lighting for public service. The first patent for an incandescent lamp was awarded to the American inventor and entrepreneur, Thomas Edison, in 1879. He used a loop of carbonized cotton thread that glowed in a vacuum for over 40 hours. Earlier, in 1860, Joseph Swan in England had patented the world's first light bulb, but it had a short lifetime and was inefficient. During the 1870s, Swan improved the vacuum inside his bulbs and formed the Swan Electric Light Company in 1881. The Edison and Swan companies were rivals in the development of the incandescent lamp, and eventually merged in England as the Edi-Swan Company.

In order to capitalize on his invention of the incandescent lamp, Edison patented his electric distribution system in 1880. In 1881, he built the world's first power station at Holborn viaduct in London, which produced 160 kW of power for lighting and electric motors. In the following year he built a similar power station in New York City to provide electric lighting for Wall Street and the banking community. There was intense rivalry between Thomas Edison's direct current system and a system using alternating current, promoted by George Westinghouse. Edison staged public events to highlight the dangers of alternating current, with live electrocutions of dogs, cats, and even an elephant! However, the alternating current system became the one generally adopted worldwide.

The first large-scale hydroelectric power station was built in 1895 on the US side of Niagara Falls, using Fourneyron turbines. The first half of the twentieth century witnessed a massive construction programme of coal power stations and hydroelectric plants. The most significant new development in the second half of the twentieth century was nuclear power. The idea of producing electricity from nuclear fission reactors was an afterthought of the Manhattan Project, a secret military enterprise by the allied powers in the Second World War to build an atomic bomb. The early fission reactors were used for producing materials for nuclear weapons. Reactors solely for electricity generation did not appear until the latter half of the 1950s.

The early years of nuclear power were heralded as a new era of cheap, inexhaustible and safe electricity. In the 1970s, the developed world became more dependent on nuclear power after a series of large jumps in oil prices following the Arab–Israeli War in 1973. The most striking example was in France, where the government decided in the interests of national energy security to commit the country to nuclear power as the principal means of generating electricity; over 75% of French electricity is currently generated from nuclear power.

A number of high-profile accidents at various nuclear plants around the world, and growing public concern about the disposal of nuclear waste, eroded public confidence in nuclear power during the late 1970s and 1980s. The worst incidents were a partial melt-down of a pressurized water reactor at Three-Mile Island (USA) in 1979 and a complete melt-down of an RMBK reactor at Chernobyl (Ukraine) in 1986. Human error and design faults were found to be significant factors in both cases. As a result there was a general improvement in nuclear safety standards worldwide and greater international support for the effective regulation of civil nuclear installations. However, the impact of these accidents virtually halted the building of new reactors. Public opinion about nuclear power is still divided but there is a growing acceptance that it may have a role in combating global warming.

Alternative energy was a fairly dormant area until the oil price shocks of the 1970s. Western governments then began to sponsor research programmes into various alternative energy technologies with the aim of reducing their dependency on oil. Funding was, however, on a much smaller scale than that for nuclear power. Wind power was the first alternative energy technology to become commercially viable, benefiting from low capital costs and tax breaks. Some other alternative technologies are now becoming competitive in niche areas. However, large-scale power generation by alternative energy sources can have a significant environmental impact. In particular, wind power has aroused public opposition in areas of outstanding natural beauty. Nonetheless, alternative energy is establishing a foothold and is likely to gain more support as the effects of global warming are realized and become more pronounced and the remaining fossil fuel reserves of the planet become uneconomic to extract.

To complete this brief historical overview, Table 1.1 compares the power scales involved in a small selection of energy-related devices, from antiquity to the present day. It is a tribute to the achievement of mankind in applying scientific knowledge for the benefit of society, but it demonstrates that modern civilization has become dependent on consuming vast amounts of energy in order to maintain a comfortable lifestyle.

Table 1.1 Power scales

Device	Power (kW)
Treadwheel (AD 0)	0.2
Tour de France cyclist (uphill)	0.5
Strong horse	0.7
Newcomen steam engine (1712)	4
Fourneyron water turbine (1832)	30
Parsons' steam turbine (1900)	10^3
Smith–Putnam wind turbine (1942)	1.3×10^3
Boeing 747 gas turbine (1969)	6×10^4
Sizewell B nuclear power station (1992)	1.2×10^6
Drax coal power station (1986)	3.9×10^6

1.2 Global energy trends

There is a strong correlation between standard of living, as measured by the gross domestic product (GDP) per capita (i.e. per head of population), and the energy consumption per capita, as shown in Fig. 1.5. The figure also shows a large spread in energy consumption per capita between different highly developed countries, indicating some scope for reducing consumption by improvements in efficiency and changes in lifestyle.

It is natural that less developed countries will seek to increase their GDP and thereby increase their energy consumption per capita. Table 1.2 compares the actual consumption of developed countries and less developed countries in 1992 with the forecasts for 2025. It shows two trends in opposite directions. In developed countries, the population size is roughly constant and it is generally assumed that there will be a significant reduction in consumption per capita due to improvements in the efficiency of energy-consuming devices. However, in less developed countries the population is expected to increase by around 50% and the consumption per capita is expected to double as their standard of living rises. The net effect is a significant increase in global energy demand. The same upward trend is demonstrated in Fig. 1.6, which shows that global energy consumption is expected to increase by around 60% between 1999 and 2020.

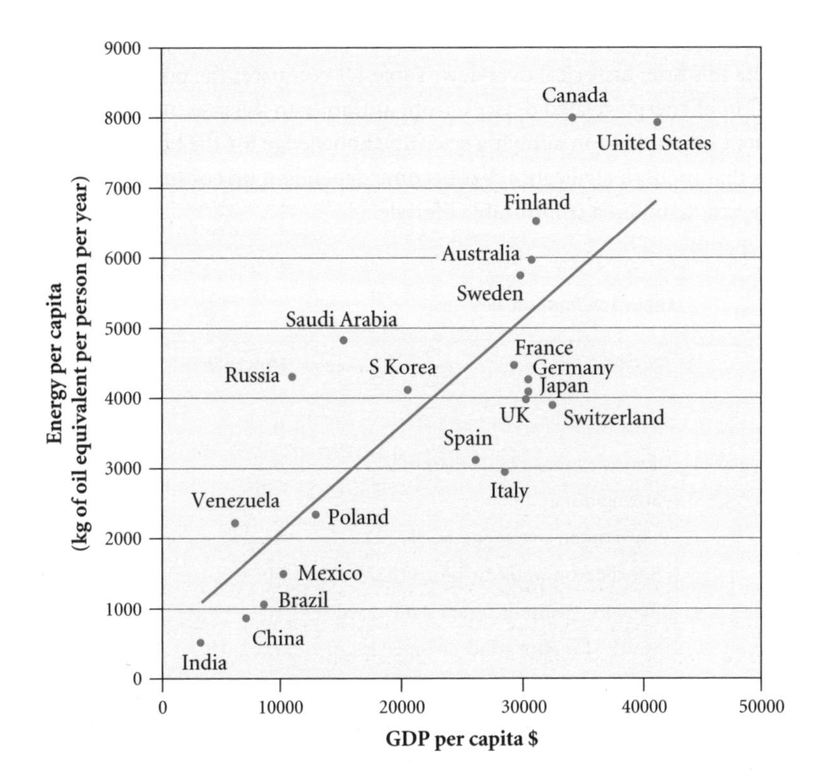

Fig. 1.5 Correlation between gross domestic product per capita and energy consumption per capita (2004).

Table 1.2 Comparison of global energy consumption in 1992 with forecast values for 2025

	Population ($\times 10^9$)	Power per capita (kW)	Total power consumption (TW)
1992			
Developed countries	1.2	7.5	9.0
Less developed countries	4.1	1.1	4.5
Total	5.3		13.5
2025			
Developed countries	1.4	3.8	5.3
Less developed countries	6.8	2.2	15.0
Total	8.2		20.3

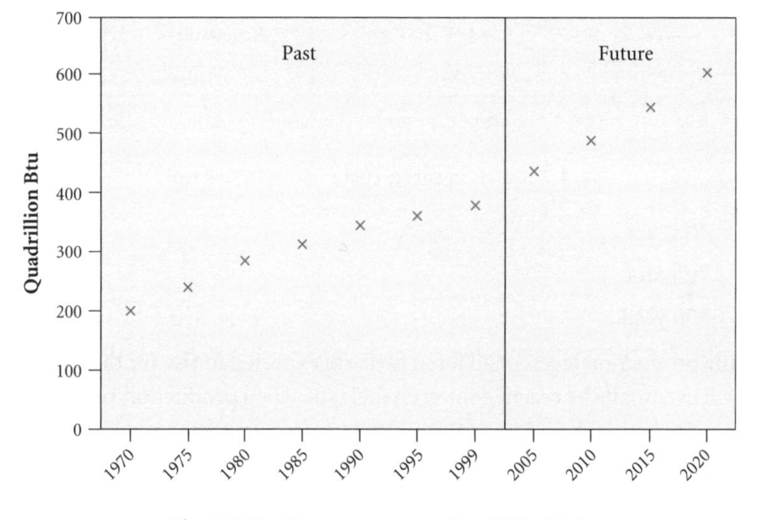

Fig. 1.6 World energy consumption 1970–2020.

The trends in energy consumption by type of primary fuel are shown in Table 1.3 and Fig. 1.7. The forecasts assume no significant change in the reliance on fossil fuels.

Oil is mainly used for transportation. The fastest growth is for natural gas, primarily for electricity generation. The expected increase in coal consumption is mainly in the developing world, particularly in China and India, which each have large coal reserves.

Oil, coal, and gas represented about 86% of all primary energy production in 2003. The number of years remaining before the remaining fossil fuel reserves are virtually exhausted depends on the rate of consumption and the discovery of new resources. An optimistic estimate (Energy Information Administration (2004), *www.eia.doe.gov*), based on the assumption that daily production levels stay at 2002 levels, gives the following approximate timescales:

Table 1.3 Primary fuel shares (% of total)

	2000	2010	2020
Oil	39.2	36.4	34.4
Gas	23.0	23.8	25.5
Coal	23.8	25.3	26.1
Nuclear	6.5	5.7	5.4
Renewables	7.6	8.9	8.7

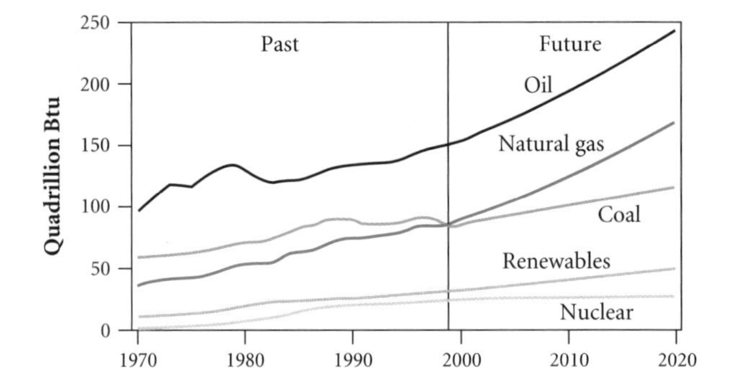

Fig. 1.7 Trends in energy consumption by fuel type.

- Oil 40 years
- Gas 70 years
- Coal 250 years

In reality, daily production levels of all fossil fuels are expected to rise for the next two decades, but a point will eventually be reached for each fuel type when production reaches a maximum, after which production will decline as the easier sources to extract are depleted and the more difficult ones are less competitive with other fuels. In 1956 the American geologist M. King Hubbert predicted that oil production in America would peak in the early 1970s by assuming that oil production would follow a bell-shaped curve. The peak actually occurred in 1970 and is now known as **Hubbert's peak**. At the time Hubbert's prediction was very controversial, but his method is now widely applied.

The growth in carbon dioxide emissions by fuel type is shown in Fig. 1.8. The mass of CO_2 produced from various sources per kilowatt-hour is shown in Table 1.4. All sources have associated CO_2 emissions, as fossil fuels are used in the construction of the generators for the carbon-free sources listed.

1.3 Global warming and the greenhouse effect

Since the Industrial Revolution there has been a sharp increase in the burning of fossil fuels. This is important for the global climate because CO_2 is what is called a **greenhouse gas**. The

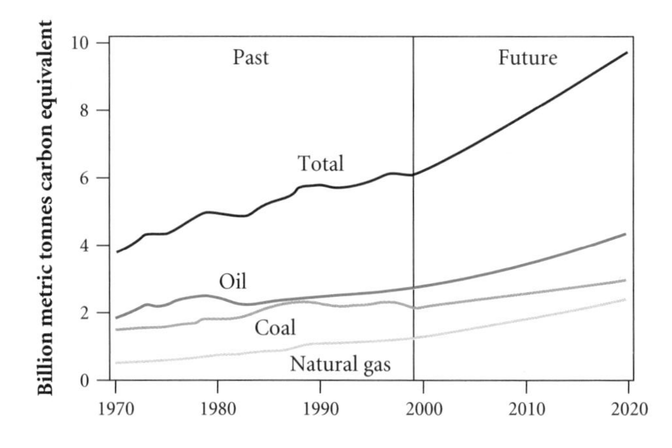

Fig. 1.8 Growth in carbon dioxide emissions by fuel type 1970–2020.

Table 1.4 CO_2 emissions from various sources (life cycle analysis)

Source	CO_2 emission (kg per kilowatt-hour)
Wood (without replanting)	1.5
Coal	0.8–1.05
Natural gas (combined cycle)	0.43
Nuclear power	0.006
Photovoltaic	0.06–0.15
Hydroelectric	0.004
Wind power	0.003–0.022

main greenhouse gases in the atmosphere are water vapour (and water droplets) and CO_2. These gases absorb infrared radiation and this affects the temperature of the Earth. We can see why this happens by first considering the temperature of the Earth with an atmosphere completely transparent to all radiation. The Sun's radiation would then impinge directly on the Earth's surface. This would heat the surface until it was at a temperature of around -19°C, at which temperature it would radiate energy back out into space at the same rate at which it was received from the Sun.

Now consider the effect of adding an atmosphere that is completely transparent to the Sun's radiation, which is mainly in the visible, but absorbs infrared radiation. Radiation emitted by a surface that is near room temperature is mainly in the infrared. Thus the atmosphere absorbs the Earth's radiation and heats up. As a result, the atmosphere radiates infrared radiation both out into space and back to the Earth's surface, which now receives more radiant energy than with a completely transparent atmosphere and gets hotter. The temperature rises until the Earth's surface emits energy at the same rate as it receives energy. This is an oversimplification of what happens in our atmosphere but illustrates the key process. The trapping of infrared radiation and the consequent temperature rise is called the **greenhouse effect**. It is actually only a small part of the temperature rise in greenhouses, which is mainly from reduced convection, but the name has stuck. We discuss a simple model of the greenhouse effect in Chapter 2 (Box 2.1).

The great French mathematician, Jean Fourier, in 1827 was the first scientist to realize that the atmosphere acts like a greenhouse. However, it was not until 1860 that the Irish scientist John Tyndall identified water vapour, carbon dioxide, and methane as the main gases in the atmosphere that absorb infrared radiation. In particular, John Tyndall realized that, without the greenhouse effect, 'the sun would rise upon an island held fast in the grip of frost'.

The greenhouse effect causes a surface temperature rise of about 35°C, so it is very important in determining and maintaining our global temperature. However, increasing the concentration of greenhouse gases substantially by burning fossil fuels has already had a noticeable impact on the global climate. Over the twentieth century, the average global temperature rose by 0.6 ± 0.2°C and the International Panel on Climate Change (IPCC) predicts that global temperatures will rise 1.4–5.8°C between 1990 and 2100, primarily due to the release of greenhouse gases from fossil fuel emissions. These predictions are based on our future reliance on fossil fuels, both for providing electricity and for transport, not changing significantly.

The average global temperatures since the mid-nineteenth century (when accurate temperature measurements started to be recorded) are shown in Fig. 1.9. There has been a steep rise since the 1970s and this rise is called **global warming**. It has been explained by physical models of the Earth's temperature that take into account the emission of greenhouse gases, in particular CO_2, due to human activity. Including these **anthropogenic** emissions gives good agreement with observations; the temperature would have been expected to have fallen slightly since the 1970s if only natural causes were included.

The main greenhouse gas is water vapour (and water droplets). The effect of carbon dioxide, methane, CFCs, and other greenhouse gases is to enhance its effect by increasing the amount of water vapour in the atmosphere. Carbon dioxide concentrations have risen from about 280 parts per million in 1750, which is just before the Industrial Revolution, to about 380 parts per

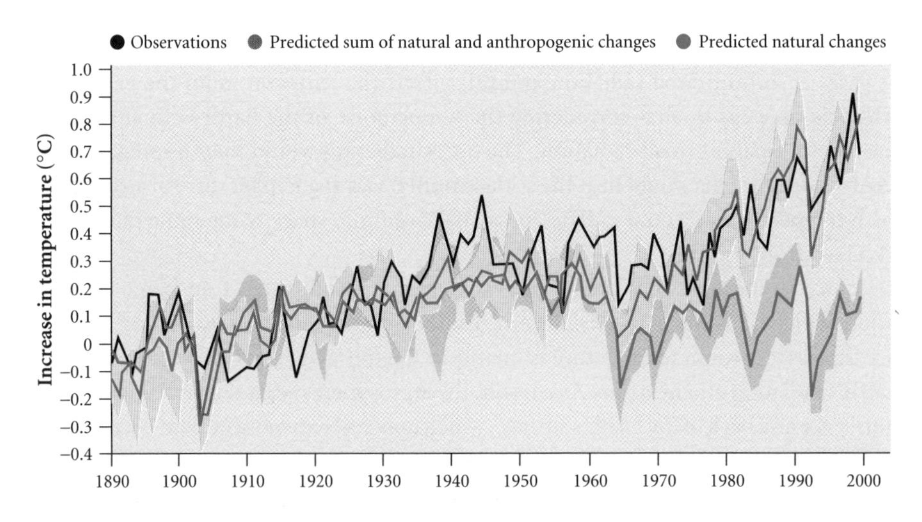

Fig. 1.9 Curves showing global temperatures 1890–2000. Measured; predicted with only natural changes; and predicted with natural plus anthropogenic changes.

million today. The characteristic timescale for an excess of water vapour in the atmosphere to disappear is a few days but for other greenhouse gases the characteristic timescales are typically 100–1000 years.

Global warming is expected to become more pronounced over the coming decades. As we discuss in Chapter 11, the impact on weather systems, ocean currents, and sea levels could be devastating in various parts of the world unless effective countermeasures are implemented and sustained over the next century. The scale of the problem requires a broad international commitment to stand any chance of success. The Kyoto Protocol was the first international agreement to commit countries to reduce or trade in emissions of carbon dioxide and other greenhouse gases, with the aim of a 5% reduction by January 2008. However, there were some notable abstainers, including the USA and Australia, and there are concerns about the effectiveness of the protocol and the likely degree of compliance of individual countries. Though the Kyoto Protocol is a step in the right direction, the world as a whole has yet to demonstrate the commitment required to combat the threat due to global warming.

1.4 Units and dimensional analysis

In order to appreciate the magnitudes of physical quantities and to be able to make comparisons, it is essential to have a sound grasp of units. To conclude this chapter, we define the most important units relating to energy. We also introduce the technique of dimensional analysis, which is employed throughout the book to derive basic physical relationships without the need for a full analysis of a system.

The SI system of units is used throughout this book but a comprehensive list of units can be found at the rear of the book, with conversions to the imperial system of units. Table 1.5 defines the key physical quantities related to energy.

Table 1.5 Energy-related units and conversion rates

Quantity	Unit	Definition
Force	newton (N)	Force required to accelerate 1 kg by 1 m s^{-2}
Energy	joule (J)	Work done by force of 1 N in moving 1 kg by 1 m
Power	watt W = J s^{-1}	1 joule per second
Energy	kilowatt-hour (kWh)	$10^3 \times 60 \times 60 = 3.6 \times 10^6$ joules ≈ 3411 Btu ≈ 859.6 kilocalories
Energy	calorie	Energy required to heat 1 g of water by 1°C ≈ 4.2 J
Energy	Btu	Energy required to heat 1 lb of water by 1°F ≈ 1.055 kJ ≈ 0.293 kWh
Energy	barrel	42 US gallons ≈ 35 imperial gallons ≈ 0.136 tonnes ≈ 159 litres
Fuel equivalence	1 tonne oil	1.5 tonnes hard coal ≈ 3.0 tonnes lignite ≈ 12 000 kWh
Power	1 horsepower	550 ft lb per second ≈ 0.746 kW

EXAMPLE 1.1

Check whether the physical dimensions of the expression for the hydrostatic pressure in a fluid, ρgh, are consistent with the physical dimensions of pressure, p, where ρ, g, and h are the density of the fluid, acceleration due to gravity, and depth, respectively.

Replacing the individual symbols in terms of their fundamental physical units, we have

$$\rho gh \equiv (\mathrm{kg\ m^{-3}})(\mathrm{m\ s^{-2}})(\mathrm{m}) = (\mathrm{kg\ m\ s^{-2}})(\mathrm{m^{-2}}) = \mathrm{N\ m^{-2}}.$$

Hence ρgh is consistent with the physical dimensions of pressure (force per unit area).

A useful development of the idea of units is **dimensional analysis**. This is a method for deriving an algebraic relationship between different physical quantities, which relies on good physical intuition in choosing the appropriate physical variables.

Each variable is expressed in terms of its fundamental units of mass M, length L, time T, . . . , raised to some arbitrary index $a, b, c, \ldots$. The unknown indices $a, b, c, \ldots$ are then determined by equating the indices of like units using simple algebra. Care is needed in choosing the relevant physical variables and it is sometimes possible to obtain more than one solution to a given problem. An independent means of checking the result obtained is therefore desirable.

EXAMPLE 1.2

Use dimensional analysis to find an expression for the hydrostatic pressure in a fluid.

Suppose the hydrostatic pressure depends on the density ρ, the acceleration due to gravity g, and depth h. We assume a general algebraic formula of the form

$$p = k\rho^a g^b h^c$$

where k is a dimensionless coefficient and the indices a, b, and c are numbers to be determined. Replacing each symbol by its fundamental physical units, we have

$$\mathrm{M^1L^{-1}T^{-2}} = (\mathrm{ML^{-3}})^a(\mathrm{LT^{-2}})^b(\mathrm{L})^c$$

or

$$\mathrm{M^1L^{-1}T^{-2}} = \mathrm{M}^a\mathrm{L}^{-3a+b+c}\mathrm{T}^{-2b}.$$

Since M, L, and T are independent physical quantities, we can equate the indices on both sides, yielding the equations

$$1 = a, \qquad -1 = -3a + b + c, \qquad -2 = -2b$$

so that $a = b = c = 1$. Hence, the expression for hydrostatic pressure is of the form $p = k\rho gh$. (The coefficient k cannot be determined from dimensional analysis because it is dimensionless.)

SUMMARY

- Global energy production is expected to increase by around 60% between 1999 and 2020.
- If production rates remain at present levels, the fossil fuels will be exhausted in about 40 years (oil), 70 years (gas), and 250 years (coal). The actual lifetimes will depend on the rate of production and new discoveries.
- The greenhouse effect is a natural phenomenon due to absorption of infrared radiation by the Earth's atmosphere, raising the temperature on the surface of Earth by about 35°C.
- The main greenhouse gas is water vapour. Carbon dioxide, methane, CFCs, and other greenhouse gases enhance the effect of water vapour by increasing its amount.
- Carbon dioxide concentrations have risen from about 280 parts per million in 1750 to about 380 parts per million today, mainly due to human activity, and there has been an associated marked rise in temperature, called global warming, since the 1970s.
- Physical expressions should be checked to make sure that the physical units are consistent.
- Dimensional analysis is a useful technique for deriving algebraic relationships between physical quantities.

FURTHER READING

Botkin D.B. and Keller, E.A. (2003). *Environmental science*. Wiley, New York. Good discussion of issues about energy and the environment.

Borowitz, S. (1999). *Farewell fossil fuels*. Plenum, New York. Interesting book about the need for alternatives to fossil fuels.

Boyle, G., Everett, R., and Ramage, J. (eds). (2003). *Energy systems and sustainability*. Oxford University Press, Oxford. Good qualitative discussion of issues.

Uglow, J. (2002). *The lunar men*. Faber and Faber, London. Interesting account of the friends who launched the Industrial Revolution.

WEB LINKS

www.iea.org International Energy Agency, advisory body on energy policy.

www.energy.gov US Department of Energy website, appraisals of energy trends and energy technologies.

www.eia.doe.gov Official energy statistics of US government.

ww.metoffice.com/research/hadleycentre Research on climate change.

www.wikipedia.org On-line encyclopaedia, articles on history of energy and energy-related topics.

www.worldenergy.org Neutral overview of global energy scene.

www.wri.org Environmental think-tank.

EXERCISES

1.1 Can global warming be combated by energy conservation alone?

1.2 Assuming that the volumes of the Greenland and Antarctic ice caps are 2.85×10^6 km^3 and 25.7×10^6 km^3, respectively, that the radius of the Earth is 6378 km, and that 70% of the Earth's surface is covered by sea, estimate the rise in sea level if both ice caps melted completely.

1.3 Describe the development of the steam engine.

1.4 Estimate the force exerted on von Guericke's hemispheres. Diameter = 51 cm; assume air pressure = 10^5 N m^{-2}.

1.5 Is the quality of life of a society better or worse without electricity?

1.6 Construct a model of an Archimedes' screw using scrap materials.

1.7 Rewrite in modern terminology Carnot's statement, 'The motive power of heat is independent of the agents employed to realise it; its quantity is fixed solely by the temperatures of the bodies between which is effected, finally, the transfer of caloric'.

1.8 Show that the physical dimensions of $\frac{1}{2}mv^2$ are consistent with the physical dimensions of work.

Table 1.6 Carbon dioxide concentrations and global temperature differences* ΔT for the period 1979–2005

Year	1979	1980	1981	1982	1983	1984	1985
CO_2 concentration (ppm)	336.53	338.34	339.96	341.09	342.07	344.04	345.10
ΔT (°C)	0.06	0.10	0.13	0.12	0.19	−0.01	−0.02
Year	**1986**	**1987**	**1988**	**1989**	**1990**	**1991**	**1992**
CO_2 concentration (ppm)	346.85	347.75	350.68	352.84	354.22	355.51	356.39
ΔT (°C)	0.02	0.17	0.16	0.10	0.25	0.20	0.06
Year	**1993**	**1994**	**1995**	**1996**	**1997**	**1998**	**1999**
CO_2 concentration (ppm)	356.98	358.19	359.82	361.82	362.98	364.90	367.87
ΔT (°C)	0.11	0.17	0.27	0.13	0.36	0.52	0.27
Year	**2000**	**2001**	**2002**	**2003**	**2004**	**2005**	
CO_2 concentration (ppm)	369.22	370.44	372.31	374.75	376.95	378.55	
ΔT (°C)	0.24	0.40	0.45	0.45	0.44	0.47	

* Temperature difference with respect to the average temperature over the period 1961–90.

1.9 Use dimensional analysis to find an expression for the power in the wind incident on a wind turbine, in terms of the air density ρ, the wind speed v, and the cross-sectional area of the wind turbine A.

1.10 Convert (a) 1 MJ into Btu, (b) 800 kg of oil equivalent per year into kW.

1.11 Derive an expression for the characteristic time taken for viscous effects in a fluid to dissipate in terms of the distance x (L), density ρ (ML^{-3}), and the coefficient of viscosity μ ($ML^{-1}T^{-1}$).

1.12 Derive a dimensionless parameter based on specific heat, c ($L^2T^{-2}\theta^{-1}$), coefficient of viscosity, μ ($ML^{-1}T^{-1}$), and thermal conductivity, k ($MLT^{-3}\theta^{-1}$), where θ represents the dimensions of absolute temperature.

1.13* Table 1.6 shows the rise in the average global near surface temperature and the atmospheric concentration of carbon dioxide for each year over the period 1979–2005. Analyse whether a statistical correlation exists between the two sets of data.

1.14* Consider an Archimedes' screw consisting of a helical tube of wavelength λ wound around a cylinder of radius a. Prove that the angle of elevation of the axis of the cylinder, θ, for the device to be able to raise water is such that

$$\tan\theta \leq \frac{2\pi a}{\lambda}.$$

1.15 Why might you expect oil production to follow a bell-shaped curve as a function of time (Hubbert's peak theory)?

2 Thermal energy

List of Topics

- Heat and temperature
- Conduction, convection, and radiation
- Laws of thermodynamics
- Basic elements of thermal power stations
- Closed thermal power cycles
- Efficiency of Carnot cycle
- Pressure, specific volume
- Internal energy
- Enthalpy
- Entropy
- Reversible process
- Irreversible process
- $T-s$ diagram
- Thermal properties of water and steam
- Disadvantages of Carnot cycle
- Rankine cycle
- Brayton cycle
- Gas turbine
- Combined cycle gas turbine
- Combined heat and power
- Fossil fuels
- Combustion
- Flue gases
- Fluidized beds
- Geothermal energy
- Aquifers and hot dry rocks

Introduction

Thermal power stations convert the heat released from burning fossil (or nuclear) fuel into electrical energy. Understanding the processes involved requires some knowledge of thermal physics. In this chapter we describe the basic laws of thermal physics, starting from the basic ideas of thermal energy and heat transfer to the laws of thermodynamics, the analysis of a closed thermal power cycle, and thermodynamic efficiency. We then consider the thermal properties of water and the thermodynamic cycles used in the various types of thermal power stations. We also describe the chemical nature of fossil fuels, the physical and chemical processes involved in combustion, and the waste products of combustion. Finally, we include a brief account of geothermal energy.

2.1 Heat and temperature

Temperature is a characteristic of the thermal energy of a body due to the internal motion of molecules. Two bodies in mutual thermal contact are said to be in thermal equilibrium if they are both at the same temperature. Temperature was originally defined in terms of the freezing point and boiling point of water, but the modern definition of temperature is based on the efficiency of an ideal fluid working in a Carnot cycle (Section 2.4), independent of the properties of any particular material.

In general, apart from when a material changes phase (e.g. from solid to liquid), the temperature of any material increases as it absorbs heat. The heat ΔQ required to raise the temperature of unit mass of a material by an amount ΔT is given by

$$\Delta Q = c\Delta T. \tag{2.1}$$

The coefficient, c, is roughly independent of temperature and is called the specific heat of the material. The original unit of thermal energy was the calorie, defined as the energy needed to change the temperature of one gram of liquid water by one degree Celsius, but thermal energy is now usually measured in joules. The energy equivalence of the two units is approximately given by

$$1 \text{ calorie} \approx 4.2 \text{ joules}. \tag{2.2}$$

During a change of phase of a material, heat is absorbed and the temperature remains constant. The heat ΔQ required to change the phase of unit mass of material is called the latent heat L. Thus

$$\Delta Q = L. \tag{2.3}$$

The latent heat of evaporation (from liquid to gas) is typically one or two orders of magnitude larger than the latent heat of fusion (from solid to liquid).

2.2 Heat transfer

There are three basic forms of heat transfer: conduction, convection, and radiation.

Conduction is the transfer of thermal energy within a body due to the random motion of molecules. The average energy of the molecules is proportional to the temperature. Consider a bar of length d and cross-sectional area A, with one end at a fixed temperature T_1 and the other at a fixed temperature T_2, where $T_1 > T_2$. The more energetic molecules at the hot end transfer kinetic energy to the less energetic molecules at the cold end. In the steady-state, the rate of flow of heat is constant along the length of the bar and is given by Fourier's law of heat conduction

$$Q = kA\frac{(T_1 - T_2)}{d} \tag{2.4}$$

where k is called the thermal conductivity.

EXAMPLE 2.1

A steel bar of length 1 m and cross-sectional area 1 cm^2 has one end at 1000°C and the other end at 0°C. If the thermal conductivity of the bar is 50 W m^{-1} K^{-1}, calculate the heat flow along the bar in the steady-state, ignoring heat losses from the surface.

From eqn (2.4) the heat flow along the bar is given by

$$Q = kA\frac{(T_1 - T_2)}{d} \approx 50 \times 10^{-4} \times \frac{10^3}{1} = 5\ \text{W}.$$

It should be noted that eqn (2.4) applies only in the steady-state. In practice, it takes time for a solid body to establish a steady-state temperature distribution. For **unsteady heat conduction**, the characteristic time to establish a steady-state is determined by the time t taken for an isotherm to diffuse a distance x, given by

$$t \approx \frac{x^2}{\kappa} \tag{2.5}$$

where $\kappa = k/\rho c$ (m^2 s^{-1}) is called the **thermal diffusivity** of the material. The validation of eqn (2.5) can be found in books on the mathematics of heat conduction (e.g. Carslaw and Jaeger 1959), but the algebraic form can be derived by dimensional analysis (Exercise 2.2).

EXAMPLE 2.2

Suppose the bar in Example 2.1 is initially at 0°C and that one end is raised to 1000°C at $t = 0$. Estimate the characteristic time for the temperature in the bar to achieve the steady-state temperature at (a) 1 cm, (b) 10 cm from the hot end. ($\rho \approx 8 \times 10^3$ kg m^{-3}, $k \approx 3.5 \times 10^2$ W m^{-1} K^{-1}, $c \approx 4 \times 10^2$ J kg^{-1}°C^{-1}.)

The thermal diffusivity of the material is given by $\kappa = \frac{k}{\rho c} \approx 1.1 \times 10^{-4}$m^2 s^{-1}. Substituting in eqn (2.5) yields: (a)$t \approx \frac{x^2}{\kappa} = \frac{(10^{-2})^2}{1.1\times10^{-4}} \approx 0.9$ s; (b) $t \approx \frac{(10^{-1})^2}{1.1\times10^{-4}} \approx 90$ s.

Convection is the transport of heat due to the bulk motion of a fluid. Consider a fluid of density ρ and temperature T moving with velocity u. The mass flow per unit area per second is ρu and the thermal energy per unit mass is cT. The rate of flow of heat per unit area by convection is the product of ρu and cT, i.e.

$$\frac{Q}{A} = (\rho u)(cT) = \rho u c T. \tag{2.6}$$

When a cold fluid flows over a hot surface (**forced convection**) the rate of heat transfer from the surface to the fluid is greater than that in the case of a stationary fluid. The temperature gradient at the surface is very large, so the layer of fluid adjacent to the wall is heated rapidly by thermal conduction. The hot fluid mixes with cold material in the bulk of the fluid and the net heat transfer is much larger than that by heat conduction alone. In forced convection, the rate of heat transfer per unit area is often expressed in the form

$$\frac{Q}{A} = Nu\frac{k(T_s - T_\infty)}{L} \tag{2.7}$$

where T_s is temperature of the surface, T_∞ is the temperature in the body of the fluid, L is a characteristic length, and Nu is a dimensionless parameter known as the **Nusselt number**.

The choice of L depends on the geometrical set-up, e.g. for heat transfer from a bar in a cross-flow it would be appropriate to take L to be the radius of the bar. The Nusselt number is a function $Nu = f(Pr, Re)$ of two other non-dimensional parameters: the Prandtl number, $Pr = c\mu/k$, and the Reynolds number, $Re = \rho uL/\mu$, where μ is the coefficient of dynamic viscosity (Section 3.5). Note that the Prandtl number depends only on the properties of the material (i.e. specific heat c, thermal conductivity k, and coefficient of viscosity μ), whereas the Reynolds number also depends on the velocity of the fluid. The numerical value of the Nusselt number for a given Pr and Re is usually obtained from empirical correlations (see Example 2.3).

EXAMPLE 2.3

For fully developed turbulent flow in a smooth pipe, the Nusselt number Nu is related to the Reynolds number Re and the Prandtl number Pr by the empirical correlation

$$Nu = \frac{\frac{1}{2}f\, Re\, Pr}{1 + 2Re^{-1/8}(Pr - 1)}$$

where $f \approx 0.08Re^{-1/4}$ is called the 'friction factor'. Estimate Nu for $Re = 10^4$ and $Pr = 1$.

Putting $Re = 10^4$ and $Pr = 1$ yields $Nu \approx 0.04\left(10^4\right)^{-1/4+1} = 40$, i.e 40 times the heat transfer due to heat conduction alone.

Radiative heat transfer is the transport of energy by electromagnetic waves. Unlike conduction and convection, heat can be transferred by radiation in a vacuum. The energy radiated per unit area per second (i.e. the power per unit area) from a surface at temperature T is given by the Stefan–Boltzmann law

$$P_e = \varepsilon\sigma T^4 \tag{2.8}$$

where ε is the **emissivity** of the surface and $\sigma \approx 5.67 \times 10^{-8}\ \mathrm{W\,m^{-2}\,K^{-4}}$ is the Stefan–Boltzmann constant. ε is a dimensionless number and varies from 0 to 1, depending on the nature of the surface.

Opaque surfaces absorb radiation from the environment. The absorptivity of a surface is the same as its emissivity. The rate of absorption per unit area is $P_a = \varepsilon\sigma T_o^4$, where T_o is the temperature of the environment. Hence the *net* rate of emission per unit area per second is given by

$$P = P_e - P_a = \varepsilon\sigma(T^4 - T_o^4). \tag{2.9}$$

A surface that absorbs *all* incident radiation is known as a **black body**. A good practical approximation to a black body is a cavity with a small pinhole that connects it with the outside environment. Since nearly all of the radiation entering the pinhole is absorbed by the surface inside the cavity before it can escape out of the pinhole, the absorptivity is very close to one and the energy distribution of the radiation emitted by the pinhole is effectively determined only by the temperature of the surface inside the cavity.

The temperature of the outer surface of the Sun determines the flux of radiation incident on the upper atmosphere of the Earth (see Example 2.4 and Chapter 6). Also, radiation is the dominant mode of heat transfer in the furnace of a fossil-fuel power plant (Section 2.11).

EXAMPLE 2.4

Estimate the solar power per square metre incident at the equator. Assume that the surface temperature of the outer surface of the Sun is 5800 K, the emissivity is unity, the radius of the Sun is 7×10^8 m, and the distance from the Sun to the Earth is 1.5×10^8 km.

The total power emitted by the Sun is the power per unit area per second, eqn (2.8), multiplied by the surface area of the Sun, i.e.

$$P_s = \sigma T^4 \times 4\pi r_s^2 \approx (5.67 \times 10^{-8}) \times (5.8 \times 10^3)^4 \times 4 \times 3.14 \times (7 \times 10^8)^2$$
$$\approx 3.9 \times 10^{26}\ \text{W}.$$

The fraction of solar power incident on 1 m^2 at the equator is the solid angle subtended from the Sun, $\Omega = 1/(4\pi d^2) \approx 1/[4 \times 3.14 \times (1.5 \times 10^{11})^2] \approx 3.5 \times 10^{-24}$, where d is the distance from the Earth to the Sun. Hence the incident solar power per unit area at the equator is $P_s\Omega \approx (3.9 \times 10^{26}) \times (3.5 \times 10^{-24}) \approx 1.37\ \text{kW m}^{-2}$.

The effect of the atmosphere on the transmission of radiation is very important in determining the temperature of the surface of the Earth. As discussed qualitatively in Chapter 1, the absorption of infrared radiation by the atmosphere gives rise to the **greenhouse effect**, which raises the Earth's temperature by about 35°C. In Box 2.1 we describe a simple model of the greenhouse effect.

Box 2.1 The greenhouse effect

As shown in Example 2.4, the solar radiation incident on the Earth's atmosphere has an intensity of 1.37 kW m^{-2}; this number is known as the **solar constant**. A fraction of this radiation, called the **albedo**, is reflected by clouds in the atmosphere and by the surface of the Earth back into outer space. The albedo is close to 30%. The radiation is absorbed by an area πR^2 of the Earth's surface (i.e. the cross-section facing the Sun), where R is the radius of the Earth. As a result, the Earth heats up until it emits as much radiation as it receives. The radiation emitted from a surface at room temperature is in the infrared. We ignore the geothermal heat flux at the surface of the Earth, which is much smaller than the incident solar flux.

We first consider what would happen if the Earth's atmosphere did not absorb any of the incident solar radiation (which is mainly in the visible part of the spectrum) nor any infrared radiation from the Earth's surface. Let A be the value of the albedo, S be the incident solar intensity, and assume the Earth's surface has an emissivity of 1 (i.e. it acts like a black body). In equilibrium when the Earth's surface is at a temperature T, we have

$$(1 - A)S\pi R^2 = 4\pi R^2 \sigma T^4,$$

noting that radiation is emitted by the whole of the Earth's surface (area $= 4\pi R^2$). Putting $A = 0.3$ and $S = 1.37$ kW m^{-2} gives $T = 255$ K $= -18°$C.

We now consider what would happen if the atmosphere absorbed all the infrared radiation emitted by the Earth, but still transmitted all of the incident solar radiation. The atmosphere would then heat up to some temperature T_a such that the energy radiated into space was equal to that it received. Assuming the same albedo as before, and assuming that the emissivity of the atmosphere is $\varepsilon = 1$, then T_a is given by

$$(1 - A)S\pi R^2 = 4\pi R^2 \sigma T_a^4,$$

which yields $T_a = T = 255$ K.

The atmosphere radiates into outer space and down towards the Earth's surface in equal amounts, as shown in Fig. 2.1. The Earth's surface therefore receives more radiation and heats up to a temperature T_E. For steady-state energy equilibrium, we require

$$(1 - A)S\pi R^2 + 4\pi R^2 \sigma T_a^4 = 4\pi R^2 \sigma T_E^4.$$

Substituting for $4\pi R^2 \sigma T_a^4$ from above, we obtain $T_E^4 = 2T^4$, so that $T_E = 303$ K $= 30°$C.

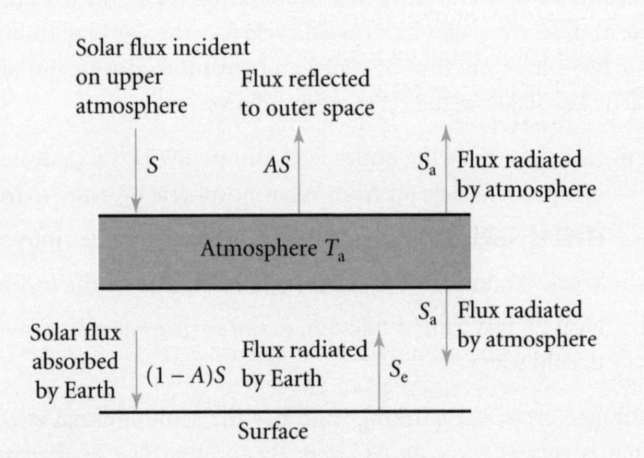

Fig. 2.1 Greenhouse effect.

The rise in surface temperature due to the absorption of infrared radiation is called the **greenhouse effect**. Whilst this model is an oversimplification it illustrates the key mechanism involved. We consider the effect of partial absorption of the solar radiation by the Earth's atmosphere in Exercise 2.4.

2.3 First law of thermodynamics and the efficiency of a thermal power plant

The maximum possible efficiency of a thermal power plant can be obtained from the laws of thermodynamics, without needing to consider the details of the fluid flow and heat transfer processes involved in the various stages of the plant.

The first law of thermodynamics is a statement of energy conservation taking thermal energy into account. Consider the system enclosed by the control volume V shown in Fig. 2.2. By energy conservation, the difference between the heat input to the system Q and the work done by the system W is equal to the change in the internal energy ΔU of the system, i.e.

$$Q - W = \Delta U. \tag{2.10}$$

(Note the convention that Q and W are positive if heat flows into the system and work is done by the system.)

2.4 Closed cycle for a steam power plant

In order to convert heat into useful work in a steam power plant, the working fluid undergoes a change of phase at different stages in a closed cycle (i.e. the working fluid is re-used), from liquid water, to a two-phase mixture of water and steam, to dry steam, and back to liquid water (Fig. 2.3). The key stages in the cycle are as follows.

1. **Compressor** (also known as the boiler feed pump). Work W_{com} done on the system to compress cold water from subatmospheric pressure to high pressure.
2. **Boiler** Heat Q_1 added to the system to convert cold water into steam.
3. **Turbine** Work W_t done by the system (i.e. by steam) on the turbines blades.
4. **Condenser** Heat Q_2 lost from the system to the environment in converting steam back to cold water.

After each complete cycle, the working fluid has the same internal energy U, so the net change in internal energy is zero, or $\Delta U = 0$. By the first law of thermodynamics (2.10) we have

$$(Q_1 - Q_2) - (W_t - W_{com}) = 0.$$

Hence the efficiency of the process is given by

$$\eta = \frac{[\text{net work output}]}{[\text{heat input}]} = \frac{W_t - W_{com}}{Q_1} = \frac{Q_1 - Q_2}{Q_1} = 1 - \frac{Q_2}{Q_1}. \tag{2.11}$$

Thus the efficiency is unity minus the ratio of the heat output in the condenser and the heat input in the boiler. In a perfect system there would be no heat loss in the condenser and all the

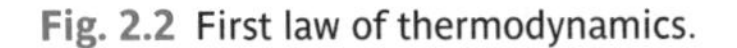

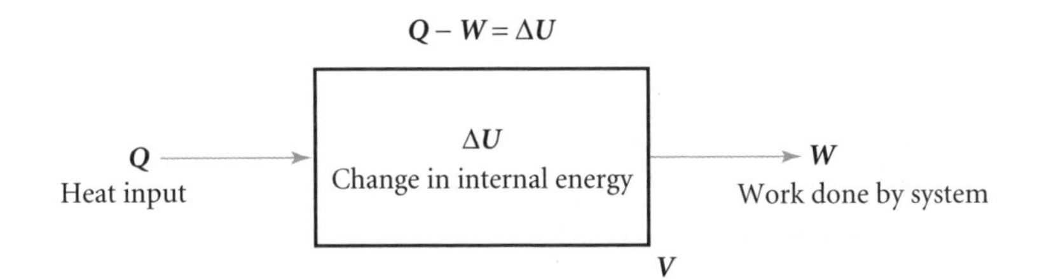

Fig. 2.2 First law of thermodynamics.

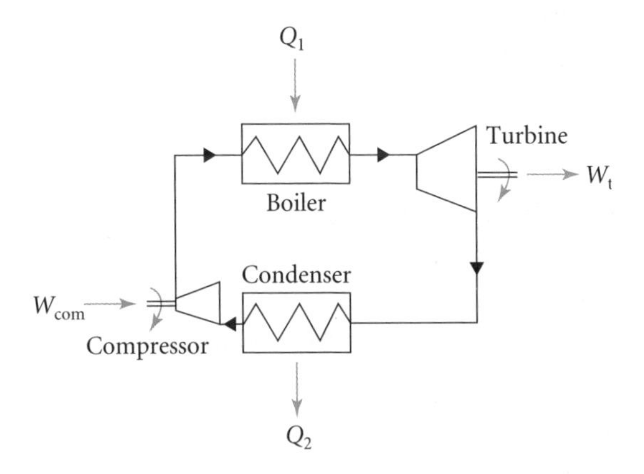

Fig. 2.3 Plant lay-out of thermal power station.

heat supplied in the boiler would be used to do useful work. However, this heat increases the disorder (entropy, Section 2.5) of the steam which is unchanged when the steam does work passing through the turbine. As a result, some heat must be lost by the working fluid to the environment, to reduce the disorder of the fluid back to its original value, and the amount depends on the temperature of the condenser. Hence $Q_2 > 0$ and $\eta < 1$. It follows that there is an upper limit to the efficiency of a thermal power plant and the thermal energy wasted heats the external environment.

From a thermodynamic point of view, the fact that the system is less than 100% efficient is a consequence of the **second law of thermodynamics**: no system operating in a closed cycle can convert all the heat absorbed from a heat reservoir into the same amount of work. Carnot proved that the maximum possible efficiency of a heat engine operating in a closed cycle between two heat reservoirs depends only on the ratio of the absolute temperatures of the reservoirs, i.e.

$$\eta_C = 1 - \frac{T_2}{T_1} \tag{2.12}$$

where T_1 and T_2 are the absolute temperatures of the upper and lower reservoirs, respectively, measured in degrees Kelvin. (See Derivation 2.1.)

Derivation 2.1 Efficiency of a Carnot cycle

Consider an ideal gas with an equation of state $pV = nRT$ operating in the closed cycle *abcda* shown in Fig. 2.4. Sections *ab* and *cd* are **isotherms** at T_1 and T_2, respectively; *bc* and *da* are **reversible adiabatics** (i.e. no heat transfer with the surroundings).

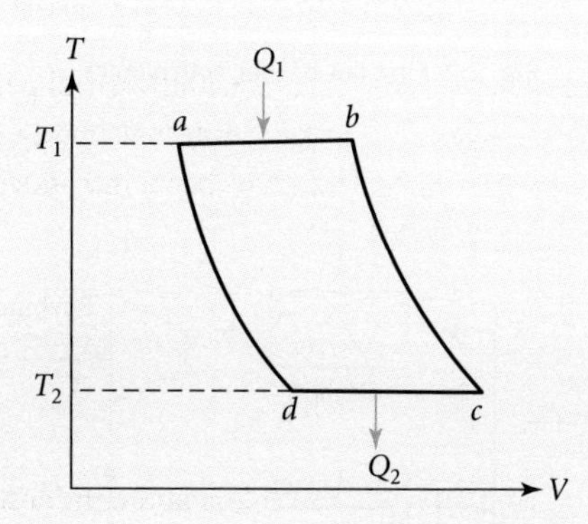

Fig. 2.4 Carnot cycle for perfect gas.

By the first law of thermodynamics (2.10) we have

$$đQ = dU + pdV = C_V\, dT + pdV \tag{2.13}$$

where C_V is the heat capacity at constant volume and đ means an inexact differential as its value depends on the path of integration.

Along isotherm *ab* we put $dT = 0$ in (2.13), so that $đQ = pdV$. The total heat input is

$$Q_1 = \int_{V_a}^{V_b} pdV = nRT_1 \int_{V_a}^{V_b} \frac{dV}{V} = nRT_1 \ln(V_b/V_a).$$

Likewise, along isotherm *cd* the total heat output is $Q_2 = nRT_2 \ln(V_c/V_d)$. Hence

$$\frac{Q_2}{Q_1} = \frac{T_2 \ln(V_c/V_d)}{T_1 \ln(V_b/V_a)}. \tag{2.14}$$

Along adiabatic *da* we put $đQ = 0$ in eqn (2.13). Hence $C_V\, dT = -pdV$, or

$$C_V \frac{dT}{T} = -nR\frac{dV}{V}.$$

Integrating along adiabatic *da* we obtain $C_V \ln(T_1/T_2) = -nR \ln(V_a/V_d)$.
Likewise, integrating along adiabatic *bc* gives $C_V \ln(T_1/T_2) = -nR \ln(V_b/V_c)$.
By inspection, $V_a/V_d = V_b/V_c$, or $V_c/V_d = V_b/V_a$, so that eqn (2.14) reduces to

$$Q_2/Q_1 = T_2/T_1.$$

Replacing Q_2/Q_1 by T_2/T_1 in eqn (2.11), yields the efficiency of the Carnot cycle as

$$\eta_C = 1 - T_2/T_1.$$

2.5 Useful thermodynamic quantities

There are six key quantities that are useful in describing the thermodynamics of a thermal power plant: temperature T, pressure p, specific volume v (volume per unit mass, i.e. the reciprocal of density), specific internal energy u, specific enthalpy h, and specific entropy s. In general, only two thermodynamic quantities are needed to completely specify the thermal state of a system.

Note. Unless otherwise stated, we assume that u, h, and s are all per unit mass and use lower case symbols, and drop the prefix *specific* in specific internal energy, etc.

The concept of **internal energy**, u, has already been introduced from the first law of thermodynamics (2.10). **Enthalpy** is defined as

$$h = u + pv. \tag{2.15}$$

Enthalpy is useful for describing:

1. heat transfer at constant pressure (e.g. in boilers and condensers), where the change in enthalpy $h_2 - h_1$ is equal to the heat input Q, i.e.

$$h_2 - h_1 = Q; \tag{2.16}$$

2. adiabatic ($Q = 0$) compression or expansion (e.g. in compressors and turbines), where the net work done on the shaft is equal to the change in enthalpy, i.e.

$$W = h_1 - h_2. \tag{2.17}$$

(See Derivation 2.2.)

The concept of **entropy** arises from the second law of thermodynamics and is a measure of the **degree of disorder** of a system. Essentially, there are two types of process whereby a system can change from one state to another: **reversible processes** and **irreversible processes**. In a reversible process, both the system and the surroundings can recover their original states. This can be achieved by changing the system so slowly that it remains in quasi-static thermal equilibrium throughout the process. In an irreversible process, however, the system and the surroundings are changed in such a way that they are unable to return to their original states (e.g. a scrambled egg). The change in entropy of a system Δs is defined as

$$\Delta s = \frac{\Delta Q_{\mathrm{rev}}}{T} \tag{2.18}$$

where ΔQ_{rev} is the heat supplied reversibly to a system at an absolute temperature T. There is therefore no change in entropy in a reversible adiabatic process.

In a reversible process the total change in entropy of a system and its surroundings is zero, whereas in an irreversible process there is a net increase in entropy. It should be realized that the concept of reversibility is an idealization that is unachievable in practice.

Derivation 2.2

1. Expansion of a gas at constant pressure

Consider a unit mass of gas contained in a thermally conducting piston tube at constant pressure p (Fig. 2.5). Heat is added slowly to the gas (by heating the outside of the tube), so that the gas remains in approximate thermal equilibrium with the surroundings.

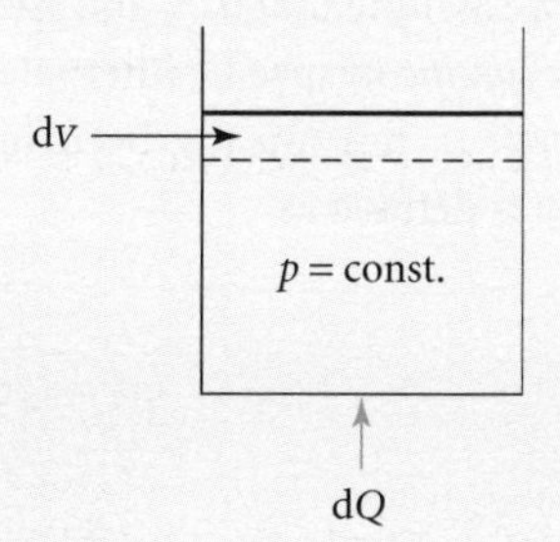

Fig. 2.5 Expansion of gas at constant pressure.

Suppose that an elemental amount of heat $đQ$ is required to expand the volume from v to $v + dv$. The work done by the gas on the surroundings is $đW = pdv$. By the first law of thermodynamics (2.10), the change in internal energy is given by $du = đQ - pdv$. Since $p = \text{const.}$, we can put $pdv = d(pv)$ and write du in the form

$$du = đQ - d(pv).$$

Integrating both sides we obtain the total change in internal energy as

$$u_2 - u_1 = Q - (p_2v_2 - p_1v_1).$$

Thus $(u_2 + p_2v_2) - (u_1 + p_1v_1) = Q$, or

$$h_2 - h_1 = Q.$$

2. Adiabatic compression or expansion

Figure 2.6 shows an adiabatic process in which fluid moves from the inlet A to the outlet B and does work by rotating a shaft immersed in the fluid. Suppose the pressure is p_1 at A and p_2 at B. The work done in moving unit mass of fluid through a volume v_1 at A is p_1v_1 and through a volume v_2 at B is $-p_2v_2$. Thus the net work done by the fluid is $(p_2v_2 - p_1v_1)$. If the amount of work done by the shaft is W_s then the total work done by the system is

$$W = W_s + (p_2v_2 - p_1v_1).$$

From the first law of thermodynamics (2.10), $Q - W = \Delta u$. Hence, for adiabatic compression or expansion $(Q = 0)$, we have $-W_s - (p_2v_2 - p_1v_1) = u_2 - u_1$. Thus the work done by the shaft is given by $W_s = (u_1 + p_1v_1) - (u_2 + p_2v_2)$ or

$$W_s = h_1 - h_2.$$

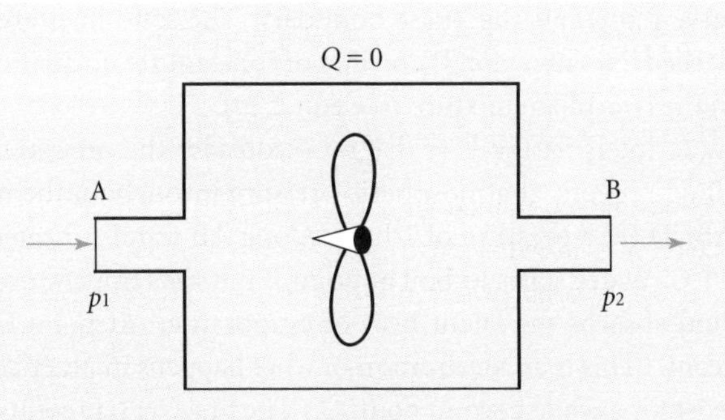

Fig. 2.6 Adiabatic compression or expansion.

2.6 Thermal properties of water and steam

Thus far, we have applied general thermodynamic considerations without specifying the physical nature of the working fluid. In a conventional thermal power plant cycle the working fluid is water and, at various stages in the cycle, the working fluid changes phase from water, to a two-phase mixture of water and steam, to dry steam and finally back to water. Some knowledge of the thermal properties of water and steam is essential in order to understand the operation of such a plant. The most convenient thermodynamic variables for describing thermal power plants are temperature T and entropy s. The T–s diagram for water and steam is shown in Fig. 2.7.

There are three distinct regions of interest:

I water;

II two-phase mixture of water and steam;

III dry steam.

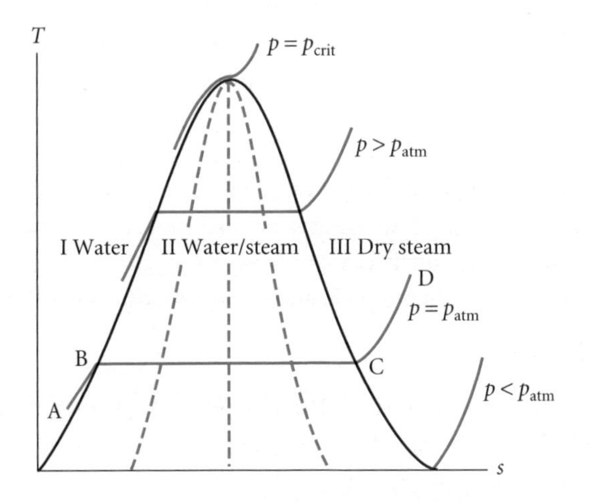

Fig. 2.7 T–s diagram for water and steam (not to scale).

The bell-shaped curve represents the phase boundary. The solid lines are isobars (constant pressure) and the dashed lines in region II are lines of constant steam quality x, i.e. the fraction by mass of steam in the two-phase mixture (see eqn 2.20).

To illustrate how to interpret the T–s diagram, consider the process of boiling water in a kettle. Since the fluid remains at atmospheric pressure throughout the heating process, we follow the isobar ABCD (at a pressure of 1 bar). Along AB water is heated from cold to the boiling point of 100°C. Water starts to boil at point B but the temperature remains at 100°C along BC as the fluid absorbs the latent heat of evaporation. At point C all the water has converted to dry steam. (This is an idealization of what happens in practice, since some water droplets may still exist for a while beyond point C.) The line CD represents superheated fluid (i.e. the temperature of the steam is above the boiling point), and the temperature of the dry steam rises at constant pressure as more heat is supplied, but the kettle should have switched off before getting this far!

The properties of water (region I) and dry steam (region III) can be obtained directly from steam tables. However, in order to use the steam tables in the two-phase region (II), some further explanation is necessary. Consider a mass of water m at B that converts into an equal mass of dry steam at C. At any point along the isobar BC the mixture of water and steam contains a mass m_f of water and m_g of steam (where the subscripts refer to *fluid* and *gas*, respectively). The total mass of the mixture is $m = m_f + m_g$. If v_f is the specific volume of liquid water at B and v_g the specific volume of dry steam at C, then the total volume of the mixture is $V = V_f + V_g = m_f v_f + m_g v_g$. Hence the specific volume v of the water–steam mixture is given by

$$v = \frac{V}{m} = \frac{m_f v_f + m_g v_g}{m} = \frac{(m - m_g)v_f}{m} + \frac{m_g v_g}{m} = \left(1 - \frac{m_g}{m}\right) v_f + \frac{m_g}{m} v_g. \tag{2.19}$$

The ratio

$$x = \frac{m_g}{m} \tag{2.20}$$

represents the proportion by mass of steam in the mixture and is called the steam quality. Likewise, $(1 - x)$ represents the proportion by mass of water in the mixture. Thus the steam quality at B is $x = 0$ (all water) and at C is $x = 1$ (all steam). Substituting for x from eqn (2.20) in eqn (2.19), gives the specific volume of the mixture as

$$v = (1 - x)v_f + x v_g. \tag{2.21}$$

The numerical values of the coefficients v_f and v_g are obtained from steam tables. Equation (2.21) can then be used to determine the specific volume of the mixture for any particular value of the steam quality, x.

Likewise the corresponding values of u, h, and s in the mixture are of the form

$$u = (1 - x)u_f + x u_g, \tag{2.22}$$

$$h = (1 - x)h_f + x h_g, \tag{2.23}$$

$$s = (1 - x)s_f + x s_g, \tag{2.24}$$

and the numerical values of the coefficients $u_f, u_g, h_f, h_g, s_f, s_g$ are also obtained from steam tables.

To illustrate how to use the T–s diagram to solve practical problems, we consider a steam power plant operating in a Carnot cycle (Example 2.5).

EXAMPLE 2.5

A steam power plant operates in the Carnot cycle shown in Fig. 2.8. The boiler is at $T_1 = 352^\circ$C, $p = 170$ bar and the condenser is at $T_2 = 30^\circ$C, $p = 0.04$ bar. Calculate: (a) the efficiency of the cycle; (b) the heat input to the boiler; (c) the work done by the turbine; (d) the heat output in the condenser; and (e) the fraction of heat used in a complete cycle, using the steam table data below.

Steam table data

		h (kJ kg^{-1})		s (kJ kg^{-1} K^{-1})	
T (°C)	p (bar)	h_f	h_g	s_f	s_g
30	0.04	126	2556	0.436	8.452
352	170	1690	2548	3.808	5.181

(a) From (2.12) the efficiency of the cycle is

$$\eta_C = 1 - \frac{T_2}{T_1} = 1 - \left(\frac{273 + 30}{273 + 352}\right) \approx 0.52\,.$$

(b) The boiler operates at constant pressure, so we can use eqn (2.16) to calculate the heat input Q_1 from the change in enthalpy along ab, i.e.

$$Q_1 = h_b - h_a = 2548\text{–}1690 = 858 \text{ kJ kg}^{-1}.$$

(c) Since bc is an adiabatic in a Carnot cycle, we can use eqn (2.17) to calculate the work done by the turbine W_t, i.e.

$$W_t = h_b - h_c = 2548 - h_c.$$

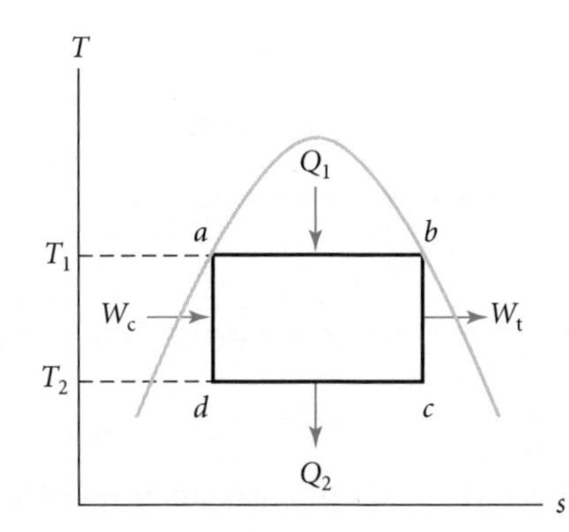

Fig. 2.8 T–s diagram for a steam power plant operating in a Carnot cycle.

To evaluate h_c we use eqn (2.23), i.e. $h_c = (1 - x_c)h_f + x_c h_g$, where x_c is the steam quality at c. In order to determine x_c we use the fact that the expansion in the turbine is adiabatic. Equating the entropy at b and c gives $s_c = s_b = 5.181$ kJ kg^{-1} K^{-1}. Using eqn (2.24), we then have the following equation for x_c:

$$s_c = (1 - x_c)s_f + x_c s_g = 5.181\,.$$

From the steam table, $s_f = 0.436$ kJ kg^{-1} K^{-1} and $s_g = 8.452$ kJ kg^{-1}K^{-1} on the isobar $p = 0.04$ bar. Hence

$$0.436(1 - x_c) + 8.452x_c = 5.181,$$

which yields $x_c \approx 0.59$.

We can now evaluate the enthalpy at c as

$$h_c = (1 - x_c)h_f + x_c h_g \approx (1\text{–}0.59)(126) + (0.59)(2556) \approx 1560 \text{ kJ kg}^{-1}.$$

The work done by the turbine is then given by

$$W_t = h_b - h_c = 2548\text{–}1560 = 988 \text{ kJ kg}^{-1}.$$

(d) Point a lies on the phase boundary, where the specific entropy is $s_a = 3.808$ kJ kg^{-1}K^{-1}. Since there is no change in entropy in the compressor, the entropy at d and a must be identical. Hence $s_d = 3.808$ kJ kg^{-1}K^{-1}. From (2.24), the steam quality at d is then given by

$$(1 - x_d)s_f + x_d s_g = 3.808\,.$$

From the steam table data, $s_f = 0.436$ kJ kg^{-1}K^{-1} and $s_g = 8.452$ kJ kg^{-1}K^{-1} on the isobar at $p = 0.04$ bar. Hence

$$0.436(1 - x_d) + 8.452x_d = 3.808, \text{ yielding } x_d \approx 0.42\,.$$

Also $h_d = (1 - x_d)h_f + x_d h_g = (1 - 0.42)(126) + (0.42)(2556) = 1147$ kJ kg^{-1}.

Since the condenser operates at constant pressure, the heat lost to the environment is equal to the change in enthalpy, so that

$$Q_2 = h_c - h_d = 1560 - 1147 = 413 \text{ kJ kg}^{-1}.$$

(e) The fraction of heat used in the complete cycle is given by

$$1 - \frac{Q_2}{Q_1} = 1 - \frac{413}{858} \approx 0.52, \text{ i.e. the Carnot efficiency derived in part (a).}$$

2.7 Disadvantages of a Carnot cycle for a steam power plant

Despite the fact that a Carnot cycle yields the maximum possible efficiency for a thermal power plant operating in a closed cycle, it suffers from a number of disadvantages that make it impractical for a real working fluid such as water.

To begin with, a Carnot cycle requires the temperature T_1 of the upper reservoir to be constant. However, from the $T-s$ diagram for water/steam (Fig. 2.7) we see that this can only be achieved by operating the boiler along an isobar in region II. It is not possible to operate the boiler in the dry steam region since the temperature rises along any given isobar in region III. Hence the upper temperature of the cycle, T_1, is constrained by the maximum temperature of the two-phase boundary.

Another problem arises in the turbine. In a Carnot cycle the turbine operates with a two-phase mixture of water and steam (region II in Fig. 2.7). The momentum of fast moving water droplets in the mixture damages the turbine blades and shortens the life of the turbine. Similarly, since the compressor is required to compress a mixture of water and steam into high pressure water, water droplets in the mixture damage the blades of the compressor.

Finally, the volume of steam in the mixture is very large, which means that the compressor needs to be very large and therefore expensive. The combined effect of these factors makes a Carnot cycle impractical for a steam power plant.

2.8 Rankine cycle for steam power plants

Fortunately, there is a thermodynamic cycle that does not suffer from the problems of the Carnot cycle. It is the **Rankine cycle**, named in honour of Thomas Rankine, one of the founders of thermodynamics. We begin by considering the simplest case of a Rankine cycle without reheat.

2.8.1 Rankine cycle without reheat

The Rankine cycle without reheat is shown in Fig. 2.9. Unlike in a Carnot cycle, all the steam in a Rankine cycle is converted into water in the condenser (*de*) before entering the compressor. The compressor increases the pressure of the water (*ef*) adiabatically before the water enters the boiler. In modern steam power plants, boilers are normally constructed in three distinct sections, each made with a different grade of steel, using cheaper steels in the lower temperature sections and more expensive steels in the higher temperature sections. In the **economizer** section (*fa*) water is heated at high pressure until it starts to boil. In the **evaporator** section (*ab*) a two-phase mixture of water and steam is heated at constant pressure until all the water has been converted into dry steam. The dry steam is then heated at constant pressure in the **superheater** section of the boiler (*bc*). Dry steam enters the turbine at high pressure and does work on the turbine blades (*cd*). On leaving the turbine, wet steam enters the condenser (at subatmospheric pressure), where it condenses on the cold outer surfaces of a large bank of condenser tubes containing cold water from the external environment. This is a simplification of the situation in a real plant. In practice there are many complicating features, e.g. a pressure drop through the boiler due to frictional losses, the incorporation of a re-circulating loop in the economizer section (to take advantage of natural circulation), and instabilities in the position of the two-phase boundaries.

In order to calculate the efficiency of a Rankine cycle without reheat it is not possible to use the Carnot formula (2.12), because the temperature of the upper reservoir of heat

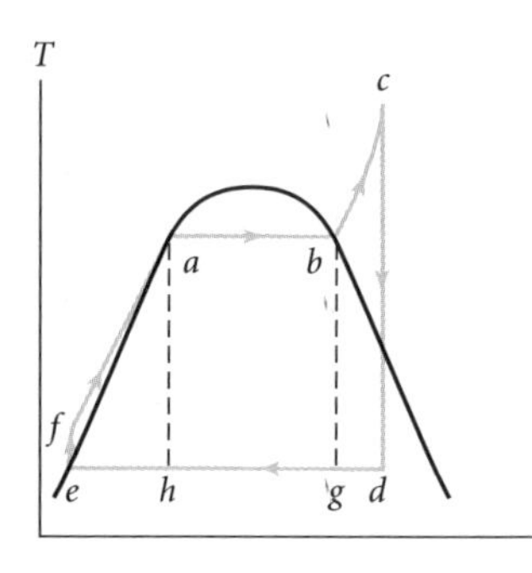

Fig. 2.9 *T–s* diagram for Rankine cycle without reheat (*abcdefa*). *abgha* represents a Carnot cycle with the same lower temperature.

is not constant. The average upper temperature in a Rankine cycle lies between the lowest temperature in the economizer and the maximum temperature in the superheater. The method of calculating the heat transfer processes in each stage of the cycle and the overall efficiency is shown in Example 2.6.

EXAMPLE 2.6

A steam power plant operates in a Rankine cycle without reheat, as shown in Fig. 2.9. The boiler and the condenser are at $p = 170$ bar and $p = 0.04$ bar, respectively. The temperature of the evaporator is 352°C, the maximum temperature in the superheater is 600°C and the temperature in the condenser is 30°C. Calculate: (a) the work done by the compressor; (b) the heat input in the boiler; (c) the work done on the turbine; (d) the heat output in the condenser; (e) the efficiency of the Rankine cycle without reheat; (f) the efficiency of the Carnot cycle operating between a reservoir at 352°C and a reservoir at 30°C.

Steam table data

		h (kJ kg^{-1})		s (kJ kg^{-1} K^{-1})	
T (°C)	p (bar)	h_f	h_g	s_f	s_g
30	0.04	126	2556	0.436	8.452
352	170	1690	2548	3.808	5.181
600	170		3564		6.603

(a) Assuming that water is incompressible, the work done by the compressor per unit mass of water is given by $W_{com} = v_f(p_f - p_e)$. Putting $v_f = 10^{-3}\text{m}^3\text{ kg}^{-1}$ and $p_f - p_e = (170 - 0.04) \times 10^5\text{ N m}^{-2}$, we have $W_{com} \approx 17\text{ kJ kg}^{-1}$.

(b) We first calculate the enthalpy at the entrance to the boiler (f). Assuming the compressor is adiabatic, the work done is equal to the change of enthalpy, i.e. $W_{com} = h_f - h_e$, or $h_f = W_{com} + h_e \approx 17 + 126 = 143\text{ kJ kg}^{-1}$.

The boiler operates at constant pressure, so the heat input is given by $Q_{in} = h_c - h_f = 3564 - 143 \approx 3421\ \mathrm{kJ\,kg^{-1}}$.

(c) The work done by the turbine is $W_t = h_c - h_d = 3564 - h_d$, where h_d is given by $h_d = h_f(1 - x_d) + h_g x_d = 126(1 - x_d) + 2556x_d$. To obtain x_d we use the fact that the expansion in the turbine is adiabatic, so that $s_d = s_c = 6.603$.
Hence $0.436(1 - x_d) + 8.452x_d = 6.603$, which yields $x_d \approx 0.77$.
Thus $h_d = 126(1 - 0.77) + 2556(0.77) \approx 1995\ \mathrm{kJ\,kg^{-1}}$,
and $W_t = 3564 - 1995 = 1569\ \mathrm{kJ\,kg^{-1}}$.

(d) The heat output in the condenser is $Q_{out} = h_d - h_e = 1995 - 126 \approx 1869\ \mathrm{kJ\,kg^{-1}}$.

(e) The efficiency of the cycle is $\eta_R = \frac{W_t - W_{com}}{Q_{in}} = \frac{1569-17}{3421} \approx 0.45$.

(f) The efficiency of the Carnot cycle is $\eta_C = 1 - \frac{T_2}{T_1} = 1 - \left(\frac{273+30}{273+352}\right) \approx 0.52$.

2.8.2 Rankine cycle with reheat

The Rankine cycle without reheat does not completely eliminate the production of water droplets with high momentum that damage the turbine blades. To overcome the problem, modern power plants tend to use the Rankine cycle with reheat, in which the steam is reheated several times before entering the condenser. Figure 2.10 shows the T–s diagram with three reheat stages. Steam is reheated after leaving the high pressure (HP) turbine before entering the intermediate pressure (IP) turbine, and is again reheated between the IP turbine and the low pressure (LP) turbine. The overall efficiency of the cycle is greater than that of a Rankine cycle without reheat and the formation of water droplets is much reduced.

In order to maximize the efficiency of a steam power plant it is desirable to operate at as high a temperature as possible in the superheater. However, above about 650°C various forms of metal fatigue become significant due to the very high temperatures and pressures that the walls of the boiler tubes have to withstand. Erosion–corrosion of the tubes due to the presence of trace chemicals in the water can also be a life-limiting factor. Replacement of boiler tubing in a power plant is a major operation that puts the plant out of service for a considerable period. The net result is that the maximum operating temperature of a steam power plant is limited to about 650°C, above which the benefits from improved efficiency are outweighed

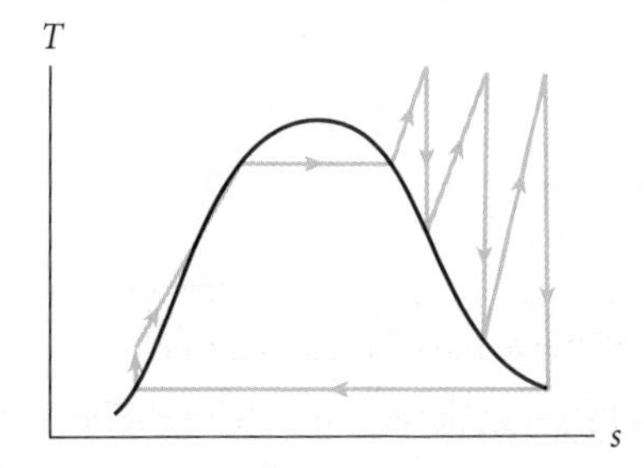

Fig. 2.10 T–s diagram for Rankine cycle with reheat.

by the cost of tube replacement and outage costs. Modern plants using Rankine cycles with reheat achieve overall efficiencies of around 40–45%.

2.9 Gas turbines and the Brayton (or Joule) cycle

As we observed in Section 2.8, the maximum temperature in a steam power station must be kept below about 650°C in order to avoid excessive metallurgical damage. In a gas turbine, however, the gaseous products of combustion are typically around 1300°C. The turbine blades are covered by a ceramic coating of low thermal conductivity so that the hot gases do not make direct contact with metal surfaces. In addition, the blade assembly is water-cooled so that the temperature of the blades is maintained below the metallurgical limit.

Gas turbines for electricity generation originally evolved from jet turbine engines. In a jet engine, the thrust arises from the combustion of gaseous fuel and the expansion of the exhaust gases. Since the working fluid does not change phase, a condenser is not involved in the process, so the overall size and cost of a gas turbine plant is less than that of an equivalent steam plant. Gas turbines operate in a Brayton (or Joule) cycle, as shown in Fig. 2.11. It is an open cycle but is equivalent to a closed cycle in the sense that the atmosphere acts as a heat exchanger that cools the air entering the combustion chamber.

Air enters the compressor (a) at atmospheric pressure and is compressed to around 10–20 bar (b). It is then mixed with fuel in the combustion chamber, producing hot combustion gases that do work on the turbine (c). The exhaust gases are then vented to atmosphere (d). Assuming the change in the compressor is an adiabatic process ($Q = 0$), the first law of thermodynamics (2.10) reduces to $-W_c = \Delta U$. Also, from (2.17) we can equate the net work done per unit mass by the compressor to the increase in enthalpy, i.e.

$$W_{com} = h_b - h_a = c_p(T_b - T_a)$$

where c_p is the specific heat at constant pressure. Similarly, assuming the turbine is adiabatic, the work done per unit mass on the turbine is

$$W_t = h_c - h_d = c_p(T_c - T_d).$$

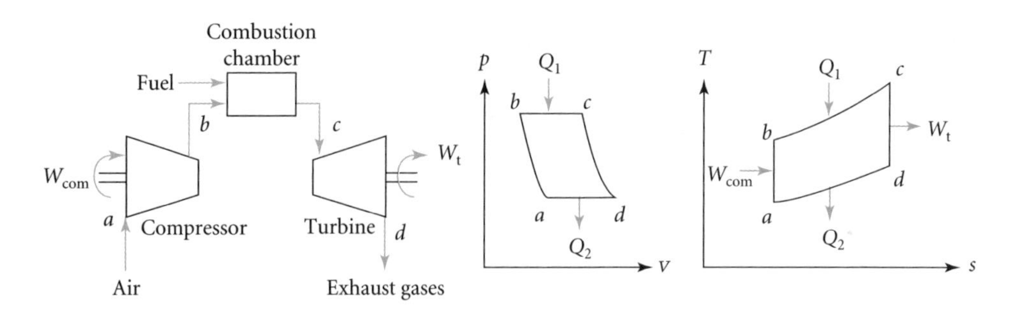

Fig. 2.11 Brayton (or Joule) cycle.

The net heat supplied is

$$Q = h_c - h_b = c_p(T_c - T_b).$$

Hence the efficiency of the cycle is given by

$$\eta = \frac{W_t - W_{com}}{Q} = \frac{(T_c - T_d) - (T_b - T_a)}{T_c - T_b}. \tag{2.25}$$

A more useful expression for the efficiency of a gas turbine is given by the formula

$$\eta = 1 - r^{-\left(\frac{\gamma-1}{\gamma}\right)} \tag{2.26}$$

where

$$r = \frac{p_b}{p_a} = \frac{p_c}{p_d} \tag{2.27}$$

is called the **pressure ratio** and $\gamma = c_p/c_v$ is the ratio of the specific heats at constant pressure and at constant volume. (See Derivation 2.3.)

Gas turbines are relatively low capital cost devices that can be started up quickly and are employed for satisfying sudden surges in electricity demand. Efficiencies of simple gas turbines are up to around 40% (Example 2.7).

Derivation 2.3 Pressure ratio and efficiency of a Brayton (or Joule) cycle

Consider the adiabatic expansion of a perfect gas. Putting $đQ = 0$ the first law of thermodynamics (2.10) becomes $du + p\,dv = 0$, or $c_v dT + p\,dv = 0$. Differentiating the equation of state for a perfect gas $pv = RT$, we have $pdv + vdp = RdT$. Putting $R = c_p - c_v$ and $v = RT/p$, and eliminating dv, we obtain

$$\frac{dT}{T} = \left(\frac{\gamma - 1}{\gamma}\right)\frac{dp}{p}$$

where $\gamma = c_p/c_v$ Integrating, we obtain $\ln T = \left(\frac{\gamma-1}{\gamma}\right)\ln p + \text{const.}$, or $T = Ap^{\frac{\gamma-1}{\gamma}}$. Hence T_b and T_c can be expressed in terms of T_a and T_d, respectively, as

$$T_a = T_b\left(\frac{p_b}{p_a}\right)^{\frac{1-\gamma}{\gamma}} = T_b r^{\frac{1-\gamma}{\gamma}} \text{ and } T_d = T_c\left(\frac{p_c}{p_d}\right)^{\frac{1-\gamma}{\gamma}} = T_c r^{\frac{1-\gamma}{\gamma}} \tag{2.28}$$

where $r = \frac{p_b}{p_a} = \frac{p_c}{p_d}$ is the pressure ratio. Substituting for T_a and T_d in eqn (2.25) and rearranging we obtain $\eta = \frac{(T_c - T_b) - (T_d - T_a)}{T_c - T_b} = 1 - r^{\frac{1-\gamma}{\gamma}}$.

EXAMPLE 2.7

Calculate the exhaust temperature and the efficiency η of an ideal gas turbine operating with a maximum temperature of 1300 K for a pressure ratio $r = 8$. (Assume $\gamma = 1.4$)

From eqn (2.28) the exhaust temperature is given by

$$T_d = T_c r^{-\left(\frac{\gamma-1}{\gamma}\right)} = 1300 \times 8^{-\left(\frac{1.4-1}{1.4}\right)} \approx 718\text{K}.$$

Equation (2.26) yields the efficiency as $\eta = 1 - r^{-\left(\frac{\gamma-1}{\gamma}\right)} = 1 - 8^{-\left(\frac{1.4-1}{1.4}\right)} \approx 0.45.$

2.10 Combined cycle gas turbine

The overall efficiency of a gas turbine can be increased by feeding the heat of the exhaust gases into a steam power plant. The combination of a Brayton cycle and a Rankine cycle is called a combined cycle gas turbine (CCGT), and is shown in Fig. 2.12. The net effect is equivalent to that of a single cycle operating between the upper temperature of a Brayton cycle and the lower temperature of a Rankine cycle. Efficiencies of up to 60% are typical in CCGT plants.

Even greater efficiencies can be achieved in a combined heat and power cycle (CHP). In a CHP plant the condenser in the steam power cycle is operated at a higher temperature than that in a conventional steam power plant, and the waste heat from the condenser is used to provide district heating in the local community. The total efficiency of CHP schemes is typically around 80%. However, the cost involved in installing the pipework and other infrastructure is high, so the application of CHP is limited to industrial complexes or to densely populated urban areas.

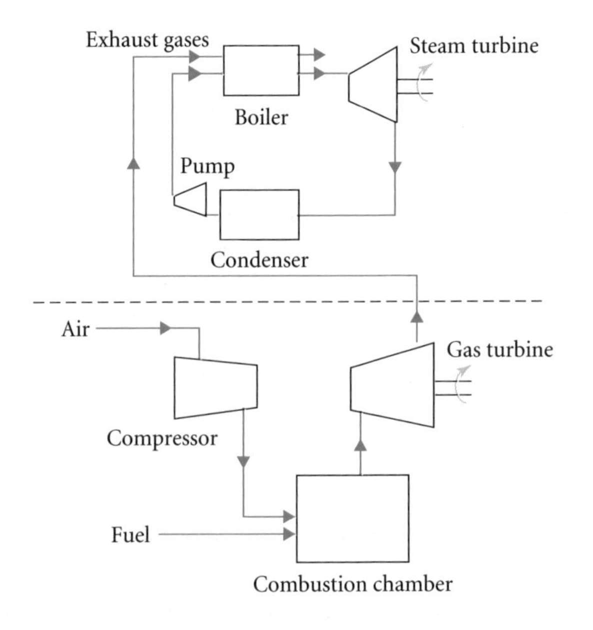

Fig. 2.12 Combined cycle gas turbine (CCGT) generation.

2.11 Fossil fuels and combustion

Fossil fuels, in solid, liquid, and gaseous forms, represent by far the largest source of energy used to generate electricity. Coal is the solid resulting from the compaction of rotting vegetation on land, whereas oil and natural gas are the remains of the compaction of microscopic marine organisms on the sea bed. Within two centuries from the start of the Industrial Revolution, mankind has consumed a significant fraction of these natural resources that took hundreds of millions of years to evolve. In the process, the atmosphere has been polluted with vast quantities of carbon dioxide, sulfur dioxide, and nitrogen oxides.

The carbon content of coal depends on its origin, varying from 55% for lignite to 90% for anthracite. Coal also contains various other combustible elements (hydrogen and sulfur) and incombustible elements (nitrogen, water, and ash-forming minerals). Oil is a complex mixture of hydrocarbons, containing carbon (~86%), hydrogen (~12%) and sulfur (~2%). Natural gas is predominantly methane (CH_4) but other forms of gaseous fuel include coal (or town) gas, producer gas, and blast furnace gas.

The combustion of methane in natural gas is the exothermic reaction

$$CH_4 + 2O_2 \rightarrow CO_2 + 2H_2O, \tag{2.29}$$

which produces 55 MJ of heat per kilogram of methane. The atomic masses of H, C, and O are in the proportion 1:12:16, so the burning of 16 kg of methane releases 44 kg of carbon dioxide.

The combustion of the hydrocarbons in oil follows similar exothermic reactions. For example, the combustion of 72 kg of pentane

$$C_5H_{12} + 8O_2 \rightarrow 5CO_2 + 6H_2O \tag{2.30}$$

releases 220 kg of carbon dioxide.

A typical 500 MW coal-fired plant consumes around 250 tonnes of coal an hour. Coal needs to be in the form of a fine powder before it can be burned. The coal lumps are ground in large coal mills and the coal powder is injected through nozzles into a combustion chamber (Fig. 2.13) where it burns in a huge fireball. The combustible material undergoes exothermic reactions with oxygen and the suspended particles of carbon and ash emit and absorb radiation. The flames are optically thick, i.e. most photons produced in the interior of the flame are re-absorbed in the flame before reaching the walls of the furnace. The outer surface of the flame is a close approximation to a black body and radiation is the dominant form of heat transfer to the boiler tubes lining the walls of the combustion chamber. The radiant heat incident on the boiler tubes is conducted through the tube walls and heats the water flowing inside. Over a long period, a solid layer of slag deposit forms on the outer surface of the boiler tubes and reduces the heat transfer; the slag is removed during plant outages for general maintenance.

The combustion of coal is a complex process, involving:

- the evaporation of water trapped inside the coal (which uses some of the energy content of the fuel);
- the production of combustible gases from the dissociation of coal (notably methane CH_4 and carbon monoxide CO), which react with oxygen and release heat;

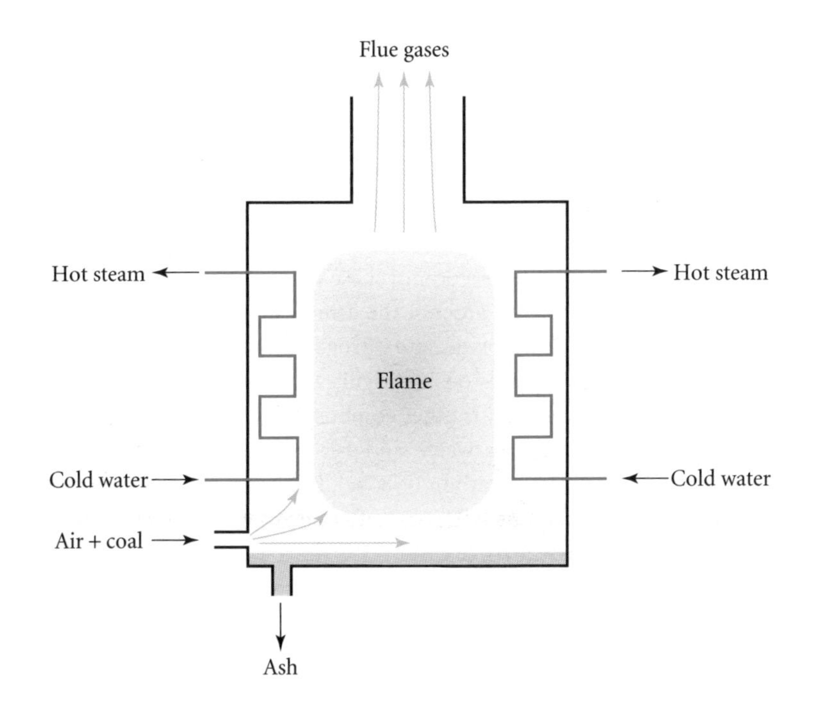

Fig. 2.13 Coal-fired combustion chamber.

- the combustion of solid carbon matter, $C + O_2 \rightarrow CO_2$; thus 12 kg of carbon releases 44 kg of CO_2.

Incomplete combustion of the carbon results in the formation of carbon monoxide

$$2C + O_2 \rightarrow 2CO. \tag{2.31}$$

Since carbon monoxide is poisonous, sufficient air must be injected into the furnace to ensure that the production of carbon monoxide is minimized, which also improves the fuel efficiency. The design of efficient coal-fired furnaces that minimize the production of environmentally harmful gases is an active area of coal technology.

Controlling the emission of sulfur dioxide from fossil fuel power stations is also an important issue since it is a major contributor to acid rain. There are various ways of tackling the problem, including:

- removing the sulfur prior to combustion by coal scrubbing and oil desulfurization;
- gasification of coal under pressure with air and steam to form gases that can be burned to produce electricity;
- flue gas desulfurization, in which the waste gases are scrubbed with a chemical absorbent (e.g. limestone).

Though these processes are beneficial to the environment, they typically add about 10% to the overall cost of electricity, so power companies need to be required or given incentives to incorporate such measures.

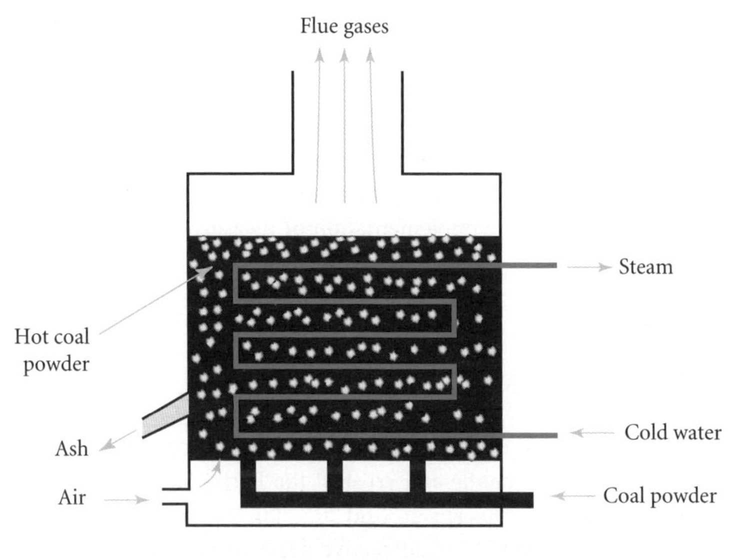

Fig. 2.14 Fluidized bed.

2.12 Fluidized beds

An alternative means of burning coal that reduces the emission of environmentally harmful gases is the fluidized bed (Fig. 2.14). The fuel in the bed is in the form of small solid particles, suspended in an upward jet of air. Turbulent mixing of the particles with the air results in more complete chemical reactions than in a conventional combustion chamber. The emission of SO_x gases is reduced by using limestone to precipitate sulfur and the generation of NO_x is also reduced because the temperature of the bed is lower than that in a combustion chamber. Heat is removed by a heat exchanger immersed in the fluidized bed.

In the early years of development, fluidized beds suffered from a major drawback due to the fact that the fine particles and gases damaged the tubing, by a process known as erosion–corrosion. The lifetime of the tubing in early prototype fluidized beds was only about a year. The problem has since been overcome by redesigning the bed to reduce the velocity of the flue gas in contact with the boiler tubing, and commercial CFD plants are now able to compete with conventional combustion units. CFD units are also used extensively for the disposal for municipal solid waste.

2.13 Carbon sequestration

Even with more efficient burning of fossil fuels, an enormous quantity of carbon dioxide would still be emitted from fossil-fired plants over the coming decades and contribute further to global warming. A novel approach to the problem is carbon sequestration (also called carbon capture), in which carbon dioxide is captured before it enters the atmosphere. Provided any later release of the carbon dioxide from the carbon dioxide stores were to occur over a long period (i.e. greater than the timescale for the depletion of the existing fossil fuel reserves of the planet), carbon capture would mitigate global warming.

One idea is to isolate the carbon dioxide from the flue gases and pump the gas underground into depleted oil fields; the carbon dioxide liquefies under pressure and can therefore be sealed. However, carbon sequestration is an expensive technology that will probably need government incentives and international agreements before it is widely employed. Policies that affect global warming and the implementation of measures such as carbon capture are discussed in Chapter 11.

2.14 Geothermal energy

The temperature in the interior of the Earth in its core is around 4000°C. As a result there is a heat flow out through the mantle and crust of the Earth, and the average temperature gradient at the Earth's surface is typically around 30°C km^{-1}. The heat flow is maintained by the generation of heat produced by the radioactive decay of isotopes of heavy nuclei and by the cooling of the core and mantle.

Geologically active parts of the world such as Iceland, California, Italy, and New Zealand are close to the interfaces between tectonic plates. Naturally-occurring steam jets (**geysers**) and hot springs up to 350°C provide a ready source of thermal energy. Even in geologically stable regions of the world, geothermal energy can still be extracted by drilling boreholes to depths of a few kilometres and flushing water through hot rock formations. Up to about 150°C geothermal energy is primarily used for district heating, industry, and agricultural purposes, but above 150°C it can be used as feed-water heating for electric power plants.

There are two basic types of rock formation that are suitable for 'mining' geothermal heat: aquifers and hot dry rocks. An **aquifer** is a layer of porous rock trapped between layers of impermeable rock, e.g. a layer of sandstone tens of metres in thickness. (Aquifers close to the surface provide vast reservoirs of rain water, which are extracted by water authorities.) Aquifers at depths of 2–3 km are typically at 60–90°C. Cold water is injected at some point in the aquifer through a borehole. The water flows through the aquifer and absorbs heat from the porous rock (see Section 2.14.1). The hot water is removed through a second borehole. In **hot dry rock** heat extraction, water is pumped at high pressure through narrow gaps in hot rock formations (see Section 2.14.2). A particularly suitable type of rock is granite, which is found in large blocks typically 10–100 m in dimension. Granite contains uranium and thorium at around 10 ppm concentrations, which release radioactive decay heat and raise the temperature of the rock. As a result, the temperature gradient in granite is higher than that in non-radioactive rock; it is thereby possible to reach high temperatures at lower depths and reduce drilling costs.

2.14.1 Heat extraction from an aquifer

In aquifer extraction heat is removed from the porous rock situated between layers of impermeable rock. Heat conduction from the impermeable rock above and below the aquifer

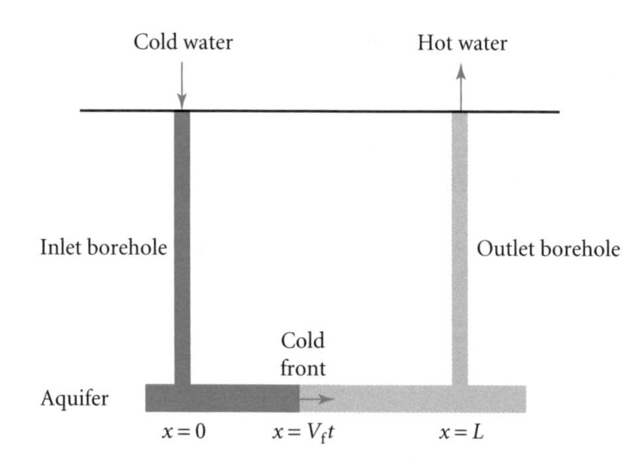

Fig. 2.15 Twin borehole system for heat extraction from an aquifer.

is usually negligible over the timescale for heat extraction from an aquifer. For simplicity we consider a simple one-dimensional fluid flow model for heat removal from an aquifer shown in Fig. 2.15. The actual flow field in the plane of the aquifer is two-dimensional but a one-dimensional model gives a useful first approximation and illustrates the main physical features.

Cold water is injected at the inlet borehole $x = 0$ and hot water is extracted at the outlet borehole $x = L$. Suppose that the aquifer is initially at temperature T_1 and the cold water at inlet is at temperature T_0. Hence the heat available per unit volume from the rock is $\rho_r c_r(T_1 - T_0)$. The power output of the system is the product of the heat per unit volume gained by the water, $\rho_w c_w(T_1 - T_0)$, and the volume flow rate Q, i.e.

$$P = \rho_w c_w (T_1 - T_0) Q. \tag{2.32}$$

As cold water flows through the aquifer it absorbs heat from the hot porous rock. A 'cold front' moves from the inlet borehole to the outlet borehole over the lifetime of the system. The speed of the cold front v_f is given by

$$v_f = \lambda v_w \tag{2.33}$$

where $\lambda = \rho_w c_w / [(1 - \varphi)\rho_r c_r]$ is a non-dimensional parameter (see Derivation 2.4) and $v_w = Q/A$ is the bulk velocity of the water in the aquifer. φ is the fraction by volume occupied by the pores, known as the **porosity**. The **lifetime** of the system is the time taken for the cold front to reach the outlet borehole and is given by

$$t_{life} = \frac{L}{v_f} = \frac{(1 - \varphi)\rho_r c_r A L}{\rho_w c_w Q}. \tag{2.34}$$

Hence, for a long lifetime it is desirable to have low porosity φ, large heat capacity per unit volume $\rho_r c_r$, large cross-sectional area of aquifer A, low volume flow rate Q, and large spacing L between the two boreholes. Since the total amount of thermal energy available from the system, $E = \rho_r c_r(T_1 - T_0)AL$, is fixed, the choice of Q and the lifetime of the system are determined by the economics of the system.

Finally, in order to obtain a given volume flow rate Q it is necessary to apply a pressure drop Δp between the boreholes. By **Darcy's law**, the volume flow rate Q through a slab of porous rock of cross-sectional area A and thickness L, is given by

$$Q = kA\frac{\Delta p}{L} \tag{2.35}$$

where k is a constant known as the **permeability**. Rearranging eqn (2.35), the pressure drop required for a given volume flow rate Q is given by

$$\Delta p = \frac{QL}{kA}. \tag{2.36}$$

Derivation 2.4 Velocity of the cold front

Suppose the 'cold front' moves a distance δx in a time interval δt (see Fig. 2.16). The volume of rock exposed is $\delta V_r = (1 - \varphi)A\delta x$, where A is the cross-sectional area of the cold front. The amount of heat removed from the element is given by $\delta h_r = \rho_r c_r (T_1 - T_0)(1 - \varphi)A\delta x$. The volume of water passing through the element in a small time interval δt is $\delta V_w = Q\delta t$ and the heat gained per unit volume by the water is $\delta h_w = \rho_w c_w (T_1 - T_0)Q\delta t$. By energy conservation, the heat lost by the rock is equal to the heat gained by the water, so that

$$\rho_r c_r (T_1 - T_o)(1 - \varphi)A\delta x = \rho_w c_w (T_1 - T_o)Q\delta t.$$

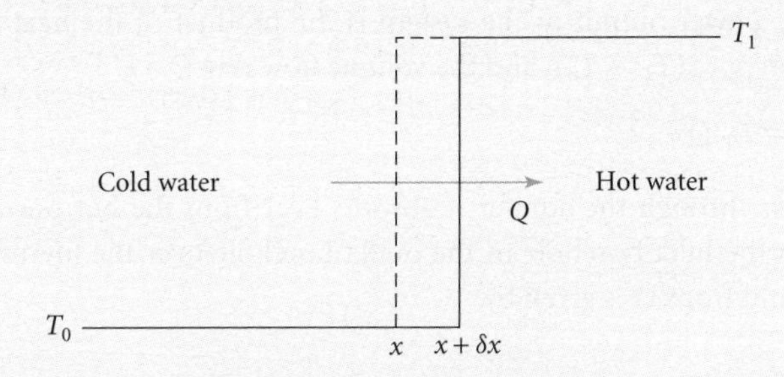

Fig. 2.16 Movement of cold front.

Rearranging, we obtain the velocity of the cold front as

$$v_f = \frac{dx}{dt} = \frac{\rho_w c_w}{(1 - \varphi)\rho_r c_r}\frac{Q}{A} = \lambda v_w.$$

EXAMPLE 2.8

A sandstone aquifer at 70°C is 20 m thick and 100 m wide. The density, specific heat, porosity and permeability are 2.3×10^3 kg m^{-3}, 1000 J kg^{-1}°C^{-1}, 0.02, and 2×10^{-9} m^3 kg^{-1}, respectively. Estimate the volume flow rate needed to generate a power output of 1 MW, the lifetime of the system, and the pressure drop required for a borehole separation of 1 km. (Assume the water at inlet is at 10°C, $\rho_w = 10^3$ kg m^{-3}, and $c_w = 4000$ J Kg^{-1}°C^{-1}.)

From eqns (2.32), (2.34), and (2.36) we have

$$Q = \frac{P}{\rho_w c_w (T_r - T_w)} = \frac{10^6}{(10^3)(4 \times 10^3)(70 - 10)} \approx 4 \times 10^{-3}\ \text{m}^3\,\text{s}^{-1}.$$

$$t_{\text{life}} = \frac{(1-\varphi)\rho_r c_r A}{\rho_w c_w Q} L \approx \frac{(1-0.02)(2.3 \times 10^3)(10^3)(20 \times 100)}{(10^3)(4 \times 10^3)(4 \times 10^{-3})} 10^3 \approx 2.8 \times 10^8 \text{s} \approx 9 \text{ years}.$$

$$\Delta p = \frac{QL}{kA} = \frac{4 \times 10^{-3} \times 10^3}{2 \times 10^{-9} \times 2 \times 10^3} = 10^6\ \text{N}\,\text{m}^{-2} = 10\ \text{bar}.$$

2.14.2 Heat extraction from hot dry rock

In hot dry rock systems, also called **enhanced geothermal systems** (EGS), water is pumped at high pressure through the narrow passages between natural fissures. In both aquifer and hot dry rock extraction it is usual to fracture the rocks around the inlet and outlet boreholes using controlled explosives in order to reduce the overall pressure drop. The water absorbs heat from the surface of the adjacent rock and the heat is transported by the fluid to the outlet borehole (Fig. 2.17). Heat is lost from the surrounding rock by unsteady heat conduction. The heat flux to the water decreases with time as the layer of cooled rock thickens. As in the case of an aquifer, a 'cold front' moves from the inlet to the outlet, but the temperature behind the front is more diffuse because some heat continues to be supplied from the walls behind the cold front.

For temperatures in the range 100–150°C it is possible to generate electricity using a binary cycle, in which a low boiling point fluid (e.g. ammonia) is heated via a heat exchanger and the vapour produced is used to drive a turbine.

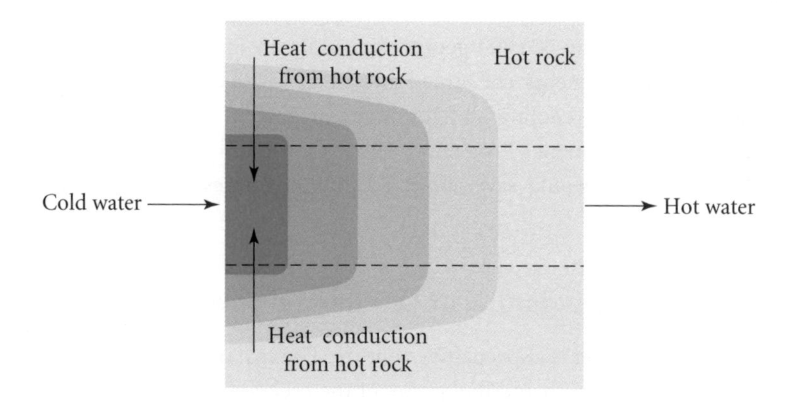

Fig. 2.17 Heat extraction from hot dry rock.

2.14.3 Geothermal heat pumps

An important use of geothermal energy is to take advantage of the fact that the temperature below ground is relatively constant, and is warmer than the surface air temperature during the winter and colder during the summer. This difference can be exploited to heat and air-condition buildings by using heat pumps. A heat pump uses the same principle of operation as a refrigerator.

In a refrigerator a fluid at high pressure and near room temperature expands and cools. The cold fluid is piped through the refrigerator where it cools the contents. The fluid is then compressed to high pressure during which process it heats up. The fluid is cooled back to near room temperature in a heat exchanger, an air-cooled coil on the back of the refrigerator, and the cycle is repeated.

We can approximate the process by a Carnot cycle where heat Q_2 is extracted at a temperature T_2, heat Q_1 is expelled at a hotter temperature T_1 (near room temperature), and work W is done by the compressor. $W = Q_1 - Q_2$ and for a Carnot cycle $Q_1/T_1 = Q_2/T_2$, so the heat extracted is given by

$$Q_2 = T_2 W/(T_1 - T_2).$$

Thus a factor $T_2/(T_1 - T_2)$ more heat is extracted than the work done by the compressor.

A heat pump works on the same principle. Heat Q is either extracted from or transferred to the building. The ratio Q/W for a heat pump is called the **coefficient of performance**, COP, and for an ideal heat pump heating a building the COP $= T_1/(T_1 - T_2)$, e.g. for $\Delta T \equiv (T_1 - T_2) = 31^\circ\text{C}$ and a ground temperature $T_2 = 6^\circ\text{C} = 279$ K, we have COP = 10. The actual COPs for heating or cooling units for buildings typically lie between 3 and 4.5. Even with these COPs, much less energy is required to heat or cool buildings compared with using direct heating units, or cooling units that exhaust heat to the hot air rather than to the colder ground.

Water or a water–antifreeze mixture is circulated through pipes that are buried in the ground at a depth of typically 100–400 ft. There they extract or transfer heat to the ground depending on whether the building is being heated or cooled. Inside the buildings ducts are used to transport the hot or cool air throughout the building. The heat pump can also be used to provide domestic hot water.

In the United States nearly 500 000 geothermal units are used for heating and cooling. Over 40% of the electrical energy consumed and nearly 40% of CO_2 emissions in the US are from space heating and cooling and for water heating in commercial and residential buildings. The US Environmental Protection Agency estimates that 100 000 domestic geothermal heat pumps units would save about 1.1 million tonnes of carbon over a 20 year period.

2.14.4 Economics and potential of geothermal power

Drilling for geothermal power is a speculative venture. The cost of drilling boreholes to depths of several kilometres is very high and the nature of the rock formation and rock temperature is unknown in advance. A hot dry rock project at Camborne in Cornwall (UK) was eventually

Table 2.1 Geothermal energy: electricity generation and direct use at end of 1999

Country	Electricity generation (MW_e)*	Direct use (MW_t)*
Iceland	170	1469
Indonesia	590	7
Italy	621	680
Japan	547	258
Mexico	750	164
New Zealand	410	308
Philippines	1863	1
USA	2228	5366
World total	7704	16 649

* MW_e, megawatts electrical; MW_t, megawatts thermal.

abandoned in 1989 because of unforeseen problems due to the rock formation. Nonetheless, geothermal power is a relatively harmless technology with little environmental impact. There is no carbon dioxide emission (except during the drilling process) but there is often a release of H_2S gas and the water already trapped in aquifers can contain toxic heavy metals. As a consequence it is normal to employ heat exchangers to keep the extracted water separate from that used for district heating or electricity generation.

Geothermal energy is a largely untapped source of carbon-free energy, and only a small fraction of the available resource (approximately 1 TW) is currently being exploited. It could provide electricity and heat in areas of the world without power and in regions where generation is by oil. Table 2.1 shows the exploitation for electricity production and for direct use at the end of 1999. It has been estimated that geothermal power could provide 5% of global electricity by 2020, but the future of geothermal energy technology depends on the economics of fossil fuels and on future policy towards tackling global warming. If tough environmental limits were to be imposed on the burning of fossil fuels then geothermal energy could make a useful contribution to the overall energy scene. Increased R&D, particularly on drilling technology, will help reduce costs and make geothermal power more economic. Combining heat pump technology with geothermal power is a fast growing technology at over 20% per annum.

SUMMARY

- For steady-state heat conduction, Fourier's law states that the heat flow is proportional to the temperature gradient, i.e. $Q = kA(T_1 - T_2)/d$.
- For unsteady heat conduction, isotherms move a characteristic distance of order $x \approx (\kappa t)^{1/2}$ in a time t.

- In forced convection the rate of heat transfer per unit area is of the form

  $$\frac{Q}{A} = Nu\frac{k(T_s - T_\infty)}{L}$$

  where Nu is a dimensionless parameter known as the **Nusselt number**.
- In radiation heat transfer the rate at which energy is radiated from a surface is given by the Stefan–Boltzmann law $P_e = \varepsilon\sigma T^4$.
- The absorption of infrared radiation by the atmosphere gives rise to the **greenhouse effect**, which raises the Earth's temperature by about 35°C.
- The first law of thermodynamics $Q - W = \Delta U$ expresses energy conservation taking thermal energy into account.
- According to the second law of thermodynamics, no system operating in a closed cycle can convert all the heat absorbed from a heat reservoir into the same amount of work.
- The maximum possible efficiency of a closed cycle operating between two heat reservoirs T_1 and T_2 (absolute temperatures) is the Carnot efficiency

  $$\eta_C = 1 - \frac{T_2}{T_1}.$$

- Enthalpy $h = u + pv$ is a useful thermodynamic quantity for describing: (a) heat transfer at constant pressure (e.g. in boilers and condensers), in which the change in enthalpy $h_2 - h_1$ is equal to the heat input Q; (b) adiabatic (i.e. no heat transfer) compression or expansion (e.g. in compressors and turbines), in which the net work done on the shaft is equal to the change in enthalpy.
- The T–s (temperature–entropy) diagram is useful for describing thermodynamic cycles in thermal power plants.
- A Carnot cycle is unsuitable for a real fluid such as water, mainly because the upper temperature is limited and water droplets in the two-phase water/steam mixture damage the blades in the compressor and in the turbine.
- A Rankine cycle overcomes the disadvantages of a Carnot cycle.
- Gas turbines use a Brayton (Joule) cycle and can operate at higher temperatures than steam turbines because the hot gases are prevented from making direct contact with the metal surfaces of the turbine blades.
- Higher efficiencies can be obtained using combined cycle gas turbines, which utilize the waste heat of a Brayton (Joule) cycle in a Rankine cycle of a steam power plant. Combined heat and power schemes are even more efficient in terms of total energy usage by providing district heating.
- Fossil fuel combustion produces carbon dioxide and other environmentally harmful gases.
- Fluidized beds produce smaller quantities of environmentally harmful gases (apart from CO_2) than conventional combustion chambers and are now commercially viable.
- Carbon sequestration is a potential means of storing carbon dioxide for long periods.
- Geothermal energy is an underutilized technology with considerable potential for supplying carbon-free energy, but carries commercial risks.

FURTHER READING

Andrews, J.G., Richardson, S.W., and White, A.A.L. (1981). Flushing geothermal heat from moderately permeable sediments. J. Geophys. Res. **86** (B10), 9439–50. Mathematical model of two-dimensional heat flow in an aquifer.

Blundell, S. and Blundell, K. (2006). *Concepts in thermal physics*, Oxford University Press, Oxford. A good textbook on thermal physics.

Carslaw, H.S. and Jaeger, J.C. (1959). *Conduction of heat in solids*, 2nd edn Clarendon Press, Oxford. Comprehensive treatment of mathematics of heat conduction.

Rogers, G. and Mayhew, Y. (1992). *Engineering thermodynamics*, 4th edn Longman, Harlow. Engineering approach to thermodynamics with many practical examples.

Rogers, G. and Mayhew, Y. (1995). *Thermodynamic and transport properties of fluids*, 5th edn. Blackwell, Oxford. Steam table data.

Zemansky, M. and Dittman, R. (1997). *Heat and thermodynamics*. McGraw-Hill, New York. Standard physics textbook on thermodynamics.

WEB LINKS

www.worldenergy.org Neutral source of data and current developments.

iga.igg.cnr.it/geo/geoenergy.php Very informative source on geothermal energy.

There are many websites that quote steam table data for specific thermal conditions.

LIST OF MAIN SYMBOLS

A	area
c	specific heat
c_p	specific heat at constant pressure
c_v	specific heat at constant volume
h	specific enthalpy
k	thermal conductivity
Nu	Nusselt number
P	power
p	pressure
Pr	Prandtl number
Q	heat or volume flow rate (i.e. in an aquifer)
r	pressure ratio
R	universal gas constant
Re	Reynolds number
s	specific entropy
t	time
T	temperature
U	internal energy
u	specific internal energy
v	specific volume
V	volume
W	work
x	steam quality
γ	ratio of specific heats
ε	emissivity
η	efficiency
κ	thermal diffusivity

μ	coefficient of dynamic viscosity	σ	Stefan–Boltzmann constant
ρ	density	φ	porosity

EXERCISES

2.1 Consider a proposal to extract heat directly from the Earth's core by drilling a cylindrical shaft of radius 1 m and depth 100 km through the Earth's mantle and filling it with copper. Estimate the power output assuming the temperature difference is 1000°C and the thermal conductivity is 3.5×10^2 W mK^{-1}.

2.2 Derive the form of eqn (2.5) using dimensional analysis. Estimate the characteristic timescale for heat to conduct through a heat shield of thickness 1 cm. ($\rho = 5 \times 10^3$ kg m^{-3}, $k \approx 10^{-1}$ W mK^{-1}, $c \approx 10^3$ J kg^{-1}°C^{-1}.)

2.3 Consider a composite wall consisting of two layers of material of thicknesses d_1 and d_2 and conductivities k_1 and k_2, respectively. If the temperature difference across the wall is ΔT, prove that the rate at which heat conducted in the steady state through the wall is of the form

$$Q = h_w \Delta T$$

and derive an expression for the heat transfer coefficient h_w.

2.4* Consider the following simple model to describe solar radiation incident on the Earth, in which the atmosphere is included. A fraction f of the incident solar flux is absorbed by the atmosphere and a fraction A is reflected, the rest being absorbed by the Earth. Assume that the only heat transfer to the atmosphere from the Earth is by radiation and that $f = 0.25$ and $A = 0.3$. Find the temperature T_E of the surface of the Earth for radiative equilibrium.

2.5 Using the empirical correlation for turbulent flow in a pipe

$$Nu = \frac{\frac{1}{2} f \, Re \, Pr}{1 + 2Re^{-1/8}(Pr - 1)}$$

where $f \approx 0.08Re^{-1/4}$, calculate the Reynolds number Re required to give a Nusselt number of $Nu = 100$ for a fluid with a Prandtl number $Pr = 3.5$.

2.6 Estimate the power radiated from a black body with a surface area 1 cm^2 at a temperature of 1000°C.

2.7 Discuss whether the caloric theory is consistent with: (a) Fourier's law of heat conduction; (b) the first law of thermodynamics.

2.8 An adiabatic compressor increases the pressure of water from 0.04 bar to 150 bar. Assuming that water is incompressible, calculate the work done per kg of water.

2.9 Using the steam table data in the table below, estimate the work done on an adiabatic turbine by 1 kg of superheated steam entering the turbine at a pressure of 200 bar and a temperature of 600°C and leaving the turbine at a pressure of 1 bar and a temperature of 100°C.

T (°C)	p (bar)	h (kJ kg^{-1})
100	1	2676
600	200	3537

2.10 Show that the internal energy u, enthalpy h, and entropy s of a two-phase mixture can be expressed in the forms

$$u = (1 - x)u_f + xu_g,$$
$$h = (1 - x)h_f + xh_g,$$
$$s = (1 - x)s_f + xs_g.$$

2.11 Consider a thermal power station operating in a Carnot cycle between an upper reservoir at $T = 400^\circ$C and $p = 180$ bar, and a lower reservoir at $T = 20^\circ$C and $p = 0.02$ bar. Using the steam table data in the table below, calculate: (a) the efficiency of the cycle; (b) the heat input in the boiler; (c) the heat output in the condenser; (d) the work done on the turbine.

		h (kJ kg^{-1})		s (kJ kg^{-1} K^{-1})	
T (°C)	p (bar)	h_f	h_g	s_f	s_g
20	0.02	84	2538	0.296	8.666
400	180	1732	2510	3.872	5.108

2.12 A power station operates in a Rankine cycle without reheat, consisting of (i) an adiabatic compressor, (ii) a three-stage boiler at 200 bar, (iii) an adiabatic turbine, (iv) a condenser at $p = 0.02$ bar, $T = 20^\circ$C. The maximum temperature of the boiler is 700°C. Using the steam table data in the table below, calculate: (a) the specific work done by the compressor; (b) the heat supplied per unit mass to the boiler; (c) the specific work obtained from the turbine; (d) the efficiency of the cycle.

			h (kJ kg^{-1})		s (kJ kg^{-1} K^{-1})	
	T (°C)	p (bar)	h_f	h_g	s_f	s_g
Water/steam mixture	20	0.02	84	2454	0.296	8.666
Dry steam	700	200		3806		6.796

2.13 Calculate the exhaust temperature and the efficiency η of an ideal gas turbine operating with a maximum temperature of 1600 K for a pressure ratio $r = 9$. (Assume $\gamma = 1.4$).

2.14 Consider an aquifer at an initial temperature of 100°C. The aquifer data are: thickness 50 m; width 100 m; density 3×10^3 kg m^{-3}; specific heat 1500 J kg^{-1}°C^{-1}; porosity 0.01; permeability 5×10^{-9} m^3 kg^{-1}. Estimate the required separation of the boreholes and the pressure drop in order to produce an output of 5 MW of heat for 10 years. Assume the water at inlet is at 5°C. ($\rho_w = 10^3$ kg m^{-3}; $c_w = 4000$ J kg^{-1}°C^{-1}.)

2.15 Derive an expression for the cost of drilling a borehole as a function of depth. Assume that the cost of drilling is independent of depth but the cost of lifting the rock material from the borehole to the surface increases linearly with depth.

2.16* Consider a two-dimensional aquifer of thickness d, with an inlet borehole at A ($x = -a$, $y = 0$) and an outlet borehole at B ($x = +a$, $y = 0$). Derive an expression for the velocity field at any point in the aquifer as a function of x and y, given that the radial velocity field due to each borehole is of the form

$$u_r = \pm \frac{Q}{2\pi r d}$$

where r is the radial distance from each borehole. Also plot the streamlines of fluid flow and derive an expression for the time taken for the cold front to move from A to B.

2.17* Two systems are in thermal contact. Energy (heat) will flow from one to the other until they are in thermal equilibrium, i.e. at the same temperature. Show that the condition that the entropy S of the combined system is a maximum is given by

$$\partial S_1/\partial U_1 = \partial S_2/\partial U_2 \equiv 1/T,$$

where T is the equilibrium temperature and $S = S_1 + S_2$, $U = U_1 + U_2$. Hence show that $\Delta Q = T\Delta S$ and deduce that in a Carnot cycle $Q_1/T_1 = Q_2/T_2$.

3 Essential fluid mechanics for energy conversion

List of Topics

- ☐ Density, pressure, viscosity
- ☐ Streamlines
- ☐ Mass continuity
- ☐ Bernoulli's equation
- ☐ Dynamics of a viscous fluid
- ☐ Laminar and turbulent flow
- ☐ Reynolds number
- ☐ Circulation
- ☐ Lift and drag forces on an aerofoil
- ☐ Euler's turbine equation

Introduction

In Chapter 2 we showed that it is possible to describe the energy transfer processes in boilers, condensers, and turbines using basic thermodynamic principles without a detailed knowledge of the fluid flow involved in each device. However, in order to understand other areas of energy conversion such as hydropower, wave power, and wind power, a basic knowledge of fluid mechanics is essential.

In this chapter we give a brief summary of the basic physical properties of fluids and derive the conservation laws of mass and energy for a fluid in which viscous effects are ignored (known as an **ideal** or **inviscid** fluid). We illustrate how the conservation laws can be applied to situations of practical interest to derive useful information about the flow. Also, we describe the effect of viscosity on the motion of a fluid around a body immersed in a fluid (e.g. a turbine blade) and show how the flow determines the forces acting on the body.

3.1 Basic physical properties of fluids

The bulk physical properties of a fluid are the following.

- **Density** (ρ). Mass per unit volume of a fluid. Unless otherwise stated, it is assumed throughout the book that the density of a fluid is constant (called **incompressible flow**; the variations in pressure arising from fluid motion (see Example 3.3) are small in comparison with atmospheric pressure). The units of density are kg m^{-3}. ($\rho_{\text{water}} \approx 10^3$ kg m^{-3} and $\rho_{\text{air}} \approx 1.2$ kg m^{-3} at $T = 20°$C and $p = 1$ atm.)
- **Pressure** (p). Force per unit area in a fluid. Pressure acts in the normal direction to the surface of a body immersed in a fluid. The unit of pressure is the pascal or N m^{-2}. (1 atmosphere $\approx$ 1 bar $= 10^5$ N m^{-2} $= 10^5$ Pa.)

- Viscosity. Force per unit area due to internal friction in a fluid arising from the relative motion between neighbouring elements in a fluid. Viscous forces act in the tangential direction to the surface of a body immersed in a flow (see Section 3.5).

3.2 Streamlines and stream-tubes

A useful concept to visualize a velocity field is to imagine a set of streamlines parallel to the direction of motion at all points in the fluid. Any element of mass in the fluid flows along a notional stream-tube bounded by neighbouring streamlines (Fig. 3.1). In practice, streamlines can be visualized by injecting small particles into the fluid. For example, smoke can be used in wind tunnels to investigate the flow over wings, turbine blades, cars, buildings, etc.

3.3 Mass continuity

One of the fundamental laws of fluid mechanics is conservation of mass (also known as mass continuity). Consider the flow along a stream-tube in a steady velocity field. Suppose that the speed of the fluid and the cross-sectional area of the stream-tube at any point are u and A, respectively. By definition, the direction of flow is parallel to the boundaries of the stream-tube, so the fluid is confined to the stream-tube and the mass flow per second is constant along the stream-tube. Hence

$$\rho u A = \text{const.} \tag{3.1}$$

Thus the speed of the fluid is inversely proportional to the cross-sectional area of the stream-tube (Example 3.1).

EXAMPLE 3.1

An incompressible ideal fluid flows at a speed of 1 m s^{-1} through a pipe of 1 m diameter in which a constriction of 0.1 m diameter has been inserted. What is the speed of the fluid inside the constriction?

Putting $\rho_1 = \rho_2$ and using eqn (3.1), we have $u_1 A_1 = u_2 A_2$, or

$$u_2 = u_1 \frac{A_1}{A_2} = (1\,\text{m}\,\text{s}^{-1}) \times \left(\frac{1}{0.1}\right)^2 = 100\,\text{m}\,\text{s}^{-1}.$$

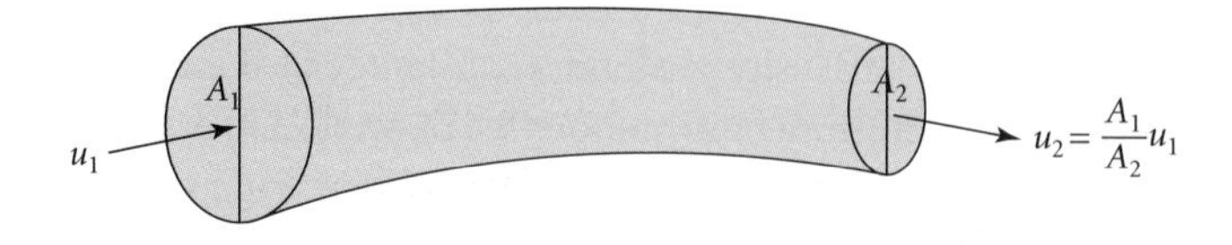

Fig. 3.1 Stream-tube.

3.4 Energy conservation in an ideal fluid: Bernoulli's equation

In many practical situations, viscous forces are much smaller than those due to gravity and pressure gradients over large regions of the flow field. We can then ignore viscosity to a good approximation and derive an equation for energy conservation in a fluid known as Bernoulli's equation (or Bernoulli's theorem). For steady flow Bernoulli's equation is of the form

$$\frac{p}{\rho} + gz + \tfrac{1}{2}u^2 = \text{const.} \tag{3.2}$$

(For a proof of Bernoulli's equation see Derivation 3.1.)

For a stationary fluid, $u = 0$ everywhere in the fluid, and eqn (3.2) reduces to

$$\frac{p}{\rho} + gz = \text{const.} \tag{3.3}$$

Equation (3.3) is the equation for hydrostatic pressure. It shows that the fluid at a given depth z is all at the same pressure p (see Example 3.2).

EXAMPLE 3.2

The atmospheric pressure on the surface of a lake is $10^5\ \text{N m}^{-2}$. Assuming the water is stationary, what is the pressure at a depth of 10 m? (Assume $\rho_{\text{water}} = 10^3\ \text{kg m}^{-3}$ and $g = 10\ \text{m s}^{-2}$.)

From eqn (3.3), we have $\frac{p_1}{\rho} + gz_1 = \frac{p_2}{\rho} + gz_2$. Putting $p_1 = 10^5\ \text{N m}^{-2}$ at $z_1 = 0$, and $z_2 = -10$ m, we have $p_2 = p_1 - \rho g(z_2 - z_1) = 10^5 - (10^3)(10)(-10) = 2 \times 10^5\ \text{N m}^{-2}$.

The significance of Bernoulli's equation is that it shows that the pressure in a moving fluid decreases as the speed increases. The practical importance of this effect is illustrated in Examples 3.3–3.5.

EXAMPLE 3.3

Assuming the pressure of stationary air is $10^5\ \text{N m}^{-2}$, calculate the percentage change due to a wind of $20\ \text{m s}^{-1}$. (Assume $\rho_{\text{air}} \approx 1.2\ \text{kg m}^{-3}$.)

From eqn (3.2) we have $\frac{p_1}{\rho} + \frac{1}{2}u_1^2 = \frac{p_2}{\rho} + \frac{1}{2}u_2^2$. The change in pressure is given by $p_2 - p_1 = \frac{1}{2}\rho(u_1^2 - u_2^2) = -\frac{1}{2}(1.2)(20)^2 = -2.4 \times 10^2\ \text{N m}^{-2}$. Hence, the percentage change in pressure is $-\frac{2.4\times10^2}{10^5} \times 100 \approx -0.24\%$.

Derivation 3.1 Bernoulli's equation for steady flow

Consider the steady flow of an ideal fluid in the control volume shown in Fig. 3.2.

The height, cross-sectional area, speed, and pressure at any point are denoted by z, A, u, and p, respectively. The increase in gravitational potential energy of a mass δm of fluid between z_1 and z_2 is $\delta m g(z_2 - z_1)$. In a small time interval δt the mass of fluid entering the control volume at P_1 is $\delta m = \rho u_1 A_1 \delta t$ and the mass exiting at P_2 is $\delta m = \rho u_2 A_2 \delta t$.

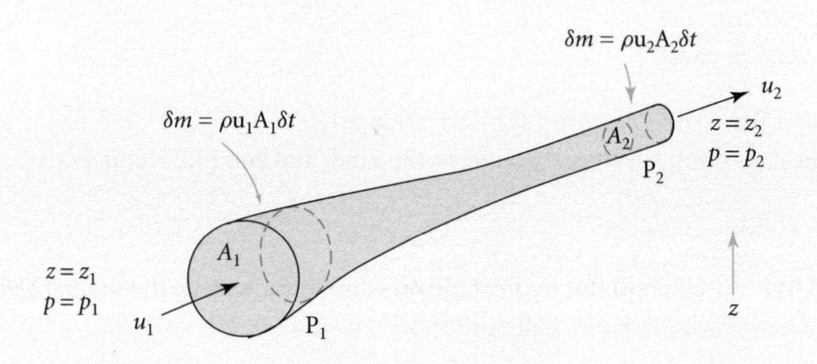

Fig. 3.2 Control volume for Bernoulli's equation.

In order for the fluid to enter the control volume it has to do work to overcome the pressure p_1 exerted by the fluid. The work done in pushing the elemental mass δm a small distance $\delta s_1 = u_1 \delta t$ at P_1 is $\delta W_1 = p_1 A_1 \delta s_1 = p_1 A_1 u_1 \delta t$. Similarly, the work done in pushing the elemental mass out of the control volume at P_2 is $\delta W_2 = -p_2 A_2 u_2 \delta t$ (note change of sign). The net work done is $\delta W_1 + \delta W_2 = p_1 A_1 u_1 \delta t - p_2 A_2 u_2 \delta t$. By energy conservation, this is equal to the increase in potential energy plus the increase in kinetic energy, so that

$$p_1 A_1 u_1 \delta t - p_2 A_2 u_2 \delta t = \delta m g(z_2 - z_1) + \tfrac{1}{2}\delta m(u_2^2 - u_1^2).$$

Putting $\delta m = \rho u_1 A_1 \delta t = \rho u_2 A_2 \delta t$ and tidying-up, we have

$$\frac{p_1}{\rho} + g z_1 + \tfrac{1}{2}u_1^2 = \frac{p_2}{\rho} + g z_2 + \tfrac{1}{2}u_2^2.$$

Finally, since points P_1 and P_2 are arbitrary it follows that

$$\frac{p}{\rho} + g z + \tfrac{1}{2}u^2 = \text{const.}$$

everywhere along the stream-tube.

EXAMPLE 3.4

A **Pitot tube** is a device for measuring the velocity in a fluid. Essentially, it consists of two tubes, (a) and (b). Each tube has one end open to the fluid and one end connected to a pressure gauge. Tube (a) has the open end facing the flow and the tube (b) has the open end normal to the flow (Fig. 3.3). Derive an expression for the velocity of the fluid in terms of the difference in pressure between the gauges.

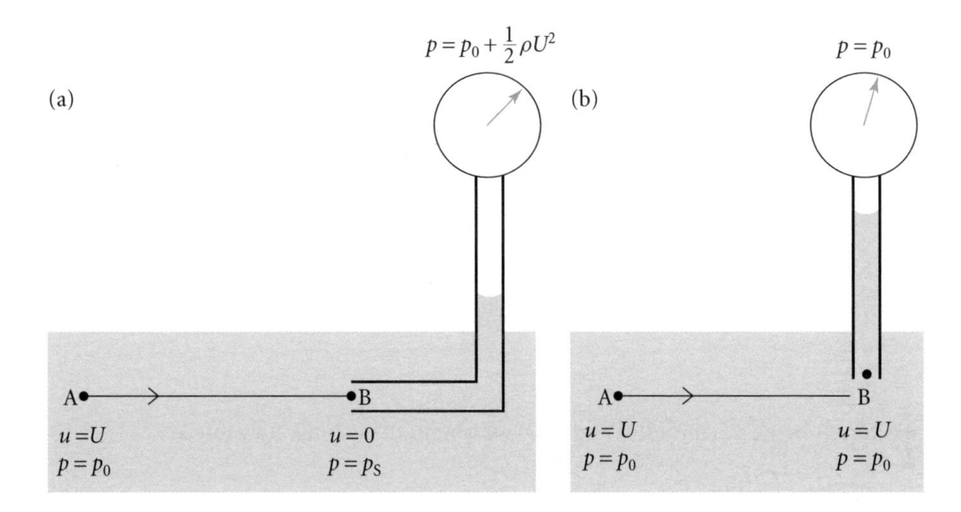

Fig. 3.3 Pitot tube. (a) Open end facing incident flow; (b) open end normal to incident flow.

Consider the fluid moving along the stream-line AB. In case (a), the fluid slows down as it approaches the stagnation point B. Putting $u = U, p = p_0$ at A and $u = 0$, $p = p_S$ at B, eqn (3.2) becomes

$$\frac{p_0}{\rho} + \tfrac{1}{2}U^2 = \frac{p_S}{\rho}. \tag{3.4}$$

Re-arranging eqn (3.4) yields the velocity in the undisturbed fluid as

$$U = \left[\frac{2(p_S - p_0)}{\rho}\right]^{1/2}. \tag{3.5}$$

In Example 3.4 and Fig. 3.3, p_S is measured by tube (a) and p_0 measured by tube (b). Note that p_S is larger than p_0 by an amount $p_S - p_0 = \frac{1}{2}\rho U^2$. The quantities $\frac{1}{2}\rho U^2, p_0$, and $p_S = p_0 + \frac{1}{2}\rho U^2$ are called the **dynamic pressure**, **static pressure** and **total pressure**, respectively.

EXAMPLE 3.5

In a **Venturi meter** an ideal fluid flows with a volume flow rate Q and pressure p_1 through a horizontal pipe of cross-sectional area A_1 (Fig. 3.4). A constriction of cross-sectional area A_2 is inserted in the pipe and the pressure is p_2 inside the constriction. Derive an expression for the volume flow rate Q in terms of p_1, p_2, A_1, and A_2.

From Bernoulli's equation (3.2) we have

$$\frac{p_1}{\rho} + \tfrac{1}{2}u_1^2 = \frac{p_2}{\rho} + \tfrac{1}{2}u_2^2. \tag{3.6}$$

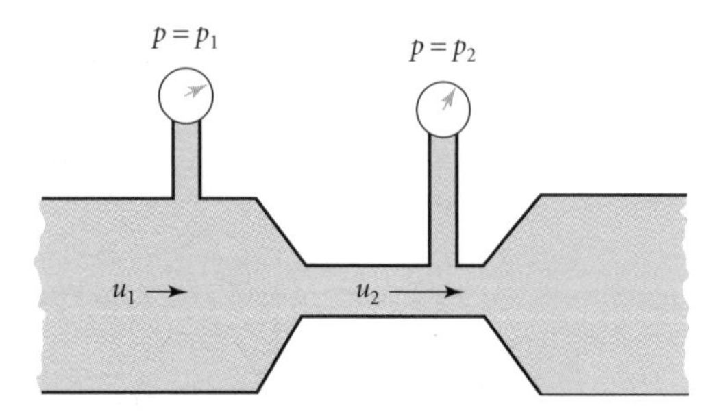

Fig. 3.4 Venturi meter.

Also, by mass continuity eqn (3.1), $\rho u_1 A_1 = \rho u_2 A_2$ or

$$u_2 = u_1 \frac{A_1}{A_2}. \tag{3.7}$$

Eliminating u_2 between eqns (3.6) and (3.7) we obtain the volume flow rate as

$$Q = A_1 u_1 = A \left[\frac{2(p_1 - p_2)}{\rho} \right]^{1/2} \tag{3.8}$$

where $A = A_1 A_2 (A_1^2 - A_2^2)^{-1/2}$.

3.5 Dynamics of a viscous fluid

In general, the motion of a viscous fluid is more complicated than that of an inviscid fluid. A simple case is that of laminar viscous flow between two parallel plates. Consider two parallel plates separated by a small distance d, with one plate moving at constant velocity U and the other plate at rest (Fig. 3.5).

There is no relative velocity between the fluid next to the plate surfaces and the plates, due to strong forces of attraction between the fluid and the surface of the plates. The velocity profile in the fluid is given by

$$u(y) = U \frac{y}{d} \qquad (0 \leq y \leq d). \tag{3.9}$$

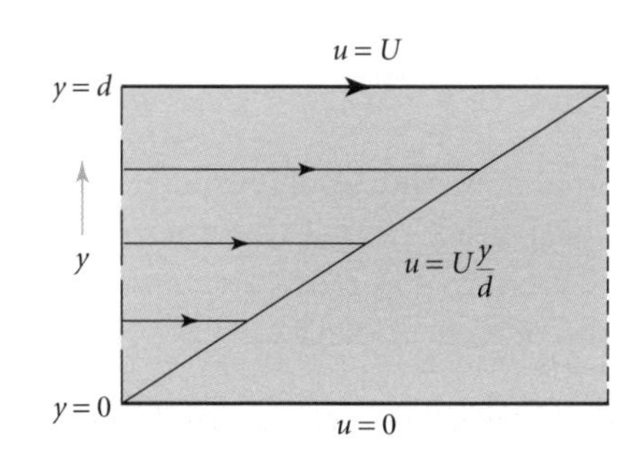

Fig. 3.5 Laminar viscous flow between parallel plates in relative motion.

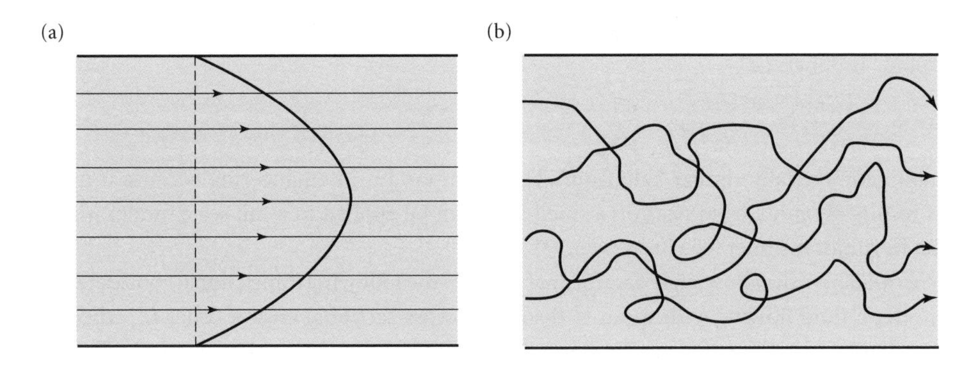

Fig. 3.6 (a) Laminar flow in a pipe. (b) Turbulent flow in a pipe.

The viscous shear force per unit area in the fluid is proportional to the velocity gradient, i.e.

$$\frac{F}{A} = -\mu\frac{du}{dy} = -\mu\frac{U}{d} \tag{3.10}$$

where the coefficient μ is known as the **coefficient of dynamic viscosity**.

In the above example, the flow arises from the viscous shear force due to the relative motion of the two plates. For viscous fluid flow along a pipe, a force needs to be applied in the axial direction (such as gravity or a pressure gradient) to overcome the viscous drag force.

A viscous fluid can exhibit two different kinds of flow regime: **laminar flow** and **turbulent flow**. In laminar flow (Fig.3.6(a)), the fluid slides along distinct stream-tubes and tends to be quite stable, but in turbulent flow the motion is disorderly and unstable (Fig. 3.6(b)).

The particular flow regime that exists in any given situation depends on the ratio of the inertial force to the viscous force. The typical magnitude of this ratio is given by the **Reynolds number**, defined as

$$Re = \frac{\rho UL}{\mu} = \frac{UL}{\nu} \tag{3.11}$$

where U, L, and $\nu = \mu/\rho$ are the characteristic speed, the characteristic length, and the kinematic viscosity (Example 3.6). *Re* is named after Osborne Reynolds, who conducted pioneering experiments on laminar and turbulent flow. He discovered that flows at small *Re* are predominantly laminar, whereas flows at large *Re* contain regions of turbulence.

EXAMPLE 3.6

Estimate the Reynolds number for: (a) treacle flowing over a plate ($\nu = 10^{-1}$ m^2 s^{-1}, $u = 10^{-2}$ m s^{-1}, $L = 0.1$ m); (b) air flowing around a jet aircraft ($\nu = 1.5 \times 10^{-5}$m^2 s^{-1}, $u = 3 \times 10^2$ m s^{-1}, $L = 10$ m).

The Reynolds numbers in each case are: (a) $Re = \frac{UL}{\nu} \approx \frac{10^{-2}\times 10^{-1}}{10^{-1}} << 1$; (b) $Re = \frac{UL}{\nu} \approx \frac{3\times 10^2 \times 10^1}{1.5\times 10^{-5}} \gg 1$. Hence in case (a) the flow is predominantly laminar but in case (b) the flow contains regions of turbulence.

Another important aspect of the Reynolds number is that two different flows with the same Reynolds number, i.e.

$$Re = \frac{\rho_1 U_1 L_1}{\mu_1} = \frac{\rho_2 U_2 L_2}{\mu_2},$$

exhibit geometrically similar behaviour. This is important in engineering because it implies that results obtained from tests on a small scale can be applied to a full scale model with the same Reynolds number (see Exercise 3.10).

We can derive the above algebraic form of Re from the following dimensional considerations. Consider a fluid flowing with speed U through a cross-sectional area of order L^2, where L is some characteristic length (e.g. the radius in the case of flow around a cylinder). The mass flowing per second is $\sim \rho U L^2$, so the inertial force (i.e. the rate of change of momentum) is $\sim \rho U L^2 \times U = \rho U^2 L^2$. Also, from eqn (3.10) the viscous force $\sim \mu A U/L = \mu U L$. Hence the ratio of the inertial force to the viscous force is of order

$$Re = \frac{[\text{inertial force}]}{[\text{viscous force}]} \approx \frac{\rho U^2 L^2}{\mu U L} = \frac{\rho U L}{\mu} = \frac{UL}{\nu}. \tag{3.12}$$

Figure 3.7 shows the flow around a cylinder for (a) an inviscid fluid and (b) a viscous fluid. For inviscid flow the velocity fields in the upstream and downstream regions are symmetrical. Hence, the corresponding pressure distribution is symmetrical, and it follows that the net force exerted by the fluid on the cylinder is zero. This startling result is in contradiction to common experience and is an example of d'Alembert's paradox.

For a body immersed in a viscous fluid, the component of velocity tangential to the surface of the fluid is zero at all points on the surface of a body. At large Reynolds numbers ($Re \gg 1$), the viscous force is negligible in the bulk of the fluid but is very significant in a viscous boundary layer close to the surface of the body. Rotational components of flow known as vorticity are generated within the boundary layer. At a certain point (known as the separation point) the boundary layer becomes detached from the surface and vorticity is discharged into the body of the fluid. The vorticity is transported downstream of the cylinder in the wake. Thus, the pressure distributions on the upstream side and the downstream side of the cylinder are not symmetrical in the case of a viscous fluid. Hence, the cylinder experiences a net force in the direction of motion, known as the drag force. There is no component of force normal

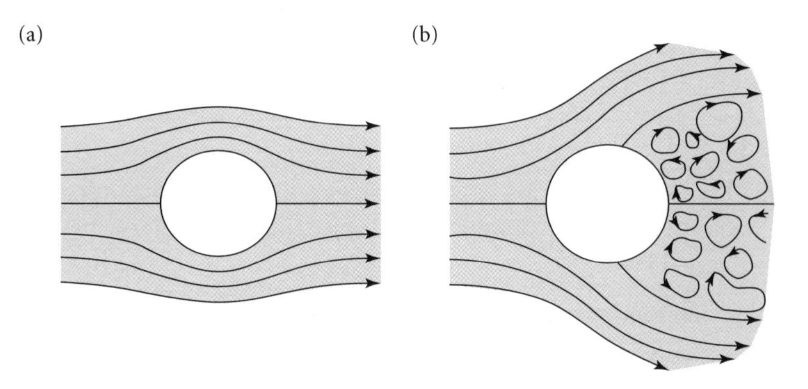

Fig. 3.7 Flow around a cylinder for (a) an inviscid fluid (b) a viscous fluid.

to the direction of the flow past the cylinder, due to the symmetry of the velocity field above and below the cylinder. However, in the case of a spinning cylinder, a force (called the lift force) arises at right angles to the direction of flow (see Section 3.6).

We can derive the algebraic form of the lift and drag forces by the following dimensional arguments. By physical intuition, it is reasonable to assume that the lift force $\mathcal{L}$ depends only on the density ρ, the speed U, and the cross-section of the body in the incident flow. Thus we assume the lift force is an algebraic function of the form

$$\mathcal{L} = \tfrac{1}{2} C_L \rho^a U^b A^c \tag{3.13}$$

where C_L is a dimensionless constant (known as the lift coefficient) and a, b, and c are unknown indices to be determined. Substituting for the physical dimensions of each quantity, we have

$$M^1L^1T^{-2} = (ML^{-3})^a(LT^{-1})^b(L^2)^c = M^aL^{-3a+b+2c}T^{-b}. \tag{3.14}$$

Equating the indices of like quantities on each side of the equation, we obtain $a = 1$, $b = 2$, and $c = 1$. Thus the lift force is of the form

$$\mathcal{L} = \tfrac{1}{2} C_L \rho U^2 A \tag{3.15}$$

and likewise the drag force $\mathcal{D}$ is of similar algebraic form

$$\mathcal{D} = \tfrac{1}{2} C_D \rho U^2 A \tag{3.16}$$

where C_D is known as the drag coefficient.

Birds are able to control the lift and drag forces by changing the shape of their wings, the ruffle of their feathers, and the angle of attack of their wings relative to the incident flow. Man has copied nature in designing the shape of an aerofoil for aircraft wings and turbine blades. For small angles of attack, the pressure distribution on the upper surface of an aerofoil is significantly lower than that on the lower surface, resulting in a net lift force on the aerofoil (Fig. 3.8).

Figure 3.9 shows the variation of the lift and drag coefficients C_L and C_D with angle of attack for a typical aerofoil. (The drag coefficient has been enlarged by a factor of 5.) The lift and drag coefficients C_L and C_D cannot be determined by dimensional analysis because they depend on non-dimensional parameters such as the Reynolds number Re, the shape of the body, and the surface roughness. In practice C_L and C_D are obtained from wind tunnel tests on model shapes or from numerical models of the flow and turbulence.

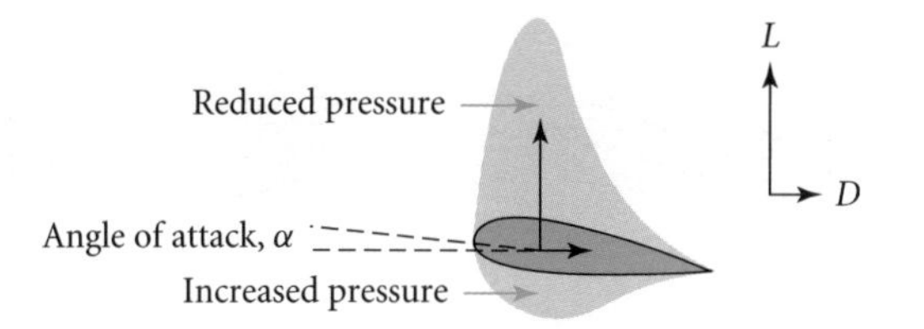

Fig. 3.8 Pressure distribution over aerofoil for small angle of attack.

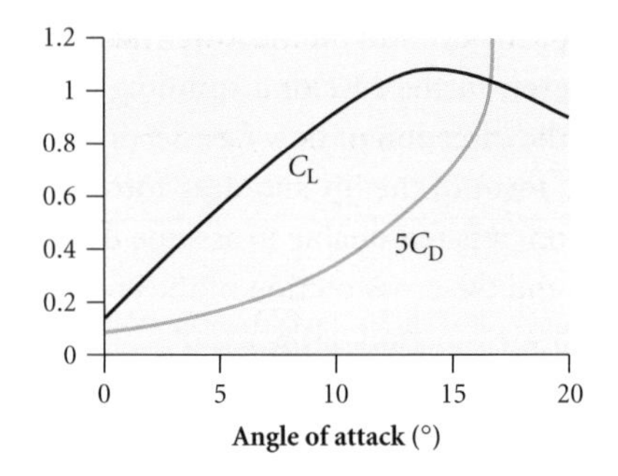

Fig. 3.9 Lift and drag coefficients.

3.6 Lift and circulation

It is possible to explain lift using inviscid fluid dynamics by introducing the concept of circulation. In order to understand circulation, it is helpful to begin by considering why a spinning ball swerves sideways as it flies through the air, an effect well known to golfers! Spinning creates an imbalance in the pressure on either side of the ball, and generates a net force at right angles to the direction of motion. This is known as the Magnus effect and is illustrated in Fig. 3.10.

We consider an inviscid fluid that is both passing over and rotating around a stationary cylinder, rather than flowing over a rotating body, since it illustrates the same effect and is more like the flow over a stationary aerofoil. Figure 3.10(a) shows the flow of a uniform stream incident on a cylinder. The velocity profile is symmetrical on the upper and lower surfaces, so the resulting pressure distribution is also symmetrical and there is no net sideways force on the cylinder. Figure 3.10(b) shows an inviscid fluid rotating around a cylinder, with a circumferential velocity u_θ that varies inversely with distance r from the centre of the cylinder, i.e.

$$u_\theta = \frac{\Gamma}{2\pi r} \tag{3.17}$$

where Γ is a constant called the circulation. Superposing the velocity profiles shown in Fig. 3.10(a) and Fig. 3.10(b) produces a velocity field in which the fluid moves faster on the upper side than on the lower side (Fig. 3.10(c)). It follows from Bernoulli's equation (3.2) that

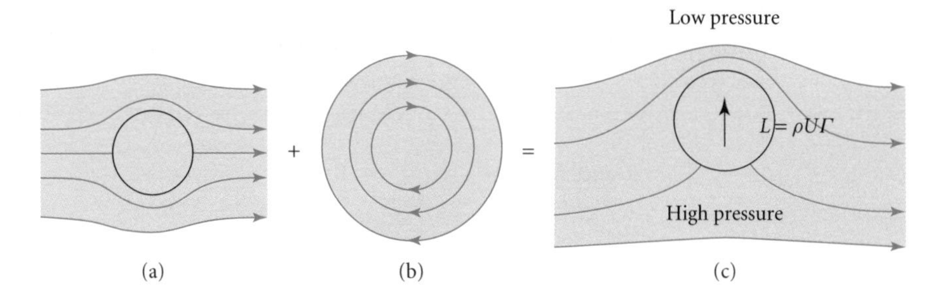

Fig. 3.10 Magnus effect. (a) Flow of a uniform stream around a cylinder; (b) rotating flow around a cylinder; (c) superposition of (a) and (b).

the pressure is lower on the upper side than on the lower side, so that a net force is exerted on the cylinder at right angles to the incident stream.

The lift force on a body (per unit length in the z-direction) is given by the **Kutta–Joukowski lift theorem**

$$L = \rho U \Gamma. \tag{3.18}$$

This expression is in fair agreement with experimental observation, despite the fact that inviscid theory allows the fluid to slip over the surface of the body. However, inviscid theory cannot account for the drag force on a body and it is necessary to consider the effects of viscosity, making the analysis more difficult. A derivation of eqn (3.18) for a cylinder is given in Derivation 3.2. For the equivalent expression for the lift on an aerofoil see Derivation 3.3.

Derivation 3.2 Magnus effect

Figure 3.10(c) shows the superposition of a uniform stream and a rotational flow around a cylinder. The velocity is enhanced on the upper side of the cylinder and reduced on the lower side. The circumferential component of velocity on the surface of the cylinder is given by (see Exercise 3.13)

$$u_\theta = -2U\sin\theta - \frac{\Gamma}{2\pi a} = -U(2\sin\theta + B) \tag{3.19}$$

where U is the speed of the uniform stream, Γ is a constant (i.e. the circulation), a is the radius of the cylinder, and B is a non-dimensional parameter given by $B = \Gamma/2\pi Ua$. For $B < 2$, there are two stagnation points on the surface of the cylinder, obtained by putting $u_\theta = 0$ in (3.19), i.e. $\sin\theta = -\frac{1}{2}B$. The fluid is stationary at these points, so from Bernoulli's theorem (3.2) the pressure is a maximum. The circulation Γ is given by the angle subtended by the stagnation points, i.e. $\Gamma = -4\pi Ua\sin\theta$.

From Bernoulli's theorem (3.2), the pressure on the surface is given by

$$p = k - \tfrac{1}{2}\rho u_\theta^2 \tag{3.20}$$

where k is a constant. Substituting for u_θ from eqn (3.19) we have

$$p = k - \tfrac{1}{2}\rho U^2(4\sin^2\theta + 4B\sin\theta + B^2) = k - \tfrac{1}{2}\rho U^2(4\sin^2\theta + B^2) - 2\rho U^2 B\sin\theta. \tag{3.21}$$

The first two terms on the right-hand side of (3.21) are symmetrical on the upper and lower sides of the cylinder. The last term $-2\rho U^2 B\sin\theta$ is negative for $0 \le \theta \le \pi$ (upper side) and positive for $\pi \le \theta \le 2\pi$ (lower side). Hence there is a net lift force in the y-direction, given by

$$\begin{aligned} F_y &= -\int_0^{2\pi} pa\sin\theta\,\mathrm{d}\theta = 2\rho U^2 Ba\int_0^{2\pi}\sin^2\theta\,\mathrm{d}\theta \\ &= \frac{\rho U\Gamma}{2\pi}\int_0^{2\pi}(1-\cos 2\theta)\,\mathrm{d}\theta = \rho U\Gamma. \end{aligned} \tag{3.22}$$

Derivation 3.3 Kutta–Joukowski theorem for lift on a turbine blade

It turns out that it is easier to analyse a cascade of turbine blades than a single blade, because the symmetry of the cascade makes it possible to simplify the calculation.

Consider the control volume $A_1B_1B_2A_2A_1$ enclosing a single blade, as shown in Fig. 3.11. The stream-lines ψ_1 and ψ_2 separate the flow passing over neighbouring blades and are a fixed vertical distance apart, h. The vertical planes A_1A_2 and B_1B_2 are chosen to be in regions of roughly uniform flow. Since there is no mass flow across a streamline it follows that $u_2 = U$. There is also no momentum transfer across the streamlines ψ_1 and ψ_2 and, by symmetry, there is no pressure gradient in the vertical direction on the streamlines ψ_1 and ψ_2. Hence the rate of change in momentum of the fluid in the vertical direction is equal to the vertical component of force on the blade.

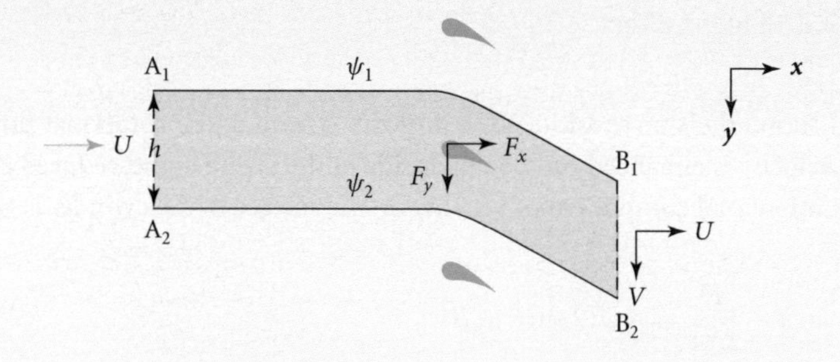

Fig. 3.11 Flow through a cascade of turbine blades.

The mass flow per second crossing B_1B_2 is ρUh, so the vertical component of force on the blade is given by

$$F_y = \rho Uhv. \tag{3.23}$$

The circulation Γ is given by the closed integral $\Gamma = \oint_C \boldsymbol{u} \cdot d\boldsymbol{s}$, where C is taken to be the contour $C = A_1B_1B_2A_2A_1$. Noting that the contribution along A_1A_2 is zero, and the contributions along the streamlines ψ_1 and ψ_2 cancel each other, the only non-zero contribution is along B_2B_1, given by

$$\Gamma = hv. \tag{3.24}$$

Hence

$$F_y = \rho U\Gamma. \tag{3.25}$$

Alternatively, putting $v = U\sin\alpha$ (where α is the angle of deflection of the incident stream), $\Gamma = hU\sin\alpha$ and

$$F_y = \rho U^2 h\sin\alpha. \tag{3.26}$$

According to inviscid flow theory for flow over an aerofoil with a sharp trailing edge (see, for example, Acheson 1990), there is only one value of the circulation such that the velocity is finite at all points on the surface of the aerofoil, given by

$$\Gamma = \pi U w \sin \alpha \qquad (3.27)$$

where w is the width (or chord) of the aerofoil. (Note the mathematical similarity between the expression for the circulation around a spinning cylinder, $\Gamma = -4\pi Ua \sin \theta$, and eqn (3.27) on putting $w = 2a$.)

Combining eqns (3.18) and (3.27) we can write the lift force on a single aerofoil of length l in the form

$$\mathcal{L} = \rho U l \Gamma = \pi \rho U^2 l w \sin \alpha. \qquad (3.28)$$

Equation (3.28) is of the same algebraic form as eqn (3.15), obtained from simple dimensional analysis. The essential difference is that eqn (3.28) gives an explicit expression for the dimensionless lift coefficient, i.e. $C_L = (2\pi lw/A)\sin \alpha = 2\pi \sin \alpha$. This predicts that C_L equals unity at an angle of 9 degrees, which is close to what is observed (see Fig. 3.9).

The physical justification for the concept of circulation arises from the way that vorticity is generated when an aerofoil starts to move from rest. In the early stages vorticity is generated around the leading edge, which is swept towards the trailing edge and then shed downstream in the wake, leaving an equal and opposite rotational flow around the aerofoil. This is why aircraft have to wait on the runway to allow time for the shed vortices generated by the previous aircraft to disperse. The effect of viscosity is therefore to produce circulation and the lift force on the aerofoil then follows from inviscid theory.

3.7 Euler's turbine equation

In most types of power generation the kinetic energy of a moving fluid is converted by a turbine into the rotational motion of a shaft. The turbine blades deflect the fluid and the rate of change of angular momentum of the fluid is equal to the net torque on the shaft.

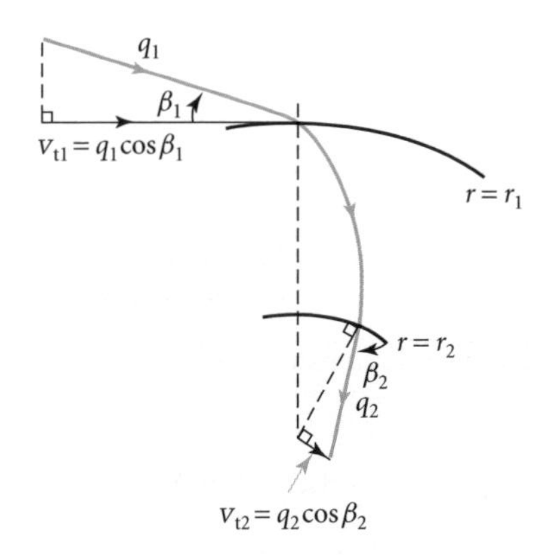

Fig. 3.12 Velocity diagram for Euler's turbine equation.

A fluid of density ρ flowing through the turbine with a volume flow rate Q has a mass flow per second given by ρQ. Suppose that the fluid enters at a radius r_1 with a circumferential velocity v_{t1} and exits at a radius r_2 with a circumferential velocity v_{t2} (Fig. 3.12).

The torque T exerted on the turbine is equal to the rate of change of angular momentum. Thus

$$T = \rho Q(r_1 v_{t1} - r_2 v_{t2}). \tag{3.29}$$

The power delivered to a turbine rotating with angular velocity ω is given by

$$P = \omega T. \tag{3.30}$$

Substituting for T from eqn (3.29) in eqn (3.30) yields the power as

$$P = \omega \rho Q(r_1 v_{t1} - r_2 v_{t2}). \tag{3.31}$$

Writing the tangential velocity in the form $v_t = q \cos\beta$, where q is the total velocity of the fluid and β is the angle between the direction of motion of the fluid and the tangent to the circle, eqn (3.31) becomes

$$P = \omega \rho Q(r_1 q_1 \cos\beta_1 - r_2 q_2 \cos\beta_2). \tag{3.32}$$

Equation (3.32) is known as Euler's turbine equation. The importance of Euler's turbine equation is that the details of the flow inside the turbine are irrelevant. All that matters is the total change in the angular momentum of the fluid between the inlet and the outlet. The maximum torque is achieved when the fluid flows out in the radial direction, i.e. when $\cos\beta_2 = 0$. Equation (3.32) then reduces to

$$P = \omega \rho Q r_1 q_1 \cos\beta_1. \tag{3.33}$$

EXAMPLE 3.7

Estimate the maximum power output of a water turbine operating at 50 Hz with $\rho = 10^3\ \mathrm{kg\,m^{-3}}$, $Q = 1\ \mathrm{m^3\,s^{-1}}$, $r_1 = 1$ m, $q_1 = 1\ \mathrm{m\,s^{-1}}$, $\cos\beta_1 = 0.5$.

Substituting in eqn (3.33) yields the power output as $P = (2\pi \times 50)(10^3)(0.5) \approx 160$ kW.

SUMMARY

- The basic physical properties that describe a fluid are density ρ (mass per unit volume), pressure p (normal force per unit area), and viscosity (shear force due to relative motion).
- Mass flow is conserved along a stream-tube, i.e.

  $$\rho u A = \text{const.}$$

- The equation for conservation of energy for steady fluid flow of an inviscid fluid is given by Bernoulli's equation

  $$\frac{p}{\rho} + gz + \tfrac{1}{2}u^2 = \text{const.}$$

- The viscous shear force per unit area is proportional to the velocity gradient. For one-dimensional flow

$$\frac{F}{A} = -\mu \frac{du}{dy}$$

where μ is the coefficient of dynamic viscosity.

- The Reynolds number,

$$Re = \frac{\rho UL}{\mu},$$

is a useful parameter for distinguishing between laminar flow and turbulent flow. Flows at $Re \ll 10^3$ are essentially laminar; flows at $Re \gg 10^3$ exhibit turbulence.

- Two flows with the same Reynolds number are dynamically similar.
- Dimensional analysis gives an algebraic expression for the forces acting on a body in a moving fluid. The drag force and the lift force have the same form

$$F = \tfrac{1}{2}C\rho u^2 A. \qquad (C = C_L \text{ for lift and } C = C_D \text{ for drag.})$$

C_L and C_D depend only on non-dimensional parameters (Reynolds number, shape of the body, the angle of attack, surface roughness).

- For an aerofoil inclined at a small angle of attack to the incident flow, the pressure on the upper surface is lower than that on the lower surface, creating an upwards lift force.
- The existence of lift can be explained using inviscid theory by introducing the concept of circulation.
- The lift force per unit length on a body is given to a good approximation by the Kutta–Joukowski theorem

$$L = \rho U\Gamma$$

where Γ is the circulation. For an aerofoil, the circulation is given by

$$\Gamma = \pi Uw \sin\alpha.$$

- Euler's turbine equation relates the power output P from a turbine to the rate of change of angular momentum of the fluid between the inlet and outlet of the turbine. The details of the flow inside the turbine are irrelevant.

FURTHER READING

Acheson, D.J. (1990). *Elementary fluid dynamics.* Clarendon Press, Oxford. Mathematical introduction to fluid mechanics.

Massey, B.S., Ward-Smith, J., and Ward-Smith, A.J. (2005). *Mechanics of fluids*, 8th edn. Taylor and Francis, London. Fluid mechanics for engineers.

Douglas, J.F., Gasiorek, J.M., and Swaffield, J.A. (2001). *Fluid mechanics.* Prentice-Hall, Englewood Cliffs, NJ. Textbook on fluid mechanics—good discussion of dimensional analysis and of turbines.

LIST OF MAIN SYMBOLS

$\mathcal{D}$	drag force	T	torque
A	area	U	characteristic speed
C_D	drag coefficient	u, v	velocity
C_L	lift coefficient	z	vertical coordinate
g	acceleration due to gravity	β	angle
$\mathcal{L}$	lift force	μ	coefficient of dynamic viscosity
L	characteristic length	ρ	density
p	pressure	ν	kinematic viscosity
q	total speed	ψ	stream function
Q	volume flow rate	ω	angular velocity
Re	Reynolds number	Γ	circulation
t	time		

EXERCISES

3.1 What happens to the speed of cars when two slow-moving lanes of cars converge into a single lane, assuming the spacing between cars remains constant?

3.2 Why does an open door swing shut when air blows through the doorway?

3.3 A Pitot tube uses two water manometers to measure the speed of flow of a river. Calculate the speed due to a difference in height of 1 cm between the manometers.

3.4 Verify that the volume flow rate through a Venturi meter is given by

$$Q = A_1 u_1 = A\left[\frac{2(p_1 - p_2)}{\rho}\right]^{1/2}, \text{ where } A = A_1 A_2 (A_1^2 - A_2^2)^{-1/2}.$$

3.5 Verify that the Reynolds number is a dimensionless parameter.

3.6 Estimate the Reynolds number Re for a body in a stream of water, for $l = 10$ mm, $u = 1\ \text{m s}^{-1}$, and $\nu = 10^{-6}\ \text{m}^2\ \text{s}^{-1}$.

3.7 Verify that all the terms appearing in Bernoulli's equation (3.2) have physical dimensions of the form L^2T^{-2}.

3.8 A fountain shoots vertically upwards with speed u_0. Use dimensional analysis to derive an algebraic expression for the maximum height h in terms of u_0 and g.

3.9 A jet of water emerges from an orifice in a dam at a depth h below the water surface. Using Bernoulli's equation, and the fact that the surface of the water in the dam and the jet are both at atmospheric pressure, show that the speed of the jet on leaving the orifice is given by $u = \sqrt{2gh}$.

3.10 It is desired to examine the flow over a model wind turbine using water instead of air. Assuming the kinematic viscosities ($\nu = \mu/\rho$) of air and water are $1.5 \times 10^{-5}\ \mathrm{m^2\,s^{-1}}$ and $10^{-6}\ \mathrm{m^2\,s^{-1}}$, respectively, and that the model is 100 times smaller that the full size turbine, what is the ratio of the speed in the water to that in air in order for the Reynolds number to be the same in both cases?

3.11* A viscous fluid flows in the x-direction between parallel plates at $y = 0$ and $y = w$, under the action of a pressure gradient $\mathrm{d}p/\mathrm{d}x = \mathrm{const}$. Given that the momentum equation for the fluid is $\mu \mathrm{d}^2u/\mathrm{d}y^2 = -\mathrm{d}p/\mathrm{d}x$, show that the velocity profile is given by $u = -(1/2\mu)(\mathrm{d}p/\mathrm{d}x)y(y - w)$.

3.12 Design an experiment to examine the Magnus effect on a rotating cylinder in a flowing stream of water to investigate how the sideways force varies with the angular velocity of the cylinder and the velocity of the stream.

3.13* A uniform stream of an ideal fluid flows around a cylinder. (a) Given that the radial and azimuthal components of velocity in polar coordinates $u_r = (1/r)\partial\psi/\partial\theta$, $u_\theta = -\partial\psi/\partial r$, show that u_r and u_θ satisfy the equation

$$\frac{1}{r}\frac{\partial}{\partial r}(ru_r) + \frac{1}{r}\frac{\partial u_\theta}{\partial \theta} = 0$$

and also show that ψ satisfies Laplace's equation

$$\frac{1}{r}\frac{\partial}{\partial r}\left(r\frac{\partial \psi}{\partial r}\right) + \frac{1}{r^2}\frac{\partial^2 \psi}{\partial \theta^2} = 0.$$

(b) Verify that $u_r \equiv 0$ and $u_\theta = \Gamma/2\pi r$ are valid solutions. (c) Show that $\psi = Ur\sin\theta(1 - a^2/r^2)$ satisfies Laplace's equation and hence show that $u_r = U(1 - a^2/r^2)\cos\theta$ and $u_\theta = -U\left(1 + a^2/r^2\right)\sin\theta$. (d) Derive expressions for the velocity components due to the superposition of (b) and (c).

3.14 Estimate the lift on an aircraft with wings of length $l = 10$ m, span $w = 2$ m, $\alpha \approx 1°$, flying at 900 $\mathrm{m\,s^{-1}}$. (Assume $\rho \approx 1.2\ \mathrm{kg\,m^{-3}}$.)

3.15 Use Euler's turbine equation to estimate the maximum power output of a water turbine operating at 50 Hz such that

$$\rho = 10^3\ \mathrm{kg\,m^{-3}}, Q = 10\ \mathrm{m^3\,s^{-1}}, r_1 = 10\ \mathrm{m}, q_1 = 5\ \mathrm{m\,s^{-1}}, \cos\beta_1 = 0.4$$

3.16 Write an article of about 500 words for a popular science magazine on the proposition that 'without viscosity, birds could not fly and fish could not swim'. (Assume that the readers have no knowledge of mathematics or fluid mechanics.)

4 Hydropower, tidal power, and wave power

List of Topics

- Natural resources
- Power from dams
- Weirs
- Water turbines
- Tides
- Tidal waves
- Tidal barrage
- Tidal resonance
- Wave energy
- Wave power devices

Introduction

In this chapter we investigate three different forms of power generation that exploit the abundance of water on Earth: hydropower; tidal power; and wave power. Hydropower taps into the natural cycle of

Solar heat → sea water evaporation → rainfall → rivers → sea.

Hydropower is an established technology that accounts for about 20% of global electricity production, making it by far the largest source of renewable energy. The energy of the water is either in the form of potential energy (reservoirs) or kinetic energy (e.g. rivers). In both cases electricity is generated by passing the water through large water turbines.

Tidal power is a special form of hydropower that exploits the bulk motion of the tides. Tidal barrage systems trap sea water in a large basin and the water is drained through low-head water turbines. In recent years, rotors have been developed that can extract the kinetic energy of underwater currents.

Wave power is a huge resource that is largely untapped. The need for wave power devices to be able to withstand violent sea conditions has been a major problem in the development of wave power technology. The energy in a surface wave is proportional to the square of the amplitude and typical ocean waves transport about 30–70 kW of power per metre width of wave-front. Large amplitude waves generated by tropical storms can travel vast distances across oceans with little attenuation before reaching distant coastlines. Most of the best sites are on the western coastlines of continents between the 40° and 60° latitudes, above and below the equator.

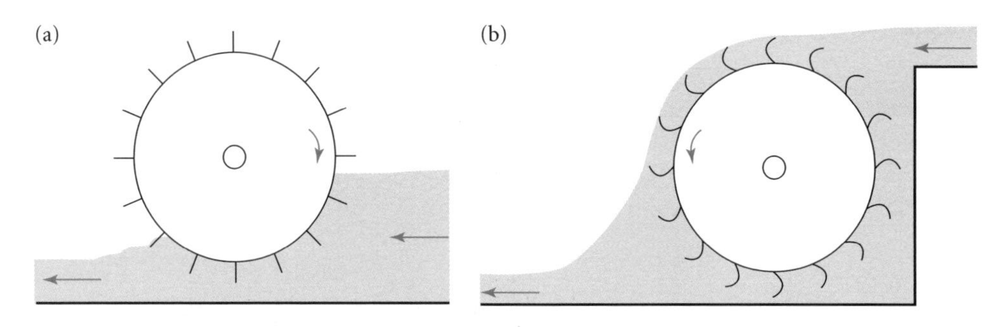

Fig. 4.1 (a) Undershot and (b) overshot waterwheels.

4.1 Hydropower

The power of water was exploited in the ancient world for irrigation, grinding corn, metal forging, and mining. Waterwheels were common in Western Europe by the end of the first millennium; over 5000 waterwheels were recorded in the *Domesday book* of 1086 shortly after the Norman conquest of England. The early waterwheels were of the undershot design (Fig. 4.1(a)) and very inefficient. The development of overshot waterwheels (Fig. 4.1(b)), and improvements in the shape of the blades to capture more of the incident kinetic energy of the stream, led to higher efficiencies.

A breakthrough occurred in 1832 with the invention of the Fourneyron turbine, a fully submerged vertical axis device that achieved efficiencies of over 80%. Fourneyron's novel idea was to employ fixed guide vanes that directed water outwards through the gaps between moving runner blades as shown in Fig. 4.2. Many designs of water turbines incorporating fixed guide vanes and runners have been developed since. Modern water turbines are typically over 90% efficient.

The main economic advantages of hydropower are low operating costs, minimal impact on the atmosphere, quick response to sudden changes in electricity demand, and long plant life—typically 40 years or more before major refurbishment. However, the capital cost of construction of dams is high and the payback period is very long. There are also serious social

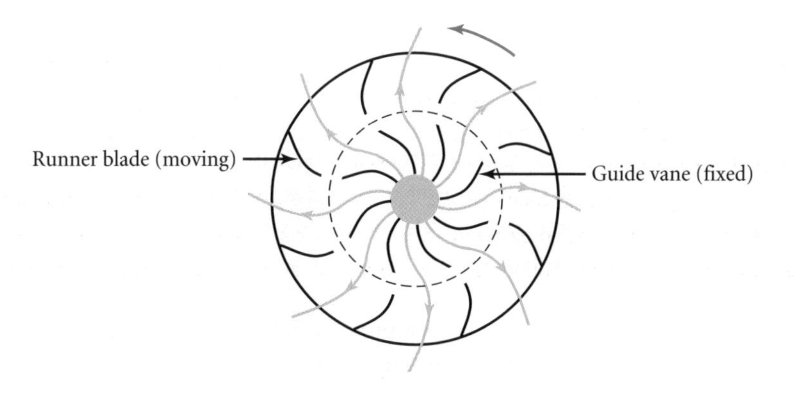

Fig. 4.2 Fourneyron water turbine.

Table 4.1 Installed hydropower

Country	Hydroelectric capacity in 2005 (GW)
USA	80
Canada	67
China	65
Brazil	58
Norway	28
Japan	27
World Total	700

Largest sites for hydropower

Country	Site	Hydroelectric capacity (GW)
China	Three Gorges*	18.2
Brazil/Paraguay	Itaipu	12.6
Venezuela	Guri	10.3
USA	Grand Coulee	6.9
Russia	Sayano–Shushenk	6.4
Russia	Krasnoyarsk	6

* Completion due in 2009.

and environmental issues to be considered when deciding about a new hydroelectric scheme, including the displacement of population, sedimentation, changes in water quality, impact on fish, and flooding.

Mountainous countries like Norway and Iceland are virtually self-sufficient in hydropower but, in countries where the resource is less abundant, hydropower is mainly used to satisfy peak-load demand. The hydroelectric capacity by country and the largest sites are shown in Table 4.1.

4.2 Power output from a dam

Consider a turbine situated at a vertical distance h (called the head) below the surface of the water in a reservoir (Fig. 4.3). The power output P is the product of the efficiency η, the potential energy per unit volume ρgh, and the volume of water flowing per second Q, i.e.

$$P = \eta \rho g h Q. \tag{4.1}$$

Note that the power output depends on the product hQ. Thus a high dam with a large h and a small Q can have the same power output as a run-of-river installation with a small h

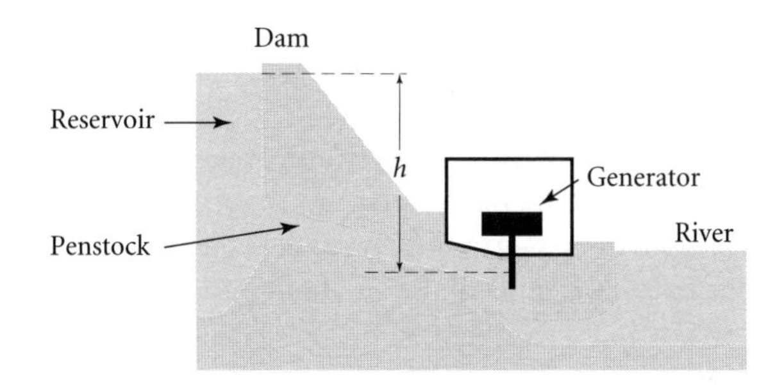

Fig. 4.3 Hydroelectric plant.

and large Q. The choice of which design of water turbine is suitable for a particular location depends on the relative magnitudes of h and Q (see Section 4.7).

EXAMPLE 4.1

Estimate the power output of a dam with a head of 50 m and volume flow rate of 20 m^3s^{-1}. (Assume $\eta = 1, \rho = 10^3$ kg m^{-3}, $g = 10$ m s^{-2}.)

From eqn (4.1) we have $P = \eta\rho g h Q \approx 1 \times 10^3 \times 10 \times 50 \times 20 \approx 10$ MW.

4.3 Measurement of volume flow rate using a weir

For power extraction from a stream it is important to be able to measure the volume flow rate of water. One particular method diverts the stream through a straight-sided channel containing an artificial barrier called a weir (Fig. 4.4). The presence of the weir forces the level of the fluid upstream of the weir to rise. The volume flow rate per unit width is related to the height of the undisturbed level of water $y_{\min}$ above the top of the weir by the formula (see Derivation 4.1)

$$Q = g^{1/2}(\tfrac{2}{3}y_{\min})^{3/2}. \tag{4.2}$$

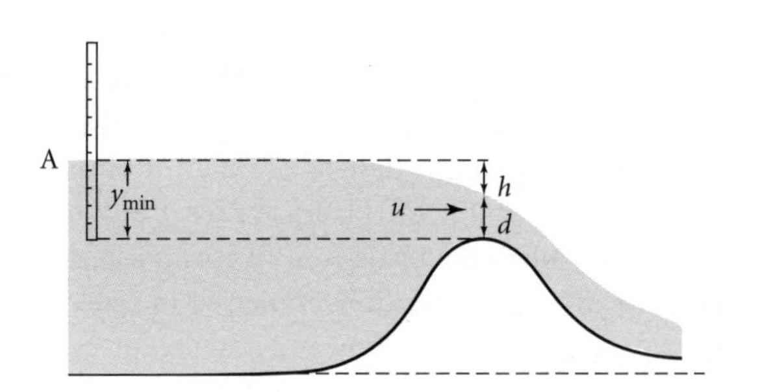

Fig. 4.4 Flow over broad-crested weir.

Derivation 4.1 Flow over a broad-crested weir

Consider a point A on the surface of the water upstream of the weir where the level is roughly horizontal (i.e. $h = 0$ in Fig. 4.4) and the velocity u_A. Towards the weir, the level drops and the speed increases. For a broad-crested weir we can ignore the vertical component of velocity and express the volume rate of water per unit width in the vicinity of the crest in the form

$$Q \approx ud \tag{4.3}$$

where d is the depth of the water near the crest. Using Bernoulli's equation (3.2), noting that the pressure on the surface is constant (atmospheric pressure), we have $\frac{1}{2}u^2 - gh \approx \frac{1}{2}u_A^2$. Hence, if the depth of the water upstream of the weir is much greater than the minimum depth over the crest of the weir, then $u_A^2 \ll u^2$ and $u \approx (2gh)^{1/2}$. Substituting for u in eqn (4.3) we obtain

$$d \approx \frac{Q}{(2gh)^{1/2}}.$$

The vertical distance from the undisturbed level to the top of the weir is $y = d + h$. Substituting for d we have

$$y = \frac{Q}{(2gh)^{1/2}} + h. \tag{4.4}$$

The first term on the right-hand side of (4.4) decreases with h but the second term increases with h. y is a minimum when $dy/dh = 0$, i.e. $-Q/(8gh^3)^{1/2} + 1 = 0$, or

$$h = \left(\frac{Q^2}{8g}\right)^{1/3}. \tag{4.5}$$

Finally, substituting for h from eqn (4.5) in eqn (4.4), yields $y_{min} = \frac{3}{2}(Q^2/g)^{1/3}$, so that

$$Q = g^{1/2}(\tfrac{2}{3}y_{min})^{3/2},$$

which is known as the Francis formula.

4.4 Water turbines

When water flows through a waterwheel the water between the blades is almost stationary. Hence the force exerted on a blade is essentially due to the difference in pressure across the blade. In a water turbine, however, the water is fast moving and the turbine extracts kinetic energy from the water. There are two basic designs of water turbines: impulse turbines and reaction turbines. In an impulse turbine, the blades are fixed to a rotating wheel and each blade rotates in air, apart from when the blade is in line with a high speed jet of water. In a reaction turbine, however, the blades are fully immersed in water and the thrust on the moving blades is due to a combination of reaction and impulse forces.

An impulse turbine called a Pelton wheel is shown in Fig. 4.5. In this example there are two symmetrical jets, and each jet imparts an impulse to the blade equal to the rate of change of

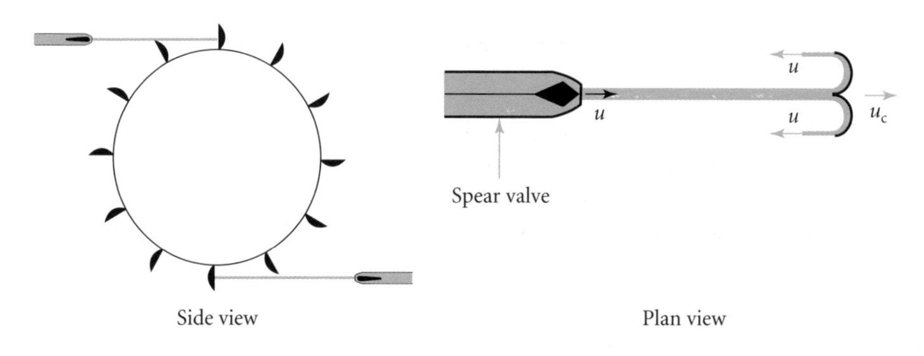

Fig. 4.5 Impulse turbine (Pelton wheel).

momentum of the jet. The speed of the jet is controlled by varying the area of the nozzle using a spear valve. Thomas Pelton went to seek his fortune in the Californian Gold Rush during the nineteenth century. By the time he arrived on the scene the easy pickings had already been taken and the remaining gold had to be extracted from rocks that needed to be crushed. Impulse turbines were being used to drive the mills to grind the rocks into small lumps. Pelton observed the motion of the turbine blades and deduced that not all the momentum of the jets was being utilized. He realized that some momentum was being lost because the water splashed in all directions on striking the blades. He redesigned the cups so that the direction of the splash was opposite to that of the incident jet. This produced a marked improvement in efficiency and Pelton thereby made his fortune.

To calculate the maximum power output from a Pelton wheel, we consider a jet moving with velocity u and the cup moving with velocity u_c. Relative to the cup, the velocity of the incident jet is $(u - u_c)$ and the velocity of the reflected jet is $-(u - u_c)$. Hence the total change in the velocity of the jet is $-2(u - u_c)$. The mass of water striking the cup per second is ρQ, so the force on the cup is given by

$$F = 2\rho Q(u - u_c). \tag{4.6}$$

The power output P of the turbine is the rate at which the force F does work on the cup in the direction of motion of the cup, i.e.

$$P = Fu_c = 2\rho Q(u - u_c)u_c \tag{4.7}$$

To derive the maximum power output we put $dP/du_c = 0$, yielding $u_c = \frac{1}{2}u$.

Substituting in eqn (4.7) then yields the maximum power as

$$P_{max} = \frac{1}{2}\rho Q u^2. \tag{4.8}$$

Thus the maximum power output is equal to the kinetic energy incident per second.

As in the Fourneyron turbine (Section 4.1), modern reaction turbines use fixed guide vanes to direct water into the channels between the blades of a runner mounted on a rotating wheel (see Fig. 4.6). However, the direction of radial flow is inward. (In the Fourneyron turbine the outward flow caused problems when the flow rate was either increased or decreased.)

The most common designs of reaction turbines are the Francis turbine and the Kaplan turbine. In a Francis turbine the runner is a spiral annulus, whereas in the Kaplan turbine it is

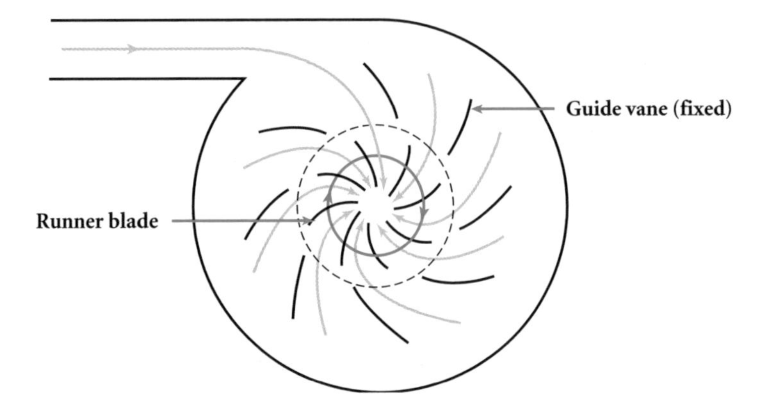

Fig. 4.6 Reaction turbine (plan view).

propeller-shaped. In both designs the kinetic energy of the water leaving the runner is small compared with the incident kinetic energy.

The term 'reaction turbine' is somewhat misleading in that it does not completely describe the nature of the thrust on the runner. The magnitude of the reaction can be quantified by applying Bernoulli's eqn (3.2) to the water entering (subscript 1) and leaving (subscript 2) the runner, i.e.

$$\frac{p_1}{\rho} + \frac{1}{2}q_1^2 = \frac{p_2}{\rho} + \frac{1}{2}q_2^2 + E \tag{4.9}$$

where E is the energy per unit mass of water transferred to the runner. Consider two cases: (a) $q_1 = q_2$, and (b) $p_1 = p_2$. In case (a), eqn (4.9) reduces to

$$E = \frac{p_1 - p_2}{\rho}, \tag{4.10}$$

i.e. the energy transferred arises from the difference in pressure between inlet and outlet. In case (b), E is given by

$$E = \tfrac{1}{2}(q_1^2 - q_2^2) \tag{4.11}$$

i.e. the energy transferred is equal to the difference in the kinetic energy between inlet and outlet. In general we define the **degree of reaction** R as

$$R = \frac{p_1 - p_2}{\rho E} = 1 - \frac{(q_1^2 - q_2^2)}{2E} \tag{4.12}$$

(see Example 4.2).

The velocity diagrams in the laboratory frame of reference for an impulse turbine and a reaction turbine are shown in Fig. 4.7(a) and (b), respectively. The symbols $\boldsymbol{u}$, $\boldsymbol{q}$ and $\boldsymbol{w}$ denote the velocity of the runner blade, the absolute velocity of the fluid, and the velocity of the fluid relative to the blade. Figure 4.7 shows the velocity triangles on the outer radius of the runner

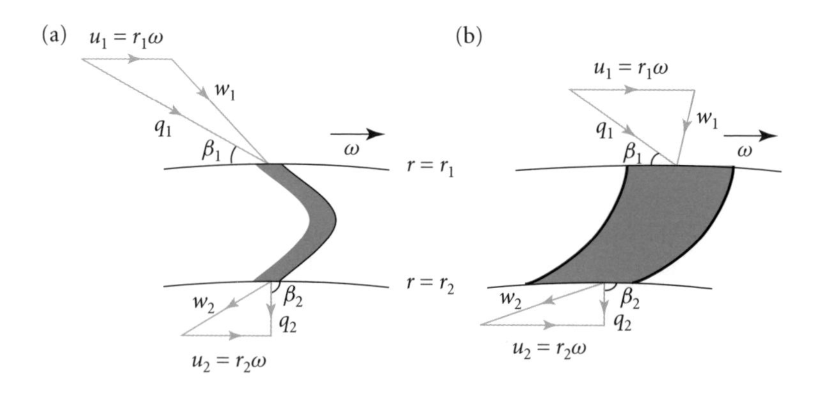

Fig. 4.7 Velocity diagrams for: (a) an impulse turbine; (b) a reaction turbine.

$r = r_1$ and the inner radius $r = r_2$. The runner rotates with angular velocity ω, so that the velocity of the blade is $u_1 = r_1\omega$ on the outer radius and $u_2 = r_2\,\omega$ on the inner radius.

The torque on the runner is

$$T = \rho Q(r_1 q_1 \cos\beta_1 - r_2 u_2 \cos\beta_2).$$

Putting $r_1 = u_1/\omega$ and $r_2 = u_2/\omega$, the work done per second is given by

$$P = T\omega = \rho Q(u_1 q_1 \cos\beta_1 - u_2 q_2 \cos\beta_2).$$

The term in brackets represents the energy per unit mass

$$E = u_1 q_1 \cos\beta_1 - u_2 q_2 \cos\beta_2. \qquad (4.13)$$

Equating the incident power due to the head of water h from eqn (4.1) to the power output of the turbine, given by Euler's turbine eqn (3.32), we have

$$\eta\rho g h Q = \rho Q(u_1 q_1 \cos\beta_1 - u_2 q_2 \cos\beta_2).$$

The term $\rho Q u_2 q_2 \cos\beta_2$ represents the rate at which kinetic energy is removed by the water leaving the runner.

We define the hydraulic efficiency as

$$\eta = \frac{u_1 q_1 \cos\beta_1 - u_2 q_2 \cos\beta_2}{gh}. \qquad (4.14)$$

The maximum efficiency is achieved when the fluid leaves the runner at right angles to the direction of motion of the blades, i.e. when $\beta_2 = \frac{\pi}{2}$ so that $\cos\beta_2 = 0$. Equation (4.14) then reduces to

$$\eta_{\max} = \frac{u_1 q_1 \cos\beta_1}{gh}. \qquad (4.15)$$

EXAMPLE 4.2

Consider a particular reaction turbine in which the areas of the entrance to the stator (the stationary part of the turbine), the entrance to the runner, and the exit to the runner are all equal. Water enters the stator radially with velocity $q_0 = 2$ m s^{-1} and leaves the stationary vanes of the stator at an angle $\beta_1 = 10^\circ$ with an absolute velocity $q_1 = 10$ m s^{-1}. The velocity of the runner at the entry radius $r = r_1$ is u_1 in the tangential direction, and is such that the velocity of the water w_1 relative to the runner is in the radial direction. On leaving the runner, the total velocity is q_2 in the radial direction. Given that the head is $h = 11$ m, calculate the degree of reaction and the hydraulic efficiency.

The volume flow rate is $q_0 A_0$ into the stator, $w_1 A_1$ into the runner, and $q_2 A_2$ out of the runner. Since $A_0 = A_1 = A_2$ it follows by mass conservation that $q_0 = w_1 = q_2$. The energy transfer per unit mass E is given by eqn (4.13). Since the total velocity q_2 leaving the runner is in the radial direction, we have $\beta_2 = \pi/2$. Putting $q_1 \cos\beta_1 = u_1$ then $E = u_1^2$. Also, the square of the total velocity is $q_1^2 = u_1^2 + w_1^2 = u_1^2 + q_2^2$, since w_1 and q_2 are equal and radial. Hence the degree of reaction is $R = 1 - \frac{(q_1^2 - q_2^2)}{2E} = 1 - \frac{u_1^2}{2u_1^2} = \frac{1}{2}$. The hydraulic efficiency is $\eta = u_1 q_1 \cos\beta_1/(gh) \approx 0.90$.

4.4.1 Choice of water turbine

The choice of water turbine depends on the site conditions, notably on the head of water h and the water flow rate Q. Figure 4.8 indicates which turbine is most suitable for any particular combination of head and flow rate. Impulse turbines are suited for large h and a low Q, e.g. fast moving mountain streams. Kaplan turbines are suited for low h and large Q (e.g. run-of-river sites) and Francis turbines are usually preferred for large Q and large h, e.g. dams. A useful parameter for choosing the most suitable turbine is the shape (or type) number S, described in Derivation 4.2.

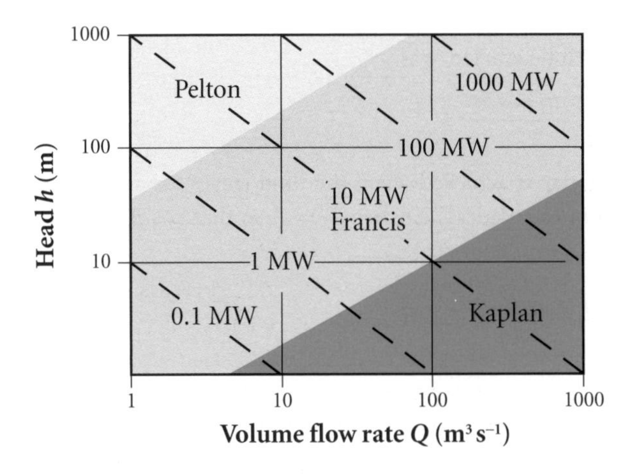

Fig. 4.8 Choice of turbine in terms of head h and volume flow rate Q.

Derivation 4.2 Shape or type number

Dimensional analysis is a useful means for choosing the appropriate type of turbine for a particular combination of h and Q. The power output P from a turbine depends on the head h, the angular velocity ω and diameter D of the turbine, and the density of water ρ. Various dimensionless parameters can be formed from these physical quantities, the power coefficient $K_{\mathrm{P}} = P/(\omega^3 D^5 \rho)$ and the head coefficient $K_{\mathrm{h}} = gh/\omega^2 D^2$ being particularly useful ones. When a turbine of a particular design is operating at its maximum efficiency, K_{P} and K_{h} will have particular values which can be used to predict the power and head in terms of the diameter D and the angular velocity ω. We can eliminate the dependence on D (which determines the size of the turbine) by forming the dimensionless ratio

$$S = \frac{K_{\mathrm{P}}^{1/2}}{K_{\mathrm{h}}^{5/4}} = \frac{\omega P^{1/2}}{\rho^{1/2}(gh)^{5/4}} \tag{4.16}$$

called the shape or type number. Substituting $P = \eta\rho g h Q$ from eqn (4.1) and assuming $\eta = 1$, eqn (4.16) becomes

$$S = \frac{\omega Q^{1/2}}{(gh)^{3/4}}. \tag{4.17}$$

Putting $Q = v_{\mathrm{w}}A$, $gh = \frac{1}{2}v_{\mathrm{w}}^2$ (where $A = \pi r^2$ is the inlet area and v_{w} is the speed of the water) and $\omega = 2\pi f = 2\pi\left(\frac{v_{\mathrm{b}}}{2\pi R}\right) = \frac{v_{\mathrm{b}}}{R}$, where v_b is the speed of the blade tip of radius R, we have

$$S = \frac{\frac{v_{\mathrm{b}}}{R}(v_{\mathrm{w}}\pi r^2)^{1/2}}{(\frac{1}{2}v_{\mathrm{w}}^2)^{3/4}} = 2^{3/2}\pi^{1/2}\frac{r}{R}\frac{v_{\mathrm{b}}}{v_{\mathrm{w}}} \approx 5\frac{r}{R}\frac{v_{\mathrm{b}}}{v_{\mathrm{w}}}. \tag{4.18}$$

For a Pelton turbine $r/R \sim 0.1, v_{\mathrm{b}}/v_{\mathrm{w}} \approx 0.5$, and $S \sim 0.25$; for a Kaplan turbine $r \sim R$ and $v_{\mathrm{b}} \sim v_{\mathrm{w}}$, so $S \sim 5$, while for a Francis turbine $S \sim 1$.

4.5 Impact, economics, and prospects of hydropower

Hydropower sites tend to have a large impact on the local population. Over 1.1 million people were displaced by the Three Gorges dam in China and it has been estimated that 30–60 million people worldwide have had to be relocated due to hydropower. Proposed hydropower plants often provoke controversy and in some countries public opposition to hydropower has stopped all construction except on small-scale projects. Also, dams sometimes collapse for various reasons, e.g. overspilling of water, inadequate spillways, foundation defects, settlement, slope instability, cracks, erosion, and freak waves from landslides in steep-sided valleys around the reservoir. As with nuclear plants, the risk of major accidents is small but the consequences can be catastrophic. Given the long lifetime of dams, even a typical failure rate as low as one per 6000 dam years means that any given dam has a probability of about 1% that it will collapse at sometime in its life. In order to reduce the environmental impact and the consequences of dam failure, the question arises as to whether it is better to build a small number of large reservoirs or a large number of small ones. Though small reservoirs tend to be more acceptable to the public than large ones, they need a much larger total reservoir area than a single large reservoir providing the same volume of stored water.

An argument in favour of hydropower is that it does not produce greenhouse gases or acid rain gases. However, water quality may be affected both upstream and downstream of a dam due to increases in the concentrations of dissolved gases and heavy metals. These effects can be mitigated by inducing mixing at different levels and oxygenating the water by auto-venting turbines. The installation of a hydropower installation can also have a major impact on fish due to changes in the habitat, water temperature, flow regime, and the loss of marine life around the turbines.

The capital cost of construction of hydropower plants is typically much larger than that for fossil fuel plants. Another cost arises at the end of the effective life of a dam, when it needs to be decommissioned. The issue as to who should pay for the cost involved in decommissioning is similar to that for nuclear plants: the plant owners, the electricity consumers, or the general public? On the positive side, production costs for hydropower are low because the resource (rainfall) is free. Also, operation and maintenance costs are minimal and lifetimes are long: typically 40–100 years. The efficiency of a hydroelectric plant tends to decrease with age due to the build-up of sedimentation trapped in the reservoir. This can be a life-limiting factor because the cost of flushing and dredging is usually prohibitive.

The economic case for any hydropower scheme depends critically on how future costs are discounted (see Chapter 11). Discounting reduces the benefit of long-term income, disadvantaging hydropower compared with quick payback schemes such as CCGT generation (see Chapter 2). Hydropower schemes therefore tend to be funded by governmental bodies seeking to improve the long-term economic infrastructure of a region rather than by private capital.

Despite the strong upward trend in global energy demand, the prospects for hydropower are patchy. In the developed world the competitive power market has tilted the balance away from capitally intensive projects towards plants with rapid payback of capital. As long as relatively cheap fossil fuels are available, the growth of hydropower is likely to be limited to parts of the world where water is abundant and labour costs for construction are low. However, it is a source of carbon-free energy and its importance would be enhanced by restrictions on carbon emissions aimed at tackling global warming.

4.6 Tides

There are two high tides and two low tides around the Earth at any instant. One high tide is on the longitude closest to the Moon and the other on the longitude furthest from the Moon. The low tides are on the longitudes at 90° to the longitudes where the high tides are situated. On any given longitude the interval between high tides is approximately 12 hours 25 minutes (see Exercise 4.8). The difference in height between a high tide and a low tide is called the tidal range. The mid-ocean tidal range is typically about 0.5–1.0 metres but is somewhat larger on the continental shelves. In the restricted passages between islands and straits the tidal range can be significantly enhanced, up to as much as 12 m in the Bristol Channel (UK) and 13 m in the Bay of Fundy (Nova Scotia). Tidal power has the advantage

Table 4.2 Tidal potential of some large tidal range sites

Country	Site	Mean tidal range (m)	Basin area (km^2)	Capacity (GW)
Argentina	Golfo Nuevo	3.7	2376	6.6
Canada	Cobequid	12.4	240	5.3
India	Gulf of Khambat	7.0	1970	7.0
Russia	Mezen	6.7	2640	15.0
Russia	Penzhinsk	11.4	20530	87.4
UK	Severn	7	520	8.6

over other forms of alternative energy of being predictable. For conventional tidal power generation it is necessary to construct huge tidal basins in order to produce useful amounts of electricity. However, in recent years, an alternative technology for exploiting strong tidal currents has been under development using underwater rotors, analogous to wind turbines. Table 4.2 shows the potential of some large tidal range sites in various locations around the world.

4.6.1 Physical cause of tides

The main cause of tides is the effect of the Moon. The effect of the Sun is about half that of the Moon but increases or decreases the size of the lunar tide according to the positions of the Sun and the Moon relative to the Earth. The daily rotation of the Earth about its own axis only affects the location of the high tides. In the following explanation we ignore the effect of the Sun (see Exercise 4.9).

For simplicity we assume that the Earth is covered by water. Consider a unit mass of water situated at some point P as shown in Fig. 4.9. The gravitational potential due to the Moon is given by

$$\phi = -\frac{Gm}{s}$$

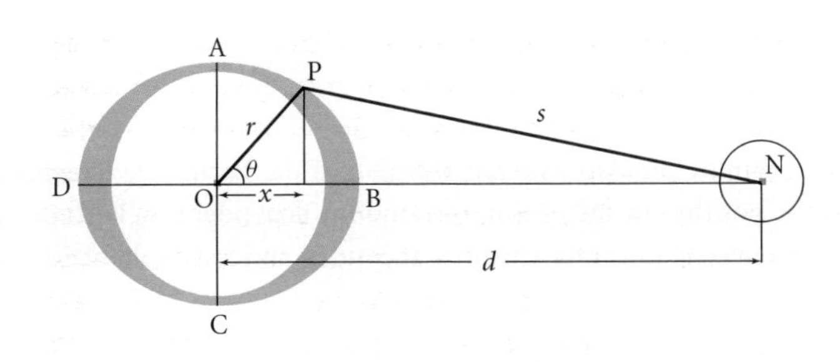

Fig. 4.9 Tidal effects due to the Moon (not to scale).

where G is the gravitational constant, m is the mass of the Moon, and s is the distance from P to the centre of the Moon. For $d \gg r$ we can expand $1/s$ as follows:

$$\frac{1}{s} = \frac{1}{[d^2 + r^2 - 2dr\cos\theta]^{\frac{1}{2}}} = \frac{1}{d}\left[1 + \left(-\frac{2r}{d}\cos\theta + \frac{r^2}{d^2}\right)\right]^{-\frac{1}{2}}$$

$$= \frac{1}{d}\left[1 + \frac{r}{d}\cos\theta + \frac{r^2}{d^2}\left(\tfrac{3}{2}\cos^2\theta - \tfrac{1}{2}\right) + \cdots\right].$$

The first term in the expansion does not yield a force and can be ignored. The second term corresponds to a constant force, Gm/d^2, directed towards N, which acts on the Earth as a whole and is balanced by the centrifugal force due to the rotation of the Earth–Moon system. The third term describes the variation of the Moon's potential around the Earth. The surface profile of the water is an equipotential surface due to the combined effects of the Moon and the Earth. The potential of unit mass of water due to the Earth's gravitation is gh, where h is the height of the water above its equilibrium level and $g = GM/r^2$ is the acceleration due to gravity at the Earth's surface, where M is the mass of the Earth. Hence, the height of the tide $h(\theta)$ is given by

$$gh(\theta) - \frac{Gmr^2}{d^3}\left(\frac{3}{2}\cos^2\theta - \frac{1}{2}\right) = 0$$

or

$$h(\theta) = h_{\max}\left(\frac{3}{2}\cos^2\theta - \frac{1}{2}\right) \tag{4.19}$$

where

$$h_{\max} = \frac{mr^4}{Md^3} \tag{4.20}$$

is the maximum height of the tide, which occurs at points B and D ($\theta = 0$ and $\theta = \pi$). Putting $m/M = 0.0123, d = 384\,400$ km, and $r = 6378$ km we obtain $h_{\max} \approx 0.36$ m, which is roughly in line with the observed mean tidal height.

4.6.2 **Tidal waves**

There are two tidal bulges around the Earth at any instant. A formula for the speed of a tidal wave in a sea of uniform depth h_0 is obtained from shallow water theory (see Derivation 4.3) as

$$c = \sqrt{gh_0}. \tag{4.21}$$

The tidal bulges cannot keep up with the rotation of the Earth (see Exercise 4.12), so the tides lag behind the position of the Moon, the amount dependent on latitude. The presence of continents and bays significantly disturbs the tides and can enhance their range (see Section 4.9).

Derivation 4.3 Shallow water theory

We consider a wave such that the wavelength λ is much greater than the mean depth of the sea h_0. We also assume that the amplitude of the wave is small compared with the depth, in which case the vertical acceleration is small compared with the acceleration due to gravity, g. Hence the pressure below the surface is roughly hydrostatic and given by

$$p = p_0 + \rho g(h - y) \tag{4.22}$$

where p_0 is atmospheric pressure and $y = h(x, t)$ is the wave profile on the free surface (Fig. 4.10). Differentiating eqn (4.22) with respect to x we have $\partial p/\partial x = -\rho g \partial h/\partial x$. Neglecting second-order terms, the equation of motion in the x-direction is of the form

$$\partial u/\partial t = -g\partial h/\partial x. \tag{4.23}$$

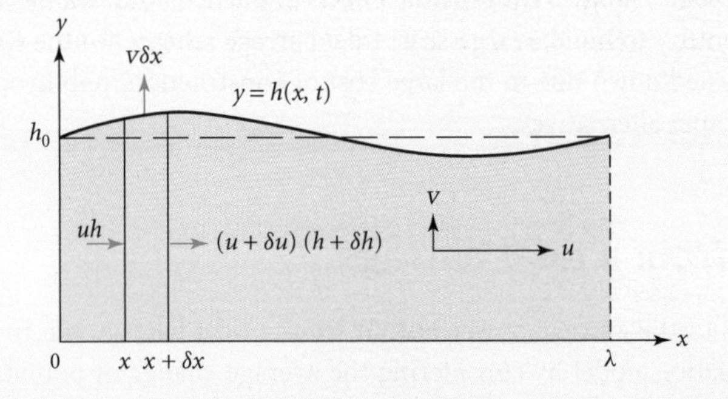

Fig. 4.10 Shallow water wave.

Since $h(x, t)$ is independent of y it follows from (4.23) that u is also independent of y. This allows us to derive an equation of mass conservation in terms of u and h. Consider a slice of fluid between the planes x and $x+\delta x$. The volume flowing per second is uh across x and $(u + \delta u)(h + \delta h)$ across $x+\delta x$. By mass conservation, the difference in the volume flowing per second from x to $x+\delta x$ is equal to the volume displaced per second $v\delta x$ in the vertical direction. Hence

$$uh = (u + \delta u)(h + \delta h) + v\delta x.$$

Putting $v = \partial h/\partial t$, $\delta u \approx (\partial u/\partial x)\delta x$, $\delta h \approx (\partial h/\partial x)\delta x$, and noting that $h\partial u/\partial x \gg u\partial h/\partial x$ and $h \approx h_0$, yields the mass continuity equation as

$$-\frac{\partial u}{\partial x} = \frac{1}{h_0}\frac{\partial h}{\partial t}. \tag{4.24}$$

Eliminating u between eqns (4.23) and (4.24) we obtain the wave equation

$$\frac{\partial^2 h}{\partial x^2} = \frac{1}{c^2}\frac{\partial^2 h}{\partial t^2} \tag{4.25}$$

for the height profile $h(x, t)$ of the wave, where $c = \sqrt{gh_0}$ is the wave speed.

4.7 Tidal power

The earliest exploitation of tidal power was in tidal mills, created by building a barrage across the mouth of a river estuary. Sea water was trapped in a tidal basin on the rising tide and released at low tide through a waterwheel, providing power to turn a stone mill to grind corn. Tidal barrages for electricity generation use large low-head turbines and can operate for a greater fraction of the day. An important issue is whether it is better to use conventional turbines that are efficient but operate only when the water is flowing in one particular direction or less efficient turbines that can operate in both directions (i.e. for the incoming and the outgoing tides).

The first large-scale tidal power plant in the world was built in 1966 at La Rance in France. It generates 240 MW using 24 low-head Kaplan turbines. A number of small tidal power plants have also been built more recently in order to gain operational experience and to investigate the long-term ecological and environmental effects of particular locations. Various proposals during the last century to build a large-scale tidal barrage scheme for the River Severn in the UK have been turned down due to the large cost of construction, public opposition and the availability of cheaper alternatives.

4.8 Power from a tidal barrage

A rough estimate of the average power output from a tidal barrage can be obtained from a simple energy balance model by considering the average change of potential energy during the draining process. Consider a tidal basin of area A as shown in Fig. 4.11. The total mass of water in the tidal basin above the low water level is $m = \rho Ah$, where h is the tidal range. The height of the centre of gravity is $\frac{1}{2}h$, so the work done in raising the water is $mg(\frac{1}{2}h) = \frac{1}{2}\rho gAh^2$. Hence the average power output (see Example 4.3) is

$$P_{\text{ave}} = \frac{\rho gAh^2}{2T} \tag{4.26}$$

where T is the time interval between tides, i.e. the tidal period. In practice, the power varies with time according to the difference in water levels across the barrage and the volume of water

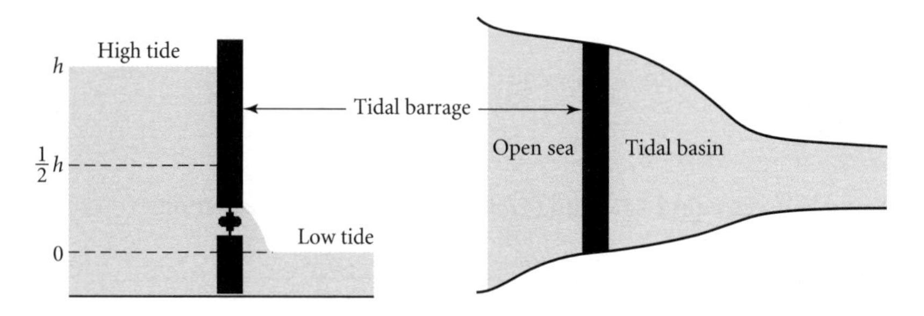

Fig. 4.11 Tidal barrage.

allowed to flow through the turbines. Also, the operating company would seek to optimize revenue by generating electricity during periods of peak-load demand when electricity prices are highest.

EXAMPLE 4.3

Estimate the average power output of a tidal basin with a tidal range of 7 m and an area of tidal basin of 520 km^2 (i.e. Severn barrage).

Substituting in eqn (4.26), noting that the tidal period is $T \approx 4.5 \times 10^4$ s ($T \approx 12.5$ h), the average power output is

$$P_{ave} = \frac{\rho g A h^2}{2T} \approx \frac{10^3 \times 10 \times 520 \times 10^6 \times 7^2}{2 \times 4.5 \times 10^4} \approx 2.8\,\text{GW}.$$

4.9 Tidal resonance

The tidal range varies in different oceans of the world due to an effect known as tidal resonance. For example, the Atlantic Ocean has a width of about 4000 km and an average depth of about 4000 m, so the speed of a shallow water wave eqn (4.21) is about $c = \sqrt{gh_0} \approx \sqrt{10 \times 4000} \approx 200$ m s^{-1}. The tidal frequency is about $2 \times 10^{-5} s^{-1}$, so the wavelength is $\lambda = c/f \approx 200/(2 \times 10^{-5})$ m $\approx 10^4$ km. This is about twice the width of the Atlantic and so resonance occurs; the time taken for the shallow water wave to make the round trip, reflecting off both shores, is about the same as the tidal period, so the amplitude builds up.

The wave amplitude also increases on the continental shelf and can reach about 3 m at the shores. River estuaries can also exhibit large tidal resonance if the length and depth of the estuary are favourable. From eqn (4.21) the time taken for a wave to propagate the length of the channel and back to the inlet is given by $t = 2L/c = 2L/\sqrt{gh_0}$ (see Example 4.4). If this time is equal to half the time between successive tides then the tidal range is doubled (see Derivation 4.4).

Derivation 4.4 Tidal resonance in a uniform channel

For simplicity, consider a uniform channel of length L such that the end at $x = 0$ is open to the sea and the other end of the channel at $x = L$ is a vertical wall. Suppose that the height of the incident tidal wave varies with time as $h_i(t) = a\cos(\omega t)$. We consider a travelling wave of the form

$$h_i(x, t) = a\cos(kx - \omega t).$$

From the mass continuity eqn (4.24), we have

$$-\frac{\partial u_i}{\partial x} = \frac{1}{h_0}\frac{\partial h_i}{\partial t} = \frac{\omega a}{h_0}\sin(kx - \omega t).$$

Integrating with respect to x yields the velocity in the horizontal direction as

$$u_i(x,t) = \frac{\omega a}{h_0 k} \cos(kx - \omega t).$$

In order to satisfy the boundary condition $u = 0$ at $x = L$ (since there cannot be any flow across the barrier) we superimpose a reflected wave of the form

$$u_r(x,t) = \frac{\omega a}{h_0 k} \cos(kx + \omega t).$$

The total velocity at $x = L$ is given by $u(L,t) = u_i(L,t) + u_r(L,t)$ $= \frac{\omega a}{h_0 k}[\cos(kL - \omega t) + \cos(kL + \omega t)] = \frac{2\,\omega a}{h_0 k} \cos(kL)\cos(\omega t) = 0$. Hence $kL = \frac{\pi}{2}(2n+1)$, and the lowest mode of oscillation ($n = 0$) is given by $kL = \frac{\pi}{2}$. Putting $k = 2\pi/\lambda$ we then obtain the minimum length of the channel as $L = \frac{\lambda}{4}$. The total height of the incident and reflected waves is

$$h(x,t) = h_i(x,t) + h_r(x,t) = a\cos(kx - \omega t) - a\cos(kx + \omega t) = 2a\sin(kx)\sin(\omega t).$$

At the end of the channel, $x = L$, the height is $h(L,t) = 2a\sin(\omega t)$, i.e. double that due to the incident wave. This causes the amplitude to build up with the result that the tidal range can be very large—in the River Severn estuary between England and Wales a range of 10–14 m is observed.

EXAMPLE 4.4

An estuary has an average depth of 20 m. Estimate the length of estuary required for tidal resonance.

Equating the time for a wave to travel the length of the channel and back again to half the tidal period we have $\frac{2L}{\sqrt{10 \times 20}} = \frac{1}{2} \times 4.5 \times 10^4$, so that $L \approx 160$ km.

4.10 Kinetic energy of tidal currents

In particular locations (e.g. between islands) there may be strong tidal currents that transport large amounts of kinetic energy. In recent years various devices for extracting the kinetic energy have been proposed. These devices are essentially underwater versions of wind turbines and obey the same physical principles as those described in Chapter 5. In the majority of designs the axis of rotation of the turbine is horizontal and the device is mounted on the seabed or suspended from a floating platform. Before installation, the tidal currents for any particular location need to be measured to depths of 20 m or more in order to determine the suitability of the site. The first generation of prototype kinetic energy absorbers have been operated in shallow water (i.e. 20–30 m) using conventional engineering components. Later generations are likely to be larger, more efficient, and use specially designed low-speed electrical generators and hydraulic transmission systems.

4.11 Ecological and environmental impact of tidal barrages

The installation of a tidal barrage has a major impact on both the environment and ecology of the estuary and the surrounding area for the following reasons.

1. The barrage acts as a major blockage to navigation and requires the installation of locks to allow navigation to pass through.
2. Fish are killed in the turbines and impeded from migrating to their spawning areas.
3. The intertidal wet/dry habitat is altered, forcing plant and animal life to adapt or move elsewhere.
4. The tidal regime may be affected downstream of a tidal barrage. For example, it has been claimed that a proposed barrage for the Bay of Fundy in Canada could increase the tidal range by 0.25 m in Boston, 1300 km away.
5. The water quality in the basin is altered since the natural flushing of silt and pollution is impeded, affecting fish and bird life.

On the positive side, there are the benefits arising from carbon-free energy, improved flood protection, new road crossings, marinas, and tourism.

4.12 Economics and prospects for tidal power

Large tidal barrages have the economic disadvantages of large capital cost, long construction times, and intermittent operation. On the other hand, they have long plant lives (over 100 years for the barrage structure and 40 years for the equipment) and low operating costs. An alternative idea is to create a closed basin in the estuary known as a **tidal lagoon**. The wall of a tidal lagoon does not extend across the whole channel so the environmental effects are lessened and the impact on fish and navigation is reduced. Also, by restricting the tidal lagoons to shallow water, the retaining wall can be low and cheap to build. The global tidal resource has been estimated as 3000 GW, but only 3% of this is in areas where the tidal range is enhanced and hence suitable for power generation. So far, large barrage schemes have not been pursued.

The economics of small tidal current devices (kinetic energy absorbers) has the attraction that they can be installed on a piecemeal basis, thereby reducing the initial capital outlay. They also have a more predictable output than wind turbines and there is no visual impact. The danger to fish is minimal because the blades rotate fairly slowly (typically about 20 revolutions per minute). The potential for tidal stream generation around the UK has been estimated as possibly 10 GW, a contribution of about a quarter to the UK electricity demand.

The long-term economic potential and environmental impact of such devices will become clearer after trials on various designs, notably in the UK, Canada, Japan, Russia, Australia, and China. The engineering challenge is to design reliable and durable equipment capable of operating for many years in a harsh marine environment with low operational and maintenance costs.

4.13 Wave energy

The waves on the surface of the sea are caused mainly by the effects of wind. The streamlines of air are closer together over a crest and the air moves faster. It follows from Bernoulli's theorem (3.2) that the air pressure is reduced, so the amplitude increases and waves are generated. As a wave crest collapses the neighbouring elements of fluid are displaced and forced to rise above the equilibrium level (Fig. 4.12).

The motion of the fluid beneath the surface decays exponentially with depth. About 80% of the energy in a surface wave is contained within a quarter of a wavelength below the surface. Thus, for a typical ocean wavelength of 100 m, this layer is about 25 m deep. We now derive an expression for the speed of a surface wave using intuitive physical reasoning. The water particles follow circular trajectories, as shown in Fig. 4.12 (See Exercise 4.16).

Consider a surface wave on deep water and choose a frame of reference that moves at the wave velocity, c, so that the wave profile remains unchanged with time. Noting that the pressure on the free surface is constant (i.e. atmospheric pressure), Bernoulli's eqn (3.2) yields

$$u_c^2 - u_t^2 - 2gh = 0 \tag{4.27}$$

where u_c is the velocity of a particle at a wave crest, u_t is the velocity of a particle at a wave trough, and h is the difference in height between a crest and a trough. If r is the radius of a circular orbit and τ is the wave period then we can put

$$u_c = \frac{2\pi r}{\tau} - c, \quad u_t = -\frac{2\pi r}{\tau} - c, \quad h = 2r. \tag{4.28}$$

Substituting for u_c, u_t and h from eqn (4.28) in eqn (4.27), and putting $\lambda = c\tau$, we obtain the wave speed as

$$c = \sqrt{g\lambda/(2\pi)}. \tag{4.29}$$

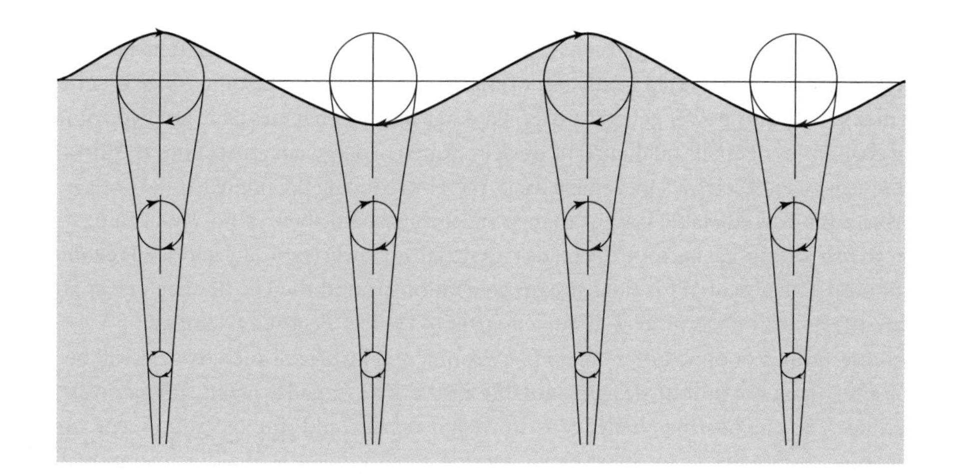

Fig. 4.12 Surface wave on deep water.

It follows from eqn (4.29) that the wave speed increases with wavelength, so that surface waves are **dispersive**. In practice the wave profile on the surface of the sea is a superposition of waves of various amplitudes, speeds, and wavelengths moving in different directions. The net displacement of the surface is therefore more irregular than that of a simple sine wave. Hence, in order for a wave power device to be an efficient absorber of wave energy in real sea conditions, it needs to be able to respond to random fluctuations in the wave profile.

The total energy E of a surface wave per unit width of wave-front per unit length in the direction of motion is given by

$$E = \tfrac{1}{2}\rho g a^2 \tag{4.30}$$

(see Derivation 4.5). The dependence of wave energy on the square of the amplitude has mixed benefits. Doubling the wave amplitude produces a fourfold increase in wave energy. However, too much wave energy poses a threat to wave power devices and measures need to be taken to ensure they are protected in severe sea conditions.

The power P per unit width in a surface wave is the product of E and the group velocity c_g, given by

$$c_g = \tfrac{1}{2}\sqrt{g\lambda/(2\pi)} \tag{4.31}$$

(see Exercise 4.11). Hence the incident power per unit width of wave-front (Example 4.5) is

$$P = \tfrac{1}{4}\rho g a^2\sqrt{g\lambda/(2\pi)}. \tag{4.32}$$

In mid-ocean conditions the typical power per metre width of wave-front is 30–70 kW m^{-1}.

EXAMPLE 4.5

Estimate the power per unit width of wave-front for a wave amplitude $a = 1$ m and wavelength of 100 m.

From eqn (4.32), the power per unit width of wave-front is

$$P = \tfrac{1}{4}\rho g a^2\sqrt{\frac{g\lambda}{2\pi}} \approx \tfrac{1}{4} \times 10^3 \times 10 \times 1^2 \times \sqrt{\frac{10\times 10^2}{2\times 3.14}} \approx 32 \text{ kW m}^{-1}.$$

Derivation 4.5 Energy in a surface wave

Consider unit width of a wave with a surface profile of the form

$$z = a\sin\left(\frac{2\pi x}{\lambda}\right)$$

as shown in Fig. 4.13. (The time-dependence is irrelevant for this derivation.) The gain in potential energy of an elemental mass $\delta m = \rho g\ \delta x\ \delta z$ of fluid in moving from $-z$ to $+z$ is $\delta V = \delta m g(2z) = 2\rho g z\ \delta x\ \delta z$. Hence the total potential energy of the elevated section of fluid is

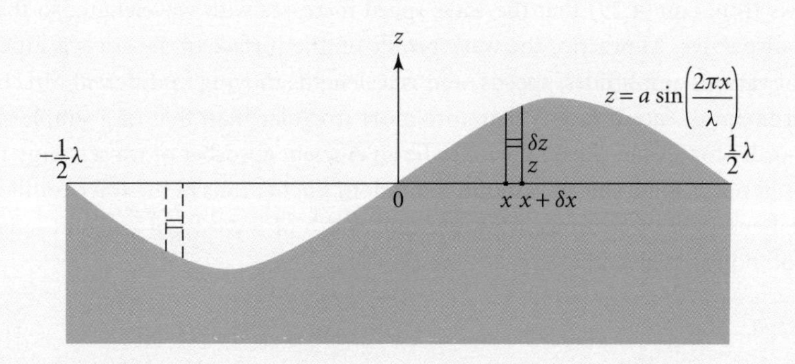

Fig. 4.13 Energy of surface wave.

$$V = 2\rho g \int_{x=0}^{x=\frac{1}{2}\lambda} \int_{z=0}^{z=a\sin(2\pi x/\lambda)} z\,\mathrm{d}z\,\mathrm{d}x = \rho g a^2 \int_{x=0}^{x=\frac{1}{2}\lambda} \sin^2(2\pi x/\lambda)\mathrm{d}x = \tfrac{1}{4}\rho g a^2 \lambda.$$

Assuming equipartition of energy, the average kinetic energy is equal to the average potential energy, so that the total energy over a whole wavelength is $E = \frac{1}{2}\rho g a^2 \lambda$, or

$$E = \tfrac{1}{2}\rho g a^2$$

per unit length in the x-direction and per unit width of wavefront.

4.14 Wave power devices

Though the first patent for a wave power device was filed as early as 1799, wave power was effectively a dormant technology until the early 1970s, when the world economy was hit by a series of large increases in oil prices. Wave power was identified as one of a number of sources of alternative energy that could potentially reduce dependency on oil. It received financial support to assess its technical potential and commercial feasibility, resulting in hundreds of inventions for wave power devices, but most of these were dismissed as either impractical or uneconomic. The main concerns were whether wave power devices could survive storms and their capital cost. During the 1980s, publicly funded research for wave power virtually disappeared as global energy markets became more competitive. However, in the late 1990s interest in wave power technology was revived due to increasing evidence of global climate change and the volatility of oil and gas prices. A second generation of wave power devices emerged, which were better designed and had greater commercial potential.

In general, the key issues affecting wave power devices are:

- survivability in violent storms;
- vulnerability of moving parts to sea water;
- capital cost of construction;
- operational costs of maintenance and repair;
- cost of connection to the electricity grid.

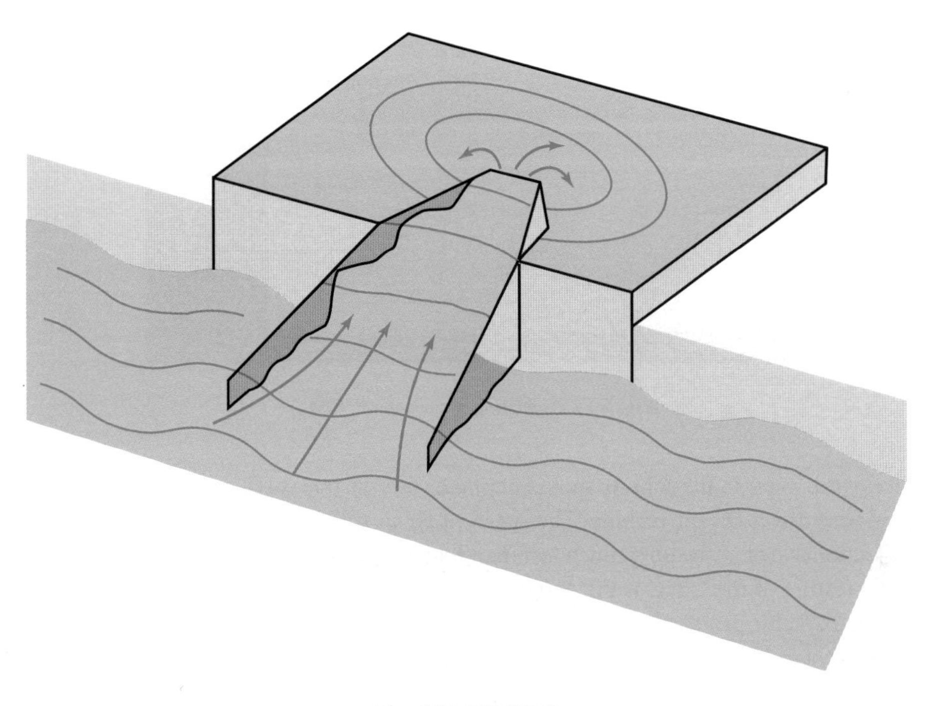

Fig. 4.14 TAPCHAN.

We now describe various types of wave power device and examine how they operate and how they address the above challenges.

4.14.1 Spill-over devices

TAPCHAN (TAPered CHANnel) is a Norwegian system in which sea waves are focused in a tapered channel on the shoreline. Tapering increases the amplitude of the waves as they propagate through the channel. The water is forced to rise up a ramp and spill over a wall into a reservoir about 3–5 m above sea level (Fig. 4.14). The potential energy of the water trapped in the reservoir is then extracted by draining the water back to the sea through a low-head Kaplan turbine. Besides the turbine, there are no moving parts and there is easy access for repairs and connections to the electricity grid. Unfortunately, shore-based TAPCHAN schemes have a relatively low power output and are only suitable for sites where there is a deep water shoreline and a low tidal range of less than about a metre. To overcome these limitations, a floating offshore version of TAPCHAN called Wave Dragon is under development, with an inlet span of around 200 m, to generate about 4 MW.

4.14.2 Oscillating water columns

The oscillating water column (OWC) uses an air turbine housed in a duct well above the water surface (Fig. 4.15). The base of the device is open to the sea, so that incident waves force the water inside the column to oscillate in the vertical direction. As a result the air above the

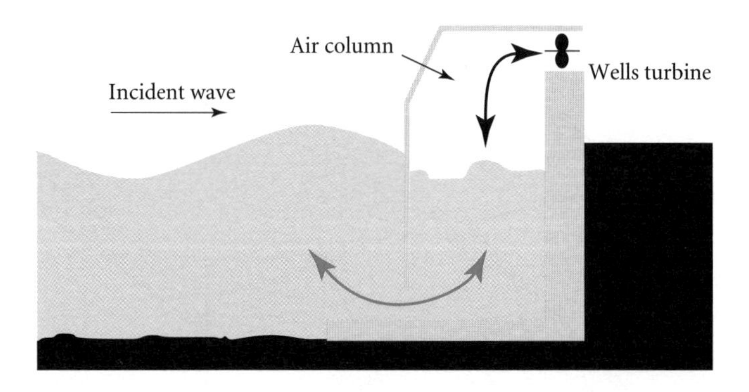

Fig. 4.15 Oscillating water column (OWC).

surface of the water in the column moves in phase with the free surface of the water inside the column and drives the air turbine. The speed of air in the duct is enhanced by making the cross-sectional area of the duct much less than that of the column.

A key feature of the OWC is the design of the air turbine, known as the Wells turbine. It has the remarkable property of spinning in the same direction irrespective of the direction of air flow in the column! Unlike conventional turbine blades, the blades in a Wells turbine are symmetrical about the direction of motion (Fig. 4.16). Relative to a blade, the direction of air flow is at a non-zero angle of attack α. The net force acting on the blade in the direction of motion is then given by

$$F = \mathcal{L}\sin\alpha - \mathcal{D}\cos\alpha \tag{4.33}$$

where $\mathcal{L}$ and $\mathcal{D}$ are the lift and drag forces acting on the blade. It is clear from the force diagram in Fig. 4.16(b) that the direction of the net force is the same, irrespective of whether the air is flowing upwards or downwards inside the air column.

The shape of the blade is designed such as to maximize the net force on the blade and the operational efficiency of a Wells turbine is around 80%. At low air velocities the turbine absorbs power from the generator in order to maintain a steady speed of rotation, whilst for

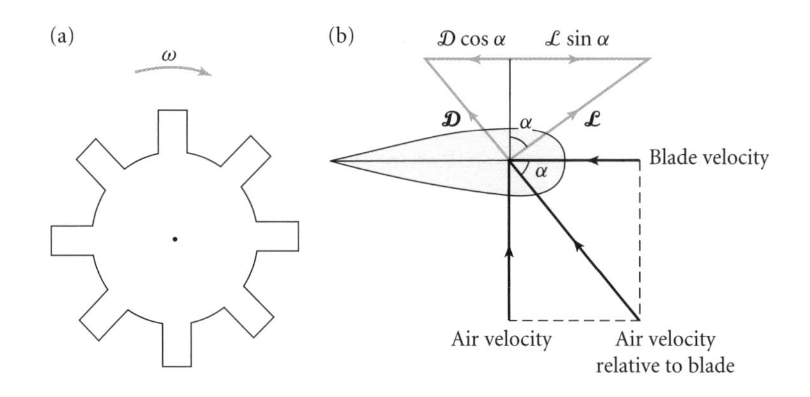

Fig. 4.16 Wells turbine. (a) Plan view of blades; (b) velocity and force triangles in frame of reference of a blade.

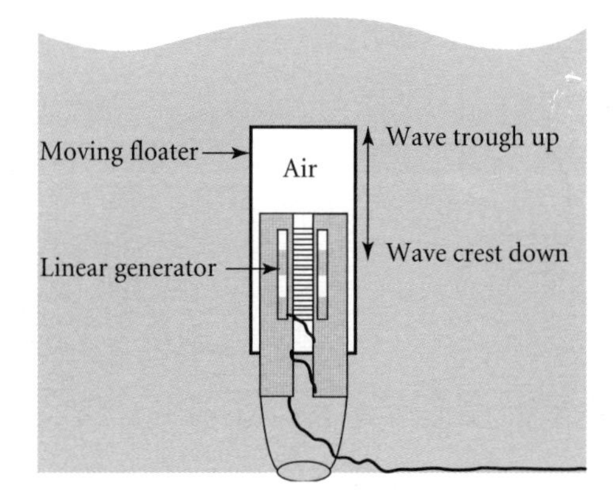

Fig. 4.17 Archimedes Wave Swing.

large air velocities the air flow around the blades is so turbulent that the net force in the direction of motion of the blade becomes erratic and the efficiency is reduced.

Two designs of shore-based OWCs are the Limpet (UK) and the Osprey (UK); generating about 0.5 and 1.5 MW, respectively. A prototype 0.5 MW Australian OWC scheme is also being developed, which uses a 40 m wide parabolic wave reflector to focus waves onto a 10 m wide shoreline OWC; the capital cost is 30% higher but the output is increased by 300%. A large floating OWC known as the Mighty Whale has been developed in Japan. It generates 110 kW but its primary role is as a wave breaker to produce calm water for fisheries and other marine activities.

4.14.3 Submerged devices

Submerged devices have the advantage of being able to survive despite rough sea conditions on the surface. They exploit the change in pressure below the surface when waves pass overhead: the pressure is increased for a wave crest but is decreased in the case of a wave trough. An example of this type of device is the Archimedes Wave Swing (AWS, Fig. 4.17). The AWS is a submerged air-filled chamber (the 'floater'), 9.5 m in diameter and 33 m in length, which oscillates in the vertical direction due to the action of the waves. The motion of the floater energizes a linear generator tethered to the sea bed. The AWS has the advantage of being a 'point' absorber, i.e. it absorbs power from waves travelling in all directions, and extracts about 50% of the incident wave power. Also, being submerged at least 6 m below the surface it can avoid damage in violent sea conditions on the surface. The device has the advantages of simplicity, no visual impact, quick replacement and cost effectiveness in terms of the power generated per kg of steel. A pre-commercial pilot project off the coast of Portugal has three AWS devices and produces 8 MW. A fully commercial AWS system could involve up to six devices per kilometre and it is estimated that the global potential for AWS is around 300 GW.

4.14.4 Floating devices

In the early 1970s public interest in wave power was stimulated by a novel device known as the Salter duck (Fig. 4.18). The device floated on water and rocked back and forth with

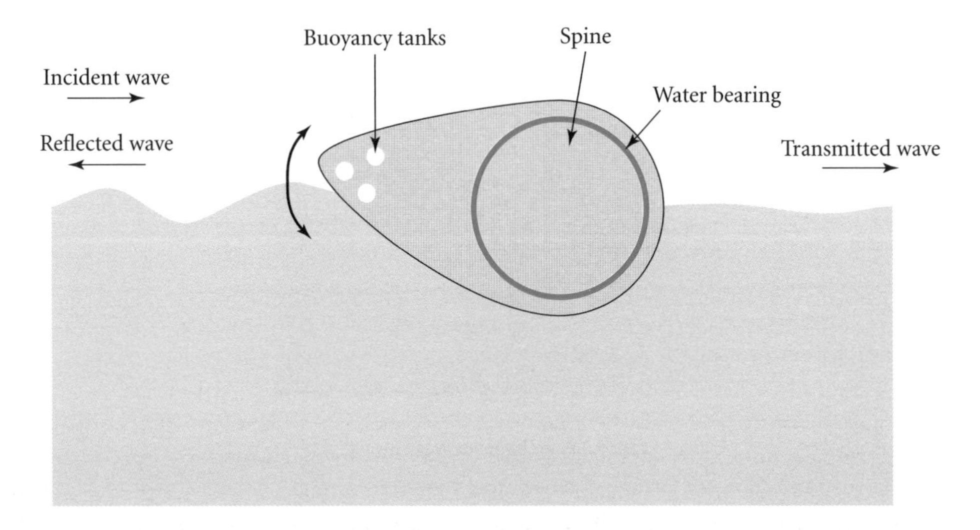

Fig. 4.18 Salter duck.

the incident waves. The shape was carefully chosen such that the surface profile followed the circular trajectories of water particles, so that most of the incident wave energy was absorbed with only minimal reflection and transmission. Efficiencies of around 90% were achieved in ideal conditions. The complete system envisaged a string of Salter ducks of several kilometres in total length parallel to a wave-front. A spinal column, of 14 m diameter, used the relative motion between each duck and the spine to provide the motive force to generate power. The device was designed to be to be used in the Atlantic Ocean for wavelengths of the order of 100 m but never got beyond small-scale trials due to lack of funding in the 1980s when governmental support for wave power in the UK was dropped in favour of wind power. Nonetheless, the Salter duck provides a useful benchmark for comparing the efficiencies of all wave power devices.

A much more recent type of floating device is the Pelamis (Fig. 4.19). It is a semi-submerged serpentine construction consisting of series of cylindrical hinged segments that are pointed towards the incident waves. As waves move along the device, the segments rock back and forth and the relative motion between adjacent segments activates hydraulic rams that pump high pressure oil through hydraulic motors and drive electrical generators. A three-segment version of Pelamis is 130 m long and 3.5 m in diameter and generates 750 kW. The combination of great length and small cross-section to the incident waves provides good protection to large amplitude waves. Three Pelamis devices are due to be installed 5 km off the coast of Portugal and it is estimated that about 30 machines per square kilometre would generate about 30 MW. In order to prevent unwanted interference effects, the devices are spaced apart by about 60–90 m.

Fig. 4.19 Pelamis.

4.15 Environmental impact, economics, and prospects of wave power

As with most forms of alternative energy, wave power does not generate harmful greenhouse gases. Opposition to shore-based sites could be an issue in areas of scenic beauty, on account of the visual impact (including the connections to the electricity transmission grid) and the noise generated by air turbines in the case of oscillating water columns. The visual impact is much less significant for offshore devices but providing cables for electricity transmission to the shore is an added cost.

The global potential of wave power is very large, with estimates of 1–10 TW. Around the UK the Department of Trade and Industry (DTI) estimated (2001) a potential of about 6 GW from wave power. The main challenges for the implementation of wave power are to reduce the capital costs of construction, to generate electricity at competitive prices, and to be able to withstand extreme conditions at sea. Wave power is generally regarded as a high risk technology. Moving to shore-based and near-shore devices reduces the vulnerability to storms but the power available is less than that further out at sea. Even the largest floating devices are vulnerable in freak storms: every 50 years in the Atlantic Ocean there is a wave with an amplitude about ten times the height of the average wave, so any device must be able to withstand a factor of a hundred times the wave energy. Measures to combat such conditions such as submerging the devices can provide an effective means of defence but add to the cost of the system. Another factor to consider is that the frequency of incident sea waves is only about 0.2 Hz, much lower than the frequency of 50–60 Hz for electricity transmission. Though this not a difficult electrical engineering problem, the challenge is to find cost-effective solutions.

Wave power is beginning to look competitive in certain niche markets, especially in remote locations where electricity supplies are expensive. However, it is likely to take one or two decades to gather sufficient operational experience for wave power to compete with other alternative energy technologies. In the long term as fossil fuel reserves become scarce, and concerns over global warming increase, forecasts of an eventual global potential for wave power to provide about 15% of total electricity production do not seem unreasonable, as part of a diverse mix of alternative energy sources.

SUMMARY

- The power output P from a dam is $P = \eta \rho g h Q$.
- The dimensionless shape number $S = \omega Q^{1/2}/(gh)^{3/4} \approx 2^{3/2}\pi^{1/2}(r/R)(v_b/v_w)$ is a useful parameter in choosing the most suitable type of turbine for a particular combination of head h and volume flow rate Q.
- The volume flow rate per unit width over a weir is given by $Q = g^{1/2}(\frac{2}{3}y_{min})^{3/2}$, where y_{min} is the height of the undisturbed level of water above the top of the weir.
- The Fourneyron water turbine employed fixed guide vanes to direct water radially outwards into the gaps between moving runner blades, and was over 80% efficient.

- In an impulse turbine the thrust arises from the momentum imparted by high speed water jets striking the cups. In a Pelton wheel the cups are shaped so that the jets splash in the opposite direction to the incident jet, in order to maximize the transfer of momentum.
- In a reaction turbine the blades are fully immersed in water. Fixed guide vanes direct the water into the gaps between the blades of a runner. The thrust is due to a combination of reaction and impulse forces.
- Combining the formula for the power output of a dam and Euler's turbine equation yields the hydraulic efficiency of a turbine as $\eta = (u_1 q_1 \cos\beta_1 - u_2 q_2 \cos\beta_2)/(gh)$.
- Hydroelectric installations have a high capital cost but low operational costs. There are environmental as well as significant social, safety, and economic issues but the electricity is almost carbon-free.
- The main cause of tides is the variation of the gravitational attraction of the Moon around the surface of the Earth. There are two tidal bulges, one facing the Moon and the other diametrically opposite.
- The speed of a tidal wave in a sea of uniform depth h_0 is given by $c = \sqrt{gh_0}$.
- The average power output of a tidal barrage is $P_{\text{ave}} = \rho g A h^2/(2T)$.
- The height of the tides can be increased by tidal resonance, due to the superposition of incident and reflected waves.
- Tidal power from barrages has limited potential, mainly due to the lack of suitable locations, the high capital cost, and the environmental impact.
- Tidal stream projects that use underwater rotors to absorb the kinetic energy of water currents are an alternative means of exploiting tidal power and have considerable potential.
- The total energy E of a surface wave per unit width of wave-front per unit length in the direction of motion is given by $E = \frac{1}{2}\rho g a^2$. About 80% of the energy is contained within a quarter of a wavelength from the surface.
- The power per unit width of wave-front is $P = \frac{1}{4}\rho g a^2 \sqrt{\frac{g\lambda}{2\pi}}$. In mid-ocean conditions the typical power per metre width of wave-front is 30–70 kW m^{-1}.
- Wave power is a vast natural resource but serious issues need to be resolved, especially survivability in storms and capital cost.
- Some shore-based wave power schemes (such as TAPCHAN and the oscillating water column) have been shown to be feasible for small-scale operation.
- Large-scale submerged and floating devices (e.g. the Archimedes Wave Swing and Pelamis) can generate much more power and are undergoing sea trials prior to commercial development.

FURTHER READING

Acheson, D. J. (1990). *Elementary fluid dynamics.* Clarendon Press, Oxford. Good account of shallow water and deep water waves.

Boyle, G. (ed.) (2004). *Renewable energy*, 2nd edn. Oxford University Press, Oxford. Good qualitative description and case studies.

Douglas, J.F., Gasiorek, J.M., and Swaffield, J.A. (2001). *Fluid mechanics.* Prentice-Hall Englewood Cliffs, NJ. Textbook on fluid mechanics—good discussion of dimensional analysis and of turbines.

Kuznetsov, N., Moz'ya, V., and Vainberg, B. (2002). *Linear water waves: a mathematical approach.* Cambridge University Press, Cambridge. Advanced textbook, including modelling of submerged devices.

WEB LINKS

www.worldenergy.org Useful data and overview of current developments.

LIST OF MAIN SYMBOLS

a	wave amplitude	R	degree of reaction
c	wave speed	S	shape factor
E	energy	t	time
g	acceleration due to gravity	u, v	velocity components
G	gravitational constant	V	potential energy
h	head	x, y, z	coordinates
k	wave number	β	angle
P	power	η	efficiency
p	pressure	λ	wavelength
q	total velocity	ρ	density
Q	volume flow rate	ω	angular velocity

? EXERCISES

4.1 Check the units to verify the expression $P = \eta\rho g h Q$ for the power output from a dam.

4.2 Estimate the power output of a dam with a head $h = 100$ m and volume flow rate $Q = 10\ \text{m}^3\ \text{s}^{-1}$. (Assume efficiency is unity, $\rho = 10^3\ \text{kg m}^{-3}, g = 9.81\ \text{m s}^{-2}$.)

4.3 Assuming the volume flow rate per unit width over a weir is of the form $Q = g^a y_{\text{min}}^b$, use dimensional analysis to determine the numerical values of a and b.

4.4 Draw a sketch of an impulse turbine consisting of four jets.

4.5 Verify that the power output of an impulse turbine is a maximum when $u_c = \frac{1}{2}u$, and that the maximum power delivered to the cup is given by $P_{\text{max}} = \frac{1}{2}\rho Q u^2$.

4.6 Explain how a rotary lawn sprinkler works.

4.7 Discuss who should pay for the cost involved in decommissioning dams when they reach the end of their life.

4.8 A turbine is required to rotate at 6 r.p.m. with a volume flow rate of 5 $m^3\ s^{-1}$ and a head of 30 m. What type of turbine would you choose?

4.9 If there are two high tides around the Earth at any instant, explain why the interval between successive high tides is 12 hours 25 min rather than 12 hours.

4.10* Compare the magnitude of the effect of Sun on the tides: (a) when the Sun and Moon are both on the same side of the Earth; (b) when the Sun and the Moon are on opposite sides of the Earth. ($m_{\text{Sun}} = 2 \times 10^{30}$ kg, $m_{\text{Moon}} = 7.4 \times 10^{22}$ kg, $d_{\text{Sun}} = 1.5 \times 10^{11}$m, $d_{\text{Moon}} = 3.8 \times 10^{8}$m.)

4.11* (a) Show by substitution that the profile $h = a \cos(kx - \omega t) + b \cos(kx + \omega t)$ satisfies the tidal wave equation $\partial^2 h/\partial x^2 = (1/c^2)\partial^2 h/\partial t^2$. (b) A uniform channel of length L is bounded at both ends by a vertical wall. Derive the height and velocity profiles of shallow water waves in the channel.

4.12 Show that the speed of a tidal bulge on the equator in the Atlantic Ocean (depth ~4000 m) is less than the speed, due to the Earth's rotation, of the seabed.

4.13 Assuming the speed c of surface waves on deep water depends only on the acceleration due to gravity g and the wavelength λ, use dimensional analysis to derive an algebraic expression of the form $c = kg^a\lambda^b$, where k is a dimensionless constant.

4.14 Calculate the speed of a surface wave on deep water of wavelength $\lambda =$ 100 m.

4.15 Given that the phase velocity and group velocity of a surface wave are $c = \sqrt{g\lambda}$ and $c_g = d\omega/dk$, respectively, where ω is the angular velocity and $k = 2\pi/\lambda$, prove that the group velocity is given by $c_g = \frac{1}{2}\sqrt{g\lambda/(2\pi)}$.

4.16* Consider a two-dimensional surface wave of amplitude a and wavelength λ such that $a \ll \lambda$ on a sea of depth $d \gg \lambda$. Assume the velocity can be expressed in the form $\mathbf{u} = \nabla\phi$, where ϕ is the velocity (called the velocity potential) satisfying Laplace's equation

$$\nabla^2\phi = \frac{\partial^2\phi}{\partial x^2} + \frac{\partial^2\phi}{\partial y^2} = 0.$$

(a) Show that travelling wave solutions exist of the form

$$\phi = Ae^{-\frac{2\pi y}{\lambda}} \sin\frac{2\pi}{\lambda}(x - ct).$$

(b) Hence show that the velocity components are given by

$$u = \frac{2\pi}{\lambda}Ae^{-\frac{2\pi y}{\lambda}} \cos\frac{2\pi}{\lambda}(x - ct) \qquad v = -\frac{2\pi}{\lambda}Ae^{-\frac{2\pi y}{\lambda}} \sin\frac{2\pi}{\lambda}(x - ct).$$

(c) Prove that particles of fluid rotate in circles of radius $r = ae^{-\frac{2\pi y}{\lambda}}$.

(d) Prove that the kinetic energy per unit width

$$E = \frac{1}{2}\rho\int_0^\lambda\int_0^\infty (u^2 + v^2)dydx = \tfrac{1}{4}\rho g a^2\lambda.$$

4.17 An oscillating water column has an air duct of cross-sectional area 1 m^2 and a water duct of cross-sectional area 10 m^2. If the average speed of the water is 1 $m\,s^{-1}$ calculate the average speed of the air.

4.18 Discuss whether it is better to build a large number of small dams or one large dam.

5 Wind power

List of Topics

- Source of wind energy
- Global wind patterns
- History of wind power
- Modern wind turbines
- Energy in the wind
- Maximum efficiency
- Turbine blade design
- Blade width and twist
- Power coefficient C_P
- Design of modern HAWT
- Fatigue in wind turbines
- Turbine control
- Turbine operation
- Wind characteristics
- Power output of turbines
- Wind farms
- Cuillagh Mountain wind farm
- Environmental impact
- Public acceptance
- Economics
- Outlook

Introduction

The international oil crises of the 1970s and recent concern over global warming have renewed interest in wind power. The wind is a carbon-free and pollution-free source of energy and wind power could produce globally 10–20% of the electrical power currently used. Wind power would therefore save a considerable amount of fuel resources. The modern wind turbine is some hundred times more powerful than the traditional windmills of the seventeenth and eighteenth centuries and wind farms already produce significant amounts of energy in some parts of the world. Twenty per cent of Denmark's electricity is currently generated by the wind.

We look firstly at the global wind resources and at the energy available in the wind. How this energy can be extracted using a wind turbine and its design are then described. We then consider the issues on the siting of turbines, generally as a wind farm, that are important for both their output and their environmental impact. We conclude the chapter with a discussion of the economics and potential of wind power.

5.1 Source of wind energy

The original source of wind energy is the radiation from the Sun, which is primarily absorbed by the land and the sea, which in turn heat the surrounding air. Materials absorb radiation differently so temperature gradients arise causing convection and pressure changes, which result in winds. A simple example is the offshore night-time wind often found on coasts, caused by the sea retaining the heat from the Sun better than the land. On a global scale, the higher intensity of solar radiation at the equator than elsewhere causes warm air to rise up from the equator and cooler air to flow in from the north and south. The direction of a wind is traditionally taken to be where it comes from, so in the Northern hemisphere the warm air rising up from the equator would give rise to a northerly wind at ground level.

An enormous amount of power resides in the winds as about 1–2% of the incident solar power of 1.37 kW/m^2 (see Example 2.4) is converted into winds. The radius of the Earth is approximately 6000 km so the cross-sectional area receiving solar radiation is about 10^{14} m^2 and the power in the winds is $\sim 10^{15}$ W. This is some 100 times the total global power usage. However, the wind is a diffuse source and it is only practical to harness a very small fraction of this amount.

Winds are variable both in time and in location, with some parts of the world exposed to frequent high winds and some to almost no wind. Places where high and low winds occur are, in particular, determined by the effect of the rotation of the Earth. Over distances of tens of kilometres the Earth's rotation has no significant effect on the direction of a wind; however, over hundreds of kilometres the effect is very noticeable. We will now explain how the Earth's rotation affects the global winds.

5.2 Global wind patterns

In a simple model, the higher intensity of solar radiation at the equator would set up a north–south convective flow of air if the Earth were not rotating. However, the Earth's rotation causes a point on the Earth to have a velocity towards the east that is highest at the equator, decreasing towards the poles. Therefore a wind moving north or south to an observer on the equator will initially have a component of velocity towards the east to an observer in space. As the wind moves away from the equator its distance to the Earth's axis decreases so its component of velocity towards the east increases. (This is a similar effect to ice skaters spinning faster when their arms are pulled in.) Air initially moving north will therefore reach a northern latitude at a point that is east of its origin. For the observer on Earth the wind appears to be accelerating towards the east and the apparent force is called the Coriolis force.

The wind speed would, in principle, reach large values by high latitudes, but by latitude 30° the flow becomes unstable. As a result the north or south motion of the wind is dissipated and such winds are thereby restricted to within the 30° latitudes. In the Northern hemisphere the sinking air near the 30° latitude gives rise to the north-east trade winds and the westerly wind belt that is the prevailing wind over Europe. A map of the resulting global wind patterns expected in this simple model, which ignores the effects of the underling configuration of oceans and continents, is shown in Fig. 5.1. Notice that there are three regions, called cells, in each hemisphere.

In practice the effects of surface friction and large-scale eddy motions have a big influence, as do seasonal variations, and only the cell nearest the equator, the Hadley cell, is clearly seen. The mid-latitude Ferrel cell is quite weak and the 'polar' cell is hardly observed. The winds are

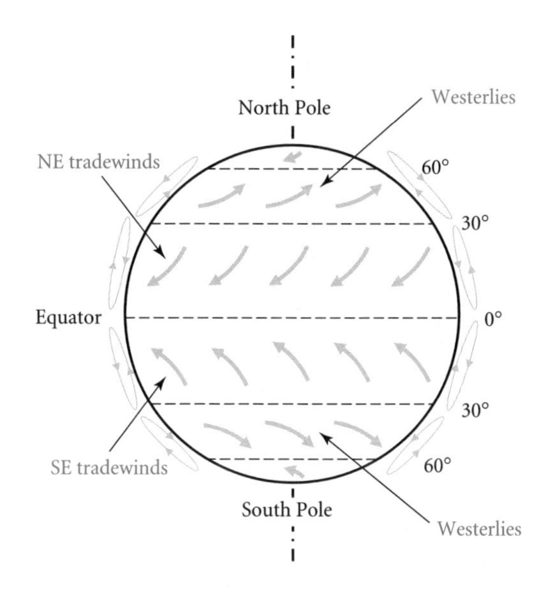

Fig. 5.1 Simplified map of global wind patterns.

generally weak in the regions between cells. However, there are many areas where the winds are strong and reliable and it is in these locations that the energy in the wind can be best exploited.

Box 5.1 History of wind power

The first recorded use of wind power was in the tenth century in the Sistan region of Persia, an area where winds can reach speeds of $\sim 45\ \mathrm{m\ s^{-1}}$, though windmills had probably been in use in the region for several centuries earlier. The windmills had a vertical axis (see Fig. 5.2(a)) and were used to pump water and grind grain. Similar windmills were used in China and may have been developed there first. In these vertical axis machines the wind pushes the sails around with a force, called a **drag** force (p. 61), dependent on the relative speed of the wind and the sail.

Vertical axis windmills using a drag force are inherently less efficient than horizontal axis windmills, which are driven by a **lift** force (p. 61). Horizontal axis windmills first appeared in Europe in England, France, and Holland around the twelfth century. Their origin is unknown (though it is possible they evolved from the horizontal axis water wheel) but such windmills spread rapidly eastward in Europe in the thirteenth century. They were used for grinding corn, pumping water, and sawing wood. The first mills were **post-mills**, where the whole mill swivelled on a post so that it could be turned manually to face the oncoming wind. In later (and larger) mills only the top with the sails and windshaft swivelled; these **tower-mills** (Fig. 5.2(b)) were introduced around the fourteenth century. From experience it was found that more power could be obtained by twisting the sails from the root to the tip of the sail.

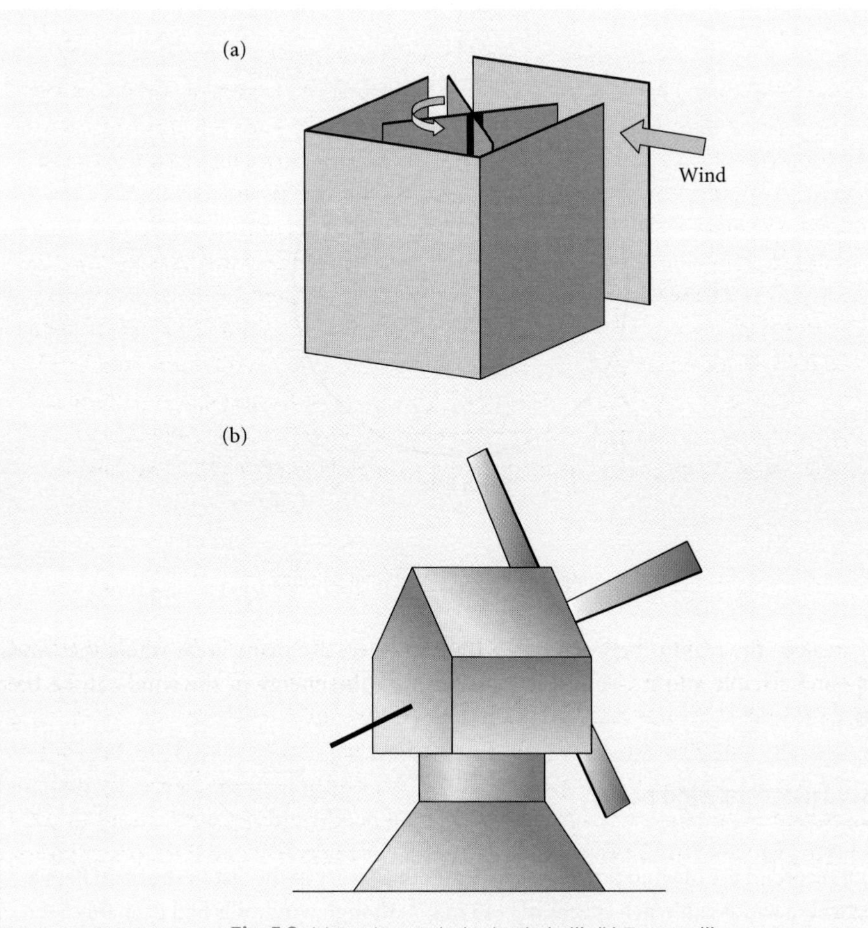

Fig. 5.2 (a) Persian vertical axis windmill. (b) Tower-mill.

The use of windmills peaked around the eighteenth century after which they were displaced by coal-powered steam engines, which were more compact and adaptable and were continuously available. However, windmills continued in regions where the land was sparsely populated, e.g. in the USA, USSR, Australia, and Argentina. In the nineteenth century small multivane windmills were developed in the USA for pumping water, where they became very common. They were eventually displaced with the development of a national electricity grid in the 1930s.

From the late nineteenth century up to the 1960s a number of wind machines were developed for generating electricity and are called wind turbines (to distinguish them from windmills). In the early twentieth century Poul La Cour built turbines using a four-bladed windmill design that produced about 25 kW. The electricity was used to produce hydrogen, which was then used for lighting. These were subsequently displaced by the introduction of diesel engines but the production of hydrogen as a fuel is now being seriously considered as hydrogen produces no CO_2 in use (see Chapter 10, section 10.12 on fuel cells). In the late 1930s a massive 1.25 MW two-bladed wind turbine was proposed by Palmer Putman and built in the USA in Vermont by the Morgan Smith Company. Though it ran successfully for a short while, a blade failure in 1945 caused the project to be terminated.

During the Second World War Denmark used wind energy when oil was not available, though this was only a temporary measure. However, following the international oil crisis in 1973 there was renewed interest in wind power as many countries began looking at sources of alternative energy. In California, the concern over the high fossil fuel costs led to a large-scale investment in wind farms, which was aided by state and federal tax incentives. The technology, though, was not then well-developed and several wind farms were unsuccessful. With the cessation of the tax incentives and the fall in oil prices in the mid-1980s, the investment in wind power in the US declined. However, in Europe, particularly in Denmark, support was maintained and the more recent alarm about global warming has stimulated considerably more interest and investment in wind turbines worldwide.

5.3 Modern wind turbines

The vast majority of current designs are horizontal-axis wind turbines (HAWTs). The turbine blades are aerofoils. These provide lift forces that drive the turbine. A modern horizontal-axis wind turbine (HAWT) is illustrated in Fig. 5.3. It consists of a tower on top of which is mounted an enclosure called the nacelle. Inside the nacelle are the bearings for the turbine shaft, the gearbox (if used), and the generator. The wind turbine blades, generally three or two, are mounted to the shaft and the nacelle is orientated by a drive mechanism, the yaw control, into the wind. The rotor is typically upwind of the tower to avoid the tower shielding the blades from the wind.

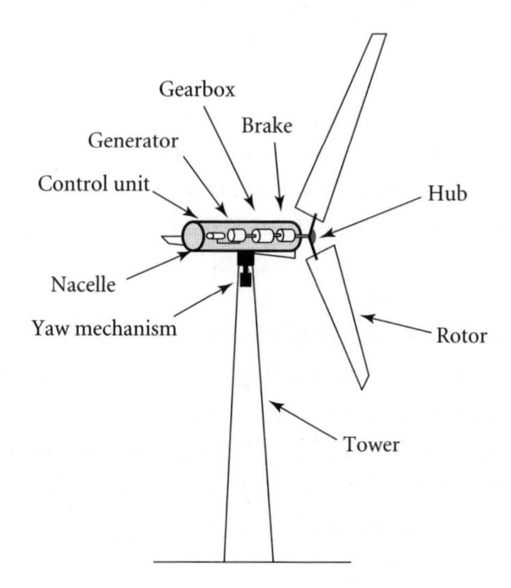

Fig. 5.3 Modern 5 MW horizontal-axis wind turbine.

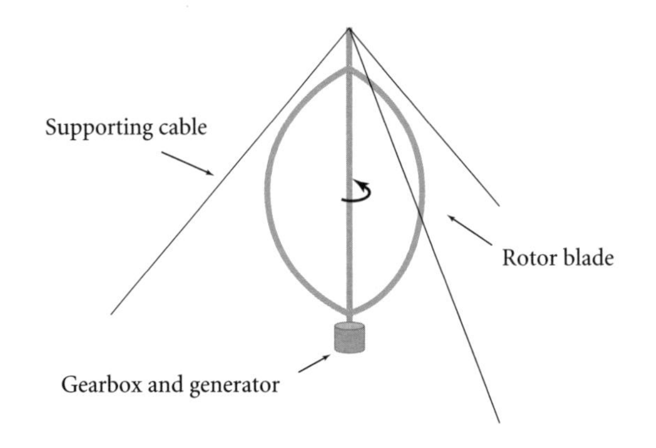

Fig. 5.4 Darrieus vertical-axis wind turbine.

There has also been some research and development of vertical-axis wind turbines (VAWTs), in particular the Darrieus design shown in Fig. 5.4. In this design the turbine is driven by lift forces generated by the wind flow over the aerofoil shaped blades. The torque is a maximum when the blades are moving across the direction of wind flow and zero when the blades are moving parallel to the direction of wind flow (see Exercise 5.9). VAWTs do not require any yaw mechanism and are easier to maintain since the gearbox and generator are situated at ground level. However, VAWTs have proven to be less cost-effective than HAWTs. In the Darrieus design (Fig. 5.4) cables are required to support the top of the rotor, and these limit the mean height of the rotors and lose the advantage of the stronger winds that occur at greater heights. Darrieus VAWTs produce greater gearbox torques than those produced by a HAWT of the same power output, and therefore need more robust construction.

We will therefore concentrate on HAWTs. We consider firstly how much energy there is in the wind and then on how HAWTs extract the energy.

5.4 Kinetic energy of wind

The energy of wind is in the form of kinetic energy. For a wind speed u and air density ρ, the energy density E (i.e. energy per unit volume) is given by

$$E = \tfrac{1}{2}\rho u^2. \tag{5.1}$$

The volume of air flowing per second through a cross-sectional area A normal to the direction of the wind is uA. Hence the kinetic energy of the volume of air flowing per second through this area is given by $P = EuA$ or

$$P = \tfrac{1}{2}A\rho u^3. \tag{5.2}$$

Thus the power in the wind P varies as the cube of the wind speed u. Hence much more power is available at higher speeds and fluctuations in wind speed can cause the output of a wind turbine to vary significantly.

EXAMPLE 5.1

Calculate the power in a wind moving with speed $u = 5\ \text{m s}^{-1}$ incident on a wind turbine with blades of 100 m diameter. How does the power change if the wind speed increases to $u = 10\ \text{m s}^{-1}$? (Assume the density of air is $1.2\ \text{kg m}^{-3}$.)

Substituting in eqn (5.2) we have

$$P = \tfrac{1}{2}A\rho u^3 = \tfrac{1}{2} \times (\pi \times 50^2) \times 1.2 \times 5^3 \approx 0.6\ \text{MW}.$$

A power of 0.6 MW is sufficient to meet the average electricity usage of about 1000 European households. Doubling the wind speed increases the power by a factor of $2^3 = 8$, so the power would increase to $8 \times 0.6 = 4.8$ MW.

5.5 Principles of a horizontal-axis wind turbine

Unfortunately, not all of the power in the wind can be extracted by a wind turbine. This is because some of the kinetic energy is carried downstream of the turbine in order to maintain air flow. This effect places a theoretical maximum efficiency of 59% for extracting power from the wind, known as the Betz limit, which is described in detail in Derivation 5.1.

As the wind flows through a turbine it slows down as part of its energy is transferred to the turbine. The airflow looks like that shown in Fig. 5.5. Upstream the speed of the wind is u_0 and it passes through an area A_0. By the time the wind reaches the turbine it has slowed to u_1 and the area of the stream-tube has increased to A_1, the area swept out by the blades of the turbine. Downstream of the turbine the wind's cross-sectional area is A_2 and its speed is u_2. The drop in speed of the wind before and after the turbine gives rise to a pressure drop across the turbine, through Bernoulli's theorem, so there is a thrust on the turbine blades.

The maximum power is generated when downstream of the turbine the wind speed is one-third of the upstream speed u_0 and at the turbine the wind speed is two-thirds of u_0, i.e. $u_2 = \frac{1}{3}u_0$ and $u_1 = \frac{2}{3}u_0$ (see Derivation 5.1). Under these conditions the power extracted,

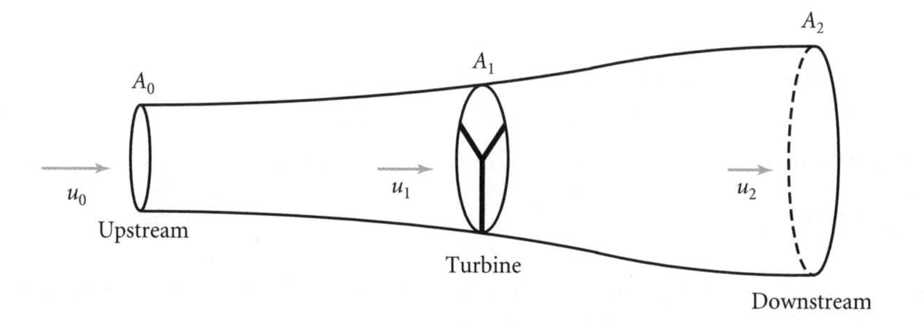

Fig. 5.5 Wind flow through a turbine.

P, is given by

$$P = \tfrac{1}{2}\rho A_1(16/27)u_0^3. \quad (5.3)$$

The power P_w in the wind passing through an area A_1 with a speed u_0 is given by eqn (5.2) as

$$P_w = \tfrac{1}{2}\rho A_1 u_0^3.$$

Thus, the fraction of the power extracted by the turbine, which is called the **power coefficient** C_P, given by

$$C_P = P/\left\{\tfrac{1}{2}\rho u_0^3 A_1\right\} \quad \text{or} \quad P = \tfrac{1}{2}C_P\rho u_0^3 A_1, \quad (5.4)$$

is $16/27 \simeq 59\%$ of the power in the incident wind passing freely through an area equal to that of the turbine, A_1. This limit for the power coefficient C_P of 16/27 of the incident wind power is called the **Betz** or **Lanchester–Betz limit**; it was first derived by Lanchester in 1915 and independently by Betz in 1921.

Derivation 5.1 Maximum extraction efficiency

We can obtain an estimate of the maximum efficiency by modelling the turbine as a thin disc (called an actuator disc) that extracts energy. Consider a stream-tube of air (see Chapter 3) shown in Fig. 5.5 that moves with speed u_1 through a wind turbine of cross-sectional area A_1. Upstream of the turbine the cross-sectional area of the stream-tube is A_0 and the air speed is u_0. Downstream of the turbine the cross-sectional area of the stream-tube is A_2 and the air speed is u_2.

Since the turbine extracts energy from the wind, the air speed decreases as it passes through the turbine and the cross-sectional area of the stream tube increases, as shown in Fig. 5.5. The thrust T exerted on the turbine by the wind is equal to the rate of change of momentum, so that

$$T = \frac{dm}{dt}(u_0 - u_2) \quad (5.5)$$

where dm/dt is the mass of wind flowing through the stream-tube per second.

The power P extracted is given by the product of the thrust and the air speed at the turbine, u_1, so that

$$P = Tu_1 = \frac{dm}{dt}(u_0 - u_2)u_1. \quad (5.6)$$

We can also express the power extracted as the rate of loss of kinetic energy of the wind, i.e.

$$P = \frac{1}{2}\frac{dm}{dt}\left(u_0^2 - u_2^2\right). \quad (5.7)$$

Comparing eqns (5.6) and (5.7), we require

$$(u_0 - u_2)u_1 = \tfrac{1}{2}\left(u_0^2 - u_2^2\right) = \tfrac{1}{2}(u_0 - u_2)(u_0 + u_2).$$

Hence

$$u_1 = \tfrac{1}{2}(u_0 + u_2), \text{ or}$$
$$u_2 = 2u_1 - u_0. \tag{5.8}$$

Also, by mass continuity (Chapter 3), the mass flow per second, $\mathrm{d}m/\mathrm{d}t$, is given by

$$\frac{\mathrm{d}m}{\mathrm{d}t} = \rho u A = \rho u_1 A_1. \tag{5.9}$$

(Note that the changes in pressure are sufficiently small that the density of air ρ is essentially constant; see Chapter 3, Example 3.3.)

Substituting for u_2 from eqn (5.8) and for $\mathrm{d}m/\mathrm{d}t$ from eqn (5.9) in eqn (5.6) yields

$$P = 2\rho u_1^2 A_1(u_0 - u_1). \tag{5.10}$$

Let $u_1 = (1-a)u_0$ where a is called the **induction factor**. Then

$$P = \tfrac{1}{2}\rho u_0^3 A_1\{4a(1-a)^2\}. \tag{5.11}$$

The power coefficient C_P, which represents the fraction of the power in the wind that is extracted by the turbine, is given by

$$C_P = P/\left\{\tfrac{1}{2}\rho u_0^3 A_1\right\} = 4a(1-a)^2. \tag{5.12}$$

Maximizing P, by setting $\mathrm{d}C_P/\mathrm{d}t$ to zero, gives the maximum power extracted P_{max} when $a = \frac{1}{3}$ equal to

$$P_{\mathrm{max}} = \tfrac{1}{2}\rho u_0^3 A_1\{16/27\}, \tag{5.13}$$

which is ~59% of the power in the incident wind passing freely through an area equal to that of the turbine, A_1. This limit for the power coefficient C_P of 16/27 of the incident wind power is called the **Betz** or **Lanchester–Betz limit**.

5.6 Wind turbine blade design

The thrust on the turbine is generated and translated into rotational energy by shaping the turbine blades as aerofoils. A wind turbine is shown in Fig. 5.6(a) and a section of a blade is shown in Fig. 5.6(b). The air speed at the turbine is u_1 and the rotational speed of the blade is v (i.e. perpendicular to the direction of air flow). The resultant velocity of the air relative to the blade, $\boldsymbol{u}_\alpha$, makes an angle φ to the direction of the blade, given by

$$\tan\varphi = u_1/v \tag{5.14}$$

The angle of attack of the wind on the blade is α. The motion of the wind over the aerofoil section gives rise to a lift force $\boldsymbol{\mathcal{L}}$ per unit length. The lift $\boldsymbol{\mathcal{L}}$ of an aerofoil is perpendicular to the direction of the air flow $\boldsymbol{u}_\alpha$, so the thrust T per unit length on the aerofoil (neglecting drag, $\boldsymbol{\mathcal{D}}$) is given by $\mathcal{L}\cos\varphi$ and the power P developed by $\mathcal{L}(\sin\varphi)v$ (force multiplied by velocity). Using eqn (5.14) to substitute for v gives

$$P = \mathcal{L}(\sin\varphi)u_1 \cot\varphi = \mathcal{L}(\cos\varphi)u_1 = Tu_1, \tag{5.15}$$

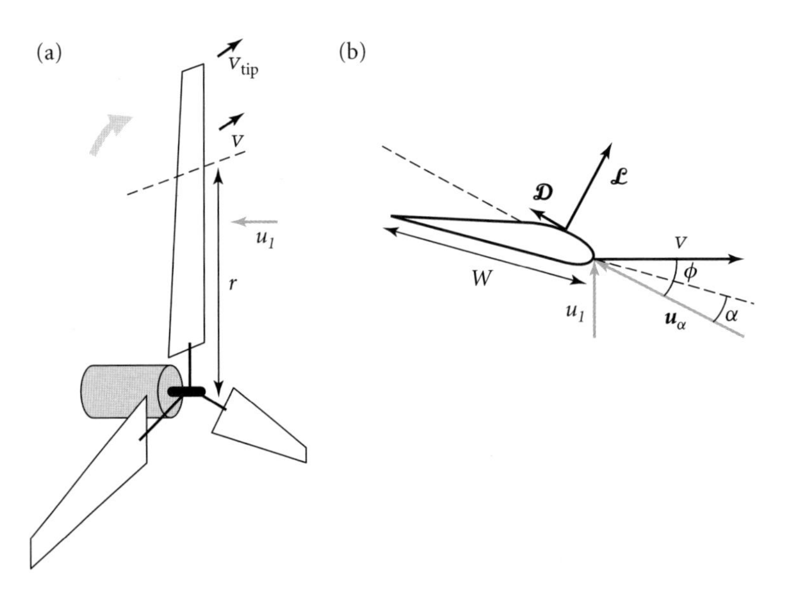

Fig. 5.6 (a) Wind incident on rotating turbine. (b) Section of turbine blade at distance r from the axis.

showing that the power developed equals that delivered by the thrust of the wind, when drag is neglected.

The speed v of the blade at a radius r is given by

$$v = \frac{r v_{tip}}{R} \tag{5.16}$$

where v_{tip} is the speed of the blade tip and R is the maximum radius of the turbine blade. An important parameter is the **tip-speed ratio** λ, defined as the ratio of the speed of the blade at the tip, v_{tip}, to the speed of the incident wind, u_0, i.e.

$$\lambda = \frac{v_{tip}}{u_0}. \tag{5.17}$$

When the Betz condition ($u_1 = 2u_0/3$) is satisfied, then using eqns (5.16) and (5.17) we can express the angle φ in terms of r, R, and λ through

$$\tan\varphi = \frac{u_1}{v} = \frac{2R}{3r\lambda}. \tag{5.18}$$

From eqn (5.18) we observe that, for a given radius r, the angle φ is only a function of the tip-speed ratio λ, which is why λ is so significant in the design of a wind turbine. The apparent angle of the wind φ increases with decreasing radius r (since $\tan\varphi \propto 1/r$). Therefore turbine blades are designed with a twist that increases with decreasing r in order for the angle of attack α to remain optimum. The blade width, W, also increases with decreasing r so that the component of lift generates the thrust required to maintain the Betz condition. The calculation of the width of a turbine blade is given in Derivation 5.2. The results for the width and angle of twist of a blade on a turbine with n blades are

$$\text{Angle of twist} = \varphi - \alpha = \tan^{-1}(2R/3r\lambda) - \alpha, \tag{5.19}$$
$$\text{Width} \approx 8\pi R(\sin\varphi)/3\lambda n. \tag{5.20}$$

EXAMPLE 5.2

A wind turbine with three blades is operating in a mean wind speed of 8 m s^{-1}. The turbine rotates at 15 rpm. Each blade is 40 m long. Estimate the width at the midpoint and tip of each blade.

The time τ for one revolution of the tip of a blade of length R is

$$\tau = 2\pi R/v_{tip}.$$

Thus the number n_{rpm} of revolution per minute (rpm) is

$$n_{rpm} = 60/\tau.$$

Therefore the tip speed v_{tip} and tip speed ratio $\lambda = v_{tip}/u_0$ are

$$v_{tip} = 2\pi R/\tau = 2\pi R n_{rpm}/60 = 2\pi(40)(15)/60 = 62.8 \text{ m s}^{-1},$$
$$\lambda = 62.8/8 = 7.85.$$

From eqn (5.18) the angle φ the wind makes to the movement of the blade at a distance r from the turbine axis is given by

$$\tan\varphi = 2R/3r\lambda.$$

So $\varphi_{tip} = \tan^{-1}(2/3\lambda) = 4.85°$. Substituting φ_{tip} into eqn (5.20) gives an estimate of the width of the blade tip

$$W_{tip} = 8\pi R(\sin\varphi_{tip})/3\lambda n = 8\pi(40)(0.0846)/\{3(7.85)(3)\} = 1.20 \text{ m}.$$

At the midpoint $\varphi_{mid} = \tan^{-1}(4/3\lambda) = 9.64°$ and $\sin\varphi_{mid} = 0.167$, so

$$W_{mid} = 2.38 \text{ m}.$$

Derivation 5.2 Blade design

A wind turbine has typically three blades. The effect of a blade on the wind flow extends over a sufficient distance in the direction of the lift that the blade's reaction on the wind occurs over a large part of the whole annular area $dA = 2\pi r\, dr$ (see Exercise 5.7). Variations in u_1 over the time for one blade to reach the position of the next blade (equal to $2\pi/n\Omega$ for n blades where Ω is the angular speed of the turbine) are modelled by taking u_1 as the average wind speed at the turbine.

The total thrust from n blades dT_n on the annular area dA equals the rate of change of momentum of the wind. From eqns (5.5) and (5.9) this is given by

$$dT_n = (\rho dA u_1)(u_0 - u_2) = \rho u_1^2 2\pi r \, dr \tag{5.21}$$

when the Betz condition ($u_2 = \frac{1}{2}u_1 = \frac{1}{3}u_0$) is satisfied.

This thrust is equal to the sum of the components of the lift $d\mathcal{L}$ from each section of blade between r and $r + dr$. (We neglect the effect of drag in this discussion.) The lift $d\mathcal{L}$ is given by (see Chapter 3, eqn (3.15))

$$d\mathcal{L} = \tfrac{1}{2} C_L \rho u_\alpha^2 W dr \tag{5.22}$$

where α is the angle of attack and W is the width (or chord) of the blade at r. From Fig. 5.6 the speed $u_\alpha = u_1/\sin\varphi$. Equating the thrust to the sum of the components of lift and substituting for u_α gives

$$\rho u_1^2 2\pi r \, dr = n d\mathcal{L} \cos\varphi = \tfrac{1}{2} n C_L \rho u_1^2 W \, dr \cos\varphi/\sin^2\varphi, \tag{5.23}$$

In order to satisfy the Betz condition (which maximizes the efficiency of the machine), the width W of the blade at a distance r from the axis must therefore be given by

$$W = 4\pi r \tan\varphi \, \sin\varphi/nC_L. \tag{5.24}$$

The tip-speed ratio $\lambda = v_{tip}/u_0$, so $\tan\varphi = u_1/v \equiv 2R/3r\lambda$. Substituting for $\tan\varphi$ yields

$$W = 8\pi R \sin\varphi/3\lambda n C_L. \tag{5.25}$$

As $\sin\varphi$ increases as r decreases, the width of the blade increases from the tip to the root of the blade (see Fig. 5.6).

The angle of attack α is chosen to give the highest lift to drag ratio and is typically a few degrees. For a three-blade wind turbine, λ is often chosen to be between 6 and 10. An aerofoil generally has C_L close to 1 at the angle of attack α_0 with the largest C_L/C_D ratio. For α_0 equal to 6 degrees and $C_L = 1$, the width W of the blade, and the shape and angle of twist of the optimum blade of length $R = 24$ m, are given in Table 5.1 for a tip-speed ratio $\lambda = 8$. These are calculated with eqn (5.25) and eqn (5.18).

Table 5.1 Width and angle of twist for an optimum turbine blade of length 24 m for $\lambda = 8$

		Angle (°)	
Radius (m)	Width (m)	Twist	Wind φ
6	2.649	12.4	18.4
12	1.377	3.5	9.5
18	0.925	0.3	6.3
24	0.696	−1.2	4.8

In practice, blade design takes into account drag as well as lift, and the effects of all the sections (blade element theory) are calculated using computer programs.

5.7 Dependence of the power coefficient C_P on the tip-speed ratio λ

In the discussion in Section 5.6 of the power extracted by a wind turbine we have neglected the effect of drag $\mathcal{D}$. This force acts in the direction of the wind at right angles to the lift $\mathcal{L}$ as shown in Fig. 5.6. Although the drag is small compared to the lift, its direction is nearly opposite to that of the blade's motion. As a result the effect of the drag is enhanced by a factor of $\sim\lambda$, the tip-speed ratio, as we will now show.

As a result of the drag the rotational force is reduced and becomes (see eqn (4.33))

$$\mathcal{L}\sin\varphi - \mathcal{D}\cos\varphi \equiv \mathcal{L}\sin\varphi(1 - g\cot\varphi) \tag{5.26}$$

where $g \equiv C_D/C_L$ is the drag to lift ratio of the blade's aerofoil shape. From eqn (5.18)

$$\cot\varphi = 3r\lambda/2R$$

so the reduction varies with radius r. We can estimate the overall reduction by taking a typical radius r as $2R/3$ so $\cot\varphi \sim \lambda$. The maximum power coefficient $C_{P_{max}}$ is therefore expected to be about

$$C_{P_{max}} \approx (1 - g\lambda)C_{P_{Betz}}. \tag{5.27}$$

For a modern wind turbine with $g{\sim}1/40$ and $\lambda \sim 10$ then the rotational force and hence C_P would be reduced by ~25%. Figure 5.7 shows a plot of the variation of C_P with λ for a turbine designed to have maximum efficiency at $\lambda \sim 10$.

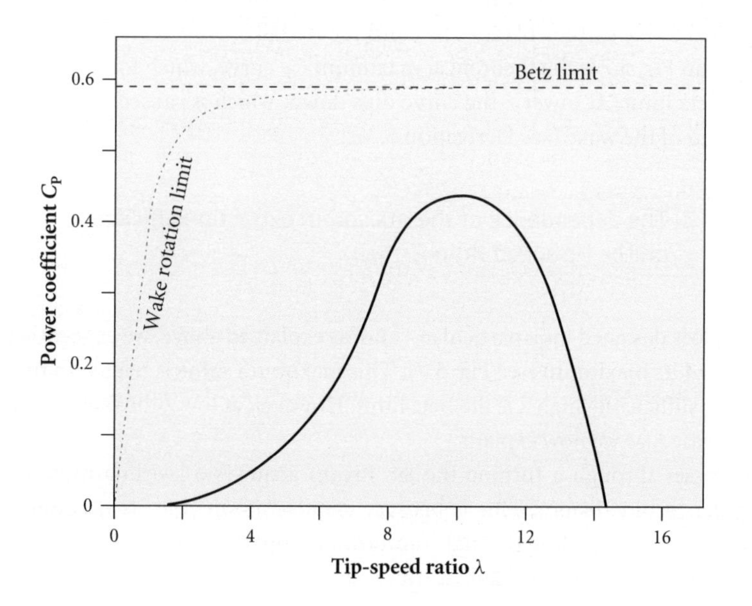

Fig. 5.7 $C_P - \lambda$ curve for a high tip-speed ratio wind turbine.

EXAMPLE 5.3

A wind turbine is operating in a mean wind speed of 10 m s^{-1}. It rotates at a speed of 20 rpm. Each blade is 45 m long and has an airfoil section with a drag to lift ratio of 1/50. The blades are optimized for the mean wind speed. Estimate the maximum power coefficient and the power output at the mean wind speed. (Assume the density of air is 1.2 kg m^{-3}.)

From Example 5.2

$$\lambda = 2\pi R n_{\mathrm{rpm}}/(60u_0) = 2\pi \times 45 \times 20/(60 \times 10) = 9.4\,.$$

Equation (5.27) gives

$$C_{P_{\max}} \approx (1 - g\lambda)C_{P_{\mathrm{Betz}}} = (1 - 9.4/50)16/27 = 0.48\,.$$

Substituting in eqn (5.4),

$$P = \tfrac{1}{2}C_P \rho u_0^3 A_{\mathrm{turbine}} = \tfrac{1}{2}(0.48)(1.2)(10^3)\pi(45)^2 = 1.83\ \mathrm{MW}.$$

The width W and twist of a turbine blade are designed for a particular λ. If the wind speed or the rotational angular velocity of the turbine alter, then φ and therefore the lift $\mathcal{L}$ changes. This changes the thrust from its optimal value and C_P decreases. So we expect C_P to fall away on either side of its maximum value as seen in Fig. 5.7. At low λ the blade stalls (see Section 5.9). Above $\lambda \simeq 13$ corresponds to the induction factor (see Derivation 5.1) $a > \frac{1}{2}$ and is where using the momentum change to derive the thrust on the turbine, eqn (5.5), is invalid as it corresponds to a negative final velocity. In this region the thrust rises rather than falls and corresponds to a turbulent wake behind the turbine blades. Although the thrust T continues to rise in this region while the speed of the wind u_1 at the turbine decreases, the power imparted to the turbine blades Tu_1 continues to fall.

Also shown in Fig. 5.7 is the theoretical maximum C_P curve, which for $\lambda \geq 4$ is close to the Lanchester–Betz limit. At lower λ the curve dips down, which is caused by the neglect of the swirling motion of the wake (see Derivation 5.3).

Derivation 5.3 The dependence of the maximum extraction efficiency on the tip-speed ratio λ

A turbine blade is designed for a particular λ and, as explained above, we expect C_P to fall away on either side of its maximum (see Fig. 5.7). This maximum value is related to the theoretical limit, which at sufficiently high λ is the Betz limit. However, at low λ the theoretical maximum for C_P is decreased, as we now explain.

As wind passes through a turbine the air stream acquires a swirling motion as a result of rotating the turbine blades. This is because angular momentum is necessarily imparted downstream since the air flow through the turbine imparts a torque to the turbine blades (see Fig. 5.8). If the speed of the blade is much greater than the speed of the wind, i.e. a high

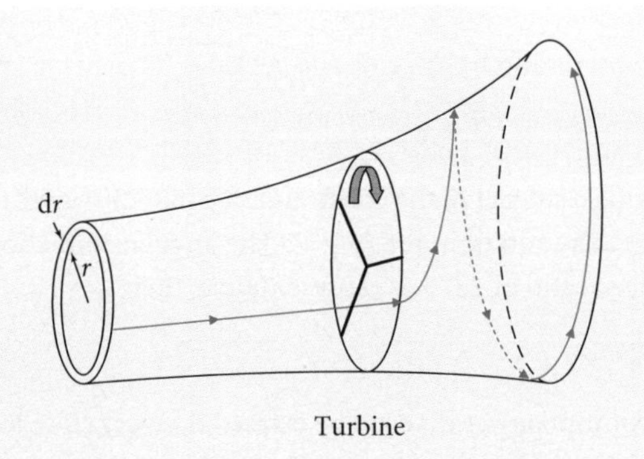

Fig. 5.8 Angular momentum of air flow after passing through turbine.

λ, then the energy associated with this rotational motion of the wind is much smaller than that associated with the momentum change of the wind in the direction of the wind (axial direction). To see this we will consider what happens to the wind when it passes through a small annular area (see Fig. 5.8).

The power extracted by the turbine from the stream-tube of wind defined by the annular area between r and $r + dr$ equals the torque dG arising from the rate of change in angular momentum of the wind multiplied by the angular speed Ω of the turbine. At a given radius this product equals the force arising from the rate of change in momentum of the wind in the direction of the blade motion multiplied by the speed of the blade.

The power extracted dP equals the loss of axial kinetic energy of the wind minus its gain in rotational kinetic energy per second. The mass flow $d\dot{m}$ equals $\rho u_1 2\pi r\, dr$. If the wind gains an angular velocity ω_1, which will be in the opposite sense to that of the blade, after passing through the turbine, the rotational kinetic energy gained per second is given by $\frac{1}{2} d\dot{m} r^2 \omega_1^2$ and the change in angular momentum per second by $d\dot{m} r^2 \omega_1$. So the torque $dG = d\dot{m} r^2 \omega_1$, and dP is given by

$$dP = dG\,\Omega = d\dot{m}\omega_1 r^2 \Omega = \tfrac{1}{2} d\dot{m}\left(u_0^2 - u_2^2\right) - \tfrac{1}{2} d\dot{m} r^2 \omega_1^2. \tag{5.28}$$

Defining $\omega = \frac{1}{2}\omega_1$ and the **angular induction factor** $a' = \omega/\Omega$,

$$dP = dP_B/(1 + a') = d\dot{m}\omega_1 r^2 \Omega \tag{5.29}$$

where dP_B is the change in linear kinetic energy per second. As in Derivation 5.1, which neglected wake rotation, dP_B equals the change in linear momentum per second multiplied by the wind speed at the turbine, eqn (5.5) leading to eqn (5.11), which can be expressed as

$$dP_B = \tfrac{1}{2} d\dot{m} 4a(1 - a)u_0^2. \tag{5.30}$$

Substituting gives

$$dP = \tfrac{1}{2} d\dot{m} 4a(1 - a)u_0^2/(1 + a'). \tag{5.31}$$

Defining the radial tip-speed ratio $\lambda_r = \Omega r/u_0$ and substituting $d\dot{m}\omega_1 r^2 \Omega$ for dP gives

$$a'(1 + a') = a(1 - a)/\lambda_r^2. \tag{5.32}$$

Substituting $d\dot{m} = \rho u_1 2\pi r\,dr$ and $u_1 = (1-a)u_0$ gives

$$dP = \tfrac{1}{2}\rho(2\pi r\,dr)4a(1-a)^2u_0^3/(1+a'). \tag{5.33}$$

Maximizing dP by varying a subject to the constraint of eqn (5.30) gives $a' << 1$ and $a \cong \frac{1}{3}$, as before when neglecting wake rotation, for $\lambda_r \geq 2$. The angular induction factor $a' \approx 2/(9\lambda_r^2)$ and, if we take a tip-speed ratio of $2\lambda/3$ as representative, then

$$dP \approx dP_B/\{1 + 1/(2\lambda^2)\}. \tag{5.34}$$

Thus for $\lambda \geq 4$ the maximum power that can be extracted is very close to the Lanchester–Betz limit, as shown in Fig. 5.7.

5.8 Design of a modern horizontal-axis wind turbine

A modern HAWT is illustrated in Fig. 5.3. Turbines are designed with a large tip-speed ratio λ of ~6–10 to give a higher power efficiency C_P. A larger λ also means a higher shaft speed and hence a lower torque. Lowering the torque allows the use of a smaller gearbox (if used) or smaller generator. Turbines with a large λ also have smaller width blades (see eqn (5.20)), so less blade material is required, which cuts costs. Ensuring that the blades are each wide enough to have sufficient strength means that there are only two or three blades on a modern large turbine. This can make starting difficult, but this problem can be overcome by using a starting motor.

Blades were originally made from wood; aluminium and steel were then employed. Nowadays though, fibre glass and other composite materials are increasingly used because of their high strength and stiffness coupled with low density. Carbon fibre and carbon/glass hybrid composites are being developed. The fatigue properties of the materials used are very important because of the very large number of revolutions, typically a few times 10^8, that a turbine makes in a 30-year design lifetime.

Fatigue causes a wire that is bent repeatedly to weaken and break—the strength of the wire decreases with the number of bends. Likewise, the rotation of the blades causes the loads experienced by the turbine to change repeatedly. These changing loads weaken the structure through fatigue, which must be allowed for in the design. The fatigue of materials is discussed more in Box 5.2.

The nacelle containing the generator and control mechanisms is mounted on a tower. Most towers have a strong and, for economy, lightweight structure. As a result the natural frequency of vibration of the tower lies below the rotational frequency of the blades. When starting or stopping the turbine the shaft speed is therefore changed quickly to avoid shaking the tower.

The size of wind turbines has increased over the last 20 years. The **rated power** is the maximum continuous power that the turbine is designed to produce. Typical specifications in 1985 were rated power 80 kW, rotor diameter 20 m, hub height 30 m; in 1995 rated power 600 kW, diameter 46 m, height 78 m; while in 2005 a modern 5 MW HAWT turbine has a hub height at 90 m and blades that sweep out a circle of diameter 115 m. The ratio of the annual energy yield to that which would be produced at the rated power is called the **capacity factor**. It is typically $\sim\frac{1}{3}$.

Box 5.2 Fatigue in wind turbines

Wind turbine fatigue requirements are particularly severe because the number of load cycles is so high. For a 30-year lifetime, an 80 m diameter turbine operating for ~80% of the time at $\lambda = 8$ in a wind speed of 10 m s^{-1} will make some 2.4×10^8 rotations. This means that the maximum stresses (force per unit area) must be lower than in other structures to avoid failure through fatigue (see Fig. 5.9).

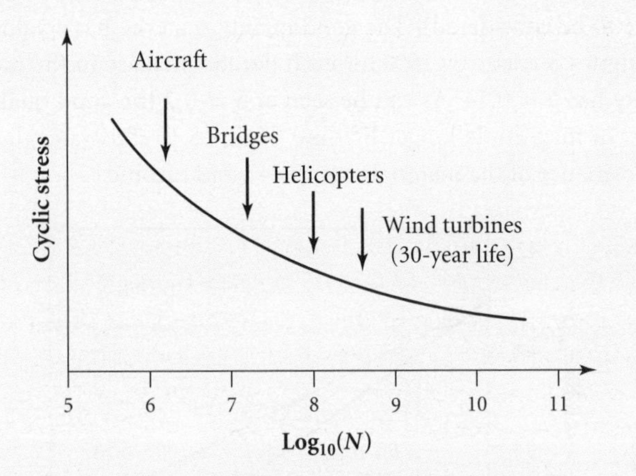

Fig. 5.9 Cyclic stress versus $\log_{10}$ (cycles to failure).

Fatigue failure is the fracture of material after it has been subjected to repeated cycles of stress changes at levels considerably below its initial static strength. The number of cycles to failure decreases as the alternating stress level increases. (The stress level can be characterized by the mean stress and its range, which is equal to $\sigma_i^{\max} - \sigma_i^{\min}$.) Fatigue involves the initiation and growth of cracks in a material under the repeated stress cycles. Discontinuities such as a sharp corner or flaws in the material are prime sites. Wind turbines installed in California in the 1980s suffered blade failures due to the fatigue not being fully understood.

Fatigue can be quantified by using the Palmer–Milner linear damage rule (often called Milner's rule). This method breaks down the cyclic stresses that a structure undergoes into the number of cycles n_i at each stress level σ_i that occur. The total damage $D_{\rm M}$ sustained by a structure is given by

$$D_{\rm M} = \sum_{1}^{s} (n_i/N_i) = n_1/N_1 + n_2/N_2 + \cdots + n_s/N_s \tag{5.35}$$

where N_i is the number of cycles to failure at the stress level σ_i. Milner's rule states that failure will occur when $D_{\rm M} = 1$, though factors of two are often found between predicted and measured lifetimes.

The fatigue strength of a material is the value of the stress level σ required to cause failure after a specified number of cycles N. The results can be expressed as an S versus

N plot (S–N plot), where S is the ratio σ/σ_0. The stress σ_0 is the static strength of the material. The data can be represented by the equation

$$\sigma/\sigma_0 = 1 - b\log_{10}(N) \tag{5.36}$$

where b is a positive constant.

Equation (5.36) predicts how the stress that can be tolerated decreases with the number of cycles. The results for two fibreglass composite materials are shown in Fig. 5.10. (The data is for tensile stresses with a ratio $R = \sigma_i^{\min}/\sigma_i^{\max} = 0.1$; data for compressive stresses would also have to be considered). The good quality material has a value of $b = 0.1$, i.e the fatigue strength decreases by 10% for each decade increase in the number of cycles. The poor quality has $b = 0.14$. As can be seen at $S = 0.2$ the good quality material has over two orders of magnitude longer lifetime. Figure 5.10 illustrates the importance of the fatigue performance of the materials used in a wind turbine.

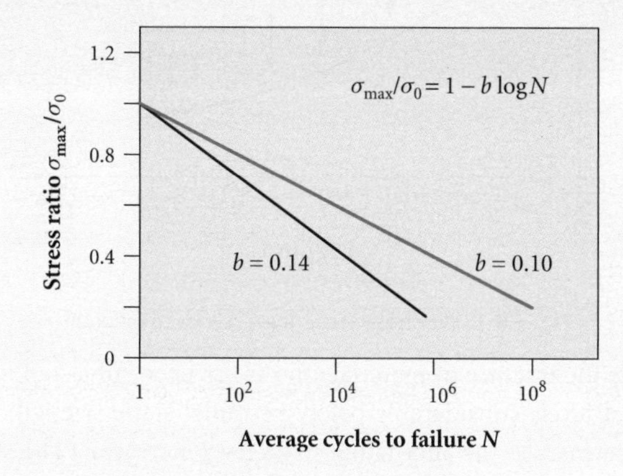

Fig. 5.10 S–N curves for two fibreglass composites.

A material with the lowest b coefficient is not necessarily the best as the static strength is also important. Increased strength could be obtained by having more fibres but with a slight increase in the value of b. Whether this change would give a better performance depends on the fatigue strength required.

Larger turbines reduce the land requirements. They also tend to operate in a greater mean wind speed, as their hubs are higher. Fewer turbines are required for the same output, so part of the maintenance and operational costs is reduced. The cost per kWh therefore decreases. The price of land and infrastructure, such as connecting to the grid, also reduces. Furthermore, many ridges only allow a single line of turbines, so more power is produced if they are larger. But, for a given amount of capital, more smaller units reduce the risk from failures. However, the reliability of turbines is now very good with turbines available for operation ~98% of the time. In practice these considerations have tended to favour the maximum size possible, which is ~5 MW (2005).

5.9 Turbine control and operation

In the generation of electrical energy from the wind, a wind turbine needs to be controlled to optimize its output. First the turbine needs to be orientated into the wind, which is accomplished by the **yaw control** mechanism. The wind provides the driving torque for the electrical generator and the current flowing in the generator produces an opposing torque, the generator torque. Ignoring friction, the wind torque equals the generator torque in steady operation. For a wind turbine with an induction generator the maximum current and opposing electrical torque are produced when the speed of the turbine is only about one per cent higher than its rated speed. The rated design speed is generally chosen to match the frequency of the electrical grid. While this (essentially) constant speed turbine is simple to control, for fixed pitch blades the aerodynamic efficiency cannot be kept optimized as the wind speed changes. For this reason most modern large wind turbines tend to be variable speed variable pitch machines and generate electricity using AC–DC–AC convertors (see Chapter 10, Box 10.1).

For a turbine running at a constant rotational speed and where the blades have a fixed twist, then as the wind speed increases so does the angle of attack α because the ratio u_1/v gets larger (see Fig. 5.6). If α becomes too large then the lift drops sharply as the aerofoil stalls. This occurs at around $\alpha \sim 10$ degrees for the aerofoil profile shown in Fig. 3.9 in Chapter 3. This effect, called 'stall' control, provides a means of limiting the torque in high winds.

The output power curve for a typical fixed pitch constant speed turbine is shown in Fig. 5.11. The turbine blade will be optimized for a tip-speed ratio λ corresponding to a wind speed close to the mean. The **rated wind speed** (u_{rated}) is such that the wind is strong enough to produce the maximum output power of the turbine generator. The tip-speed ratio λ and hence the power coefficient C_P (see Fig. 5.7) change with wind speed for a constant speed turbine. This variation in C_P is why we see that the output power at a wind speed of 14 m s^{-1} is only roughly 2.5 times that at 8 m s^{-1}, rather than $(14/8)^3 = 5.4$ times larger.

In the case of a variable speed variable pitch turbine both the speed of rotation and the pitch of the blades can be altered. This facility allows C_P (and hence the output) to be optimized

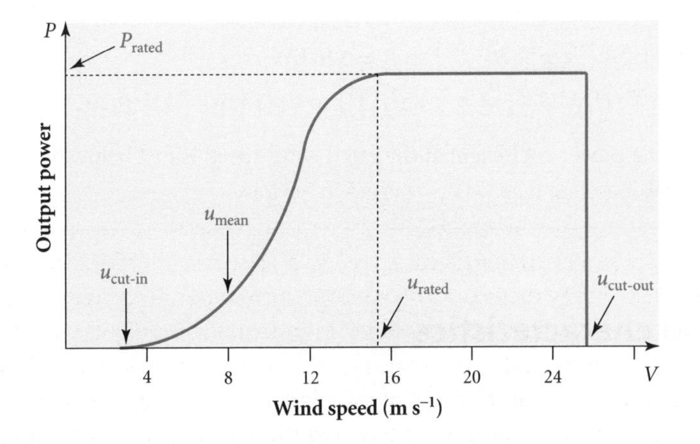

Fig. 5.11 Output power versus wind speed.

when the wind is above the minimum required to operate the turbine ($u_{\text{cut-in}}$). Generally only the speed is altered, by changing the generator torque, in the region between the cut-in and rated wind speeds to maintain the optimum tip speed ratio λ. The generator torque can be varied by changing the electrical load. Above u_{rated}, the speed of the turbine is usually maintained at a constant value and the pitch of the blade is adjusted (which alters the angle of attack and hence the lift), to reduce the wind torque and keep loads on the turbine within safe limits.

There is a balance between the maximum energy output and the capital cost of generators. To take full advantage of periods of high wind speed would require large but expensive generators. Choosing the optimum balance results in rated generator powers typically ~3 times the average power output. The ratio of the energy produced in a year to the energy that would be produced if the turbine operated at its rated power is the **capacity factor**, and is therefore typically $\sim\frac{1}{3}$.

Turbines operate typically for 65–80% of the time, depending on demand and on when the wind speed is below $u_{\text{cut-in}}$ or above $u_{\text{cut-out}}$. Above the rated wind speed the generator and wind torques can be adjusted to maintain the power at the maximum. Above the maximum safe operating wind speed ($u_{\text{cut-out}}$) the pitch of the blades can be altered to shed power by reducing the angle of attack (**feathering**), or the blades stall by design. The turbine can be stopped by application of the shaft brake.

The **specific energy output** of a wind turbine is defined as the output power per unit of swept area and is useful when comparing turbines of different size or design since the capacity factors can differ between turbines.

EXAMPLE 5.4

Calculate the average power output of a wind turbine with blades of 85 m diameter operating in wind with a mean speed of 7 m s^{-1}. At this wind speed the power coefficient is 0.45. The rated output power is 1.5 MW when the wind speed is greater than 13 m s^{-1}. What is the power coefficient at a wind speed of 13 m s^{-1}? (Assume the density of air is 1.2 kg m^{-3}.)

Substituting in eqn (5.4) we have

$$0.45 = P/\{\tfrac{1}{2}(1.2)(\tfrac{1}{4} \times \pi \times 85^2)7^3\} \text{ so } P = 526 \text{ kW},$$

$$C_P = 1.5 \times 10^6/\{\tfrac{1}{2}(1.2)(\tfrac{1}{4} \times \pi \times 85^2)13^3\} \text{ so } C_P(13 \text{ m s}^{-1}) = 0.20\,.$$

In this example the power coefficient at the rated wind speed is just below half its value at the mean wind speed.

5.10 Wind characteristics

We know from experience that the speed of the wind in any location varies considerably with time. This variation affects the amount of power in the wind and the loads felt by the turbine. In particular, there are fluctuations in the wind speed over periods of days from changes in

Table 5.2 Surface roughness (z_0) values

Terrain	z_0 (m)
Urban areas	3–0.4
Farmland	0.3–0.002
Open sea	0.001–0.0001

the weather and over periods of minutes from gusts. Averages over a ~10 min period are used to define the steady wind speed. The shorter term fluctuations about this value are quantified by the turbulence intensity I_T, defined as the ratio of the standard deviation σ_T of the wind speed to the steady wind speed. σ_T generally increases as the steady wind speed increases and I_T is found to depend in particular on the terrain and height. I_T increases with surface roughness and varies approximately as $[\ln (z/z_0)]^{-1}$, where z is the height of the turbine and z_0 characterizes the terrain (see Table 5.2). Its magnitude is important in determining the fatigue loading on a wind turbine.

The steady wind is characterized by its frequency distribution $f(u)$ and its persistence. Persistence data give, for example, the number of times the wind is expected to blow for more than an hour with a speed greater than u, while $f(u)\Delta u$ gives the percentage of time over which the wind speed is expected to be between u and $u + \Delta u$. The persistence of the wind is important in estimating the dependability of the generated wind power.

For sites that have an annual mean wind speed greater than 4.5 m s^{-1} the frequency distribution is often well described by the **Rayleigh distribution**

$$f(u) = (2u/c^2)\exp[-(u/c)^2] \tag{5.37}$$

where $c = 2\langle u\rangle/\pi^{1/2}$ and $\langle u\rangle$ is the average wind speed. For a Rayleigh distribution the power in the wind is given by

$$P = \tfrac{1}{2}\rho A\langle u^3\rangle = 0.955\rho A\langle u\rangle^3 \approx \rho A\langle u\rangle^3. \tag{5.38}$$

The Rayleigh frequency distribution for a mean wind speed of 8 m s^{-1} is shown in Fig. 5.12.

The wind speed u increases significantly with the height above the ground, with the speed zero at the surface. Its rate of change decreases with height as the frictional forces decrease, as shown in Fig. 5.13. A commonly used form to describe the dependence of u on height z is

$$u(z) = u_S(z/z_S)^{\alpha_S} \tag{5.39}$$

where z_S is the height at which u is measured to be u_S, typically 10 m, and α_S is the wind shear coefficient which is strongly dependent on the terrain. α_S also generally shows a large variation over a 24 hour period and can change from less than 0.15 during the day to greater than 0.5 at night. This diurnal variation arises because at night the surface temperature drops as the ground loses heat by radiation giving a stable atmosphere with the lower part cooler than the upper part of the atmosphere. The air is not mixed and wind shear can be high. After sunrise the ground is heated by the Sun and warms the air in contact, which then rises causing mixing and reduced wind shear.

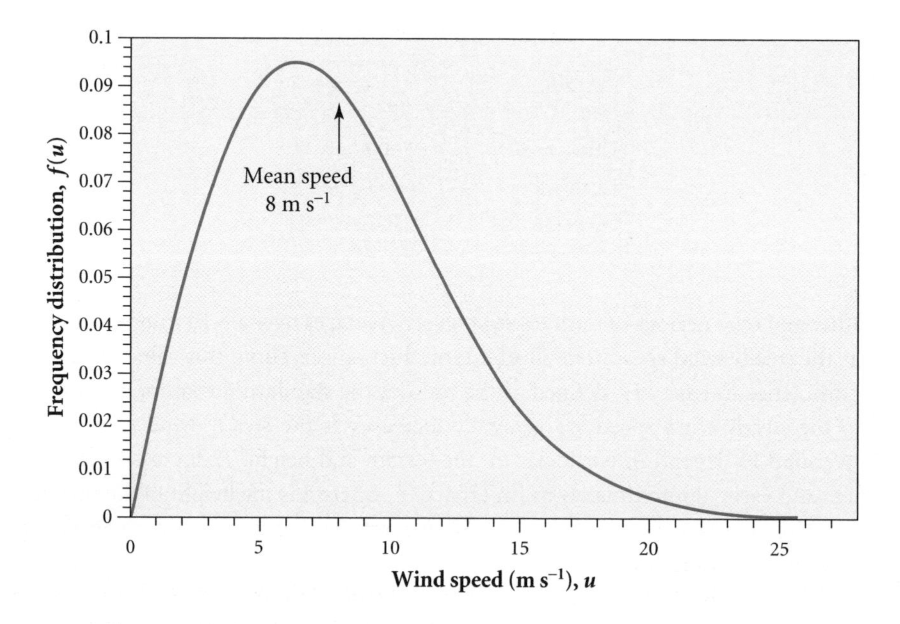

Fig. 5.12 The Rayleigh frequency distribution for a mean wind speed of 8 m s^{-1}.

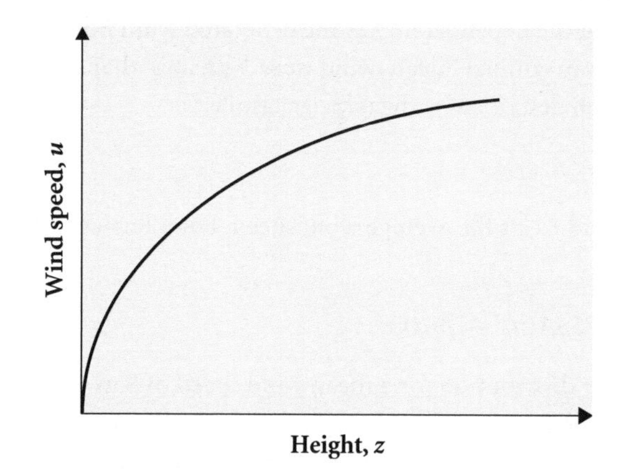

Fig. 5.13 Typical variation of wind speed with height.

The roughness of the terrain is characterized by a surface roughness parameter z_0 and some typical values are given in Table 5.2. An approximate parametrization for the dependence of the wind shear coefficient α_S on z_0 for steady wind speeds lying between 6 and 10 m s^{-1} at a height of 10 m is

$$\alpha_S = \tfrac{1}{2}(z_0/10)^{0.2}. \quad (5.40)$$

For $u_S = 8$ m s^{-1} at a height of 10 m, this relation gives $\alpha_S = 0.32, 0.13$, and 0.05 for $z_0 = 1, 0.01$, and 0.0001 m, respectively. Since a typical hub height for a 1 MW turbine is 80 m these different wind shears translate into mean hub wind speeds of 15.6, 10.5, and 8.9 m s^{-1}. We can see that, for the same wind speed at 10 m, more power is produced if a turbine is mounted higher, particularly when the wind shear is large. However, large wind shears produce more turbulence and hence more fatigue.

5.11 Power output of a wind turbine

The power in the wind P_w at a given site is given by

$$P_w = \frac{1}{2}\rho A\langle u^3\rangle = \frac{1}{2}\rho A \int \{u(z)\}^3 f(u)\,\mathrm{d}u \tag{5.41}$$

where $u(z)$ is the wind speed at the height of the turbine hub (we are neglecting the variation in wind speed over the turbine blades) and $f(u)$ is the wind frequency distribution.

The average output power P_o is given by

$$P_o = \frac{1}{2}\rho A \int C_P(\lambda)\{u(z)\}^3 f(u)\,\mathrm{d}u \tag{5.42}$$

where the tip-speed ratio $\lambda = v_{tip}/u$. For a variable pitch or speed turbine (or a variable speed and pitch turbine) C_P can be kept close to its maximum value over a range of wind speeds. For a fixed speed and fixed pitch machine then C_P changes with the wind speed (Fig. 5.7). In the wind farm example discussed in Box 5.3 the effect of C_P varying is to make the integral roughly equal to $\frac{1}{2}\langle u(z)\rangle^3$, where $\langle u(z)\rangle$ is the mean wind speed at the height of the turbine hub (see Exercise 5.14). The average output power for a fixed speed and pitch turbine is therefore given by

$$P_o \approx 0.2D^2\langle u(z)\rangle^3 \tag{5.43}$$

where D is the diameter of the turbine.

EXAMPLE 5.5

Estimate the power output of a wind farm consisting of 25 fixed pitch and constant speed 1 MW turbines. The hub height is 85 m and diameter is 55 m. The wind has an average speed of $u = 7$ m s^{-1} at a height of 10 m. The land is characterized by a surface roughness parameter $z_0 = 0.001$.

Substituting in (5.40) and (5.39) we have

$$\alpha_S = \frac{1}{7}(0.001/10)^{0.2} = 0.08, \text{ so } u_{hub} = 7(85/10)^{0.08} = 8.3 \text{ m s}^{-1}.$$

Putting u_{hub} in eqn (5.43) gives an estimate for the power output P_o for each turbine

$$P_o = 0.2(55)^2(8.3)^3 = 346 \text{ kW}.$$

So the wind farm output is estimated as 25 × 346 kW = 8.65 MW.

We now consider where the wind conditions favour the siting of wind turbines.

5.12 Wind farms

Good measurements over a period of time are required to determine where to site a turbine or a number of turbines, which is then called a **wind farm**. Experience has shown that good sites require an average wind speed greater than $6\ m\,s^{-1}$. Suitable locations include mountain gaps and passes, high altitude plains, exposed ridges, and coastal areas. Of particular importance in wind farms is the spacing required between the turbines. A spacing of between 5 and 10 rotor diameters D, dependent on conditions, is required for the reduced wind speed downwind from a turbine to have been brought up to its original speed by gaining energy from the surrounding wind that is not slowed down. Greater turbulence intensity reduces the spacing required but increases the fatigue loading. For a farm with a spacing of $8D$ (downwind) by $5D$ (crosswind), an array loss (the amount by which the output of the farm falls below the output of the turbines sited separately) of $\sim$10% might be expected. If the wind farm is on land there is a small amount of land required for the tower and service roads, but between turbines the land is available for use, e.g for grazing cattle or growing crops.

A low wind shear reduces the differential loads on turbine blades and hence the fatigue damage, which favours flat sites. Offshore sea areas can provide excellent wind conditions. These areas can be out of sight of land and so have little visual impact. Turbines can be as tall as desired, and can also operate at higher rotational speeds as noise is less of a concern out at sea. Turbines can cause electromagnetic interference and planning permission has been refused for some sites because a wind farm would affect radar coverage. Access to the electrical grid is important and some of the windiest sites are also some of the most remote. Offshore wind farms clearly need undersea cables to land, and maintenance and installation costs can be higher than for onshore farms.

The power density from a wind farm with 5 MW rated turbines, each with a capacity factor of 0.35 and diameter of 115 m, would be $\sim$5 MW per square km or $\sim$50 kW ha^{-1}, if spaced at $7D$ (downwind) $\times$ $4D$ (crosswind). The power density is roughly independent of turbine size as the output and area both depend on D^2, but larger turbines generally have a higher hub height and hence a higher wind speed. An example of a wind farm is the Cuillagh Mountain wind farm described in Box 5.3.

Besides good wind conditions we also need to consider the environmental impact that a turbine and in particular a wind farm would cause.

5.13 Environmental impact and public acceptance

Wind farms are being actively considered because they are renewable sources of energy that produce no global warming nor any pollution. For contrast the emissions from coal and gas power plants are given in Table 5.4. (There is a small amount of CO_2 emitted, $\sim$10 t (tonne) CO_2 per GWh, associated with the construction and operation of the wind farms. See Chapter 1, Table 1.4.)

Box 5.3 The Cuillagh Mountain wind farm

This farm in Ireland consists of 18 Vestas V47 wind turbines located in County Donegal at a mean height of ~320 m above sea level, where the wind speed at 30 m is 7.2 $m\,s^{-1}$. The wind turbines each have a diameter of 47 m, a hub height of 45 m, a fixed rotor speed of 28.5 rpm, 3 blades, and a rated power of 660 kW.

The output power of a turbine as a function of wind speed is shown in Table 5.3. The mean wind speed at 45 m above the ground is 8.1 $m\,s^{-1}$ and the average annual output of each turbine is 2.05 GWh equivalent to 235 kW continuous output.

Table 5.3 Output power of V47 wind turbine versus wind speed

Speed ($m\,s^{-1}$)	5	7	9	11	13	15	17	19
Power (kW)	44	166	350	538	635	657	660	660

The farm covers an area of approximately 100 ha and the array loss was calculated as 7.3%. The projected energy production from the farm was 36.9 GWh and the actual production over a 17 month period was 55.65 GWh compared with an expected output of 56.54 GWh. Individual monthly outputs varied from 74% to 140% of predicted values, but overall the output was only 1.6% less than expected showing that the wind predictions for the farm were good.

The output of the farm is expected to be highest in the winter (4.27 GWh in January) and lowest in the summer (1.86 GWh in June), which matches the variation in demand.

Primarily, wind turbines are fuel savers rather than energy suppliers. A Brookhaven National Laboratory study on the benefits of nuclear power in offsetting air pollution deaths: gas 150, coal 220, oil 140, and wood 57 per gigawatt-year of generation, would also apply to wind power.

These are all considerable environmental gains, which should not be forgotten when considering any negative aspects. One of the main concerns is the visual impact of a wind

Table 5.4 Emissions from coal and gas power plants

	Emission (kg/MWh) from	
Emission type	**Coal plant**	**Gas (CCGT) plant**
SO_2	6.10	—
NO_x	3.47	0.092
Particulates	0.64	—
CO_2	~900	~400

farm. To gain high winds the turbines are often sited on ridges, so they are very visible. If they are in a region of natural beauty they are felt by many, but not all, to be a blot on the landscape. Clearly if the site is remote there will be less concern. This planning consideration does favour offshore installations, which can be out of sight. Where a wind farm is noticeable experience has suggested that it is more acceptable if there are a few large rather than many small turbines. Making the local population aware of the power available and any job benefits, and also involving them in the planning process, can make planning permission easier.

Another concern that has been raised is their threat to birds—over a 2-year period 183 birds were killed in the Altamont wind farm in California, while in Spain over 18 wind farms there was an average of 0.13 birds killed per turbine in 2003. To put these figures in perspective, there are estimated each year to be over 57 million birds killed by cars, over 97 million killed by flying into plate glass windows, and in the UK 55 million birds killed by cats. So, while care should be taken to reduce bird mortality from turbines, the relative risk should be borne in mind.

The space that wind farms take has also been raised as an issue, but it should be noted that the land between turbines can be used for grazing or for growing crops. To get an idea of how much land is required we will work out the area needed to supply 10% of the UK's electricity.

EXAMPLE 5.6

Estimate the area of land required by wind farms to provide 10% of the UK's electricity demand.

The current (2004) UK electricity energy demand is ~350 TWh per year. The number of hours in a year is 8760, so the energy demand would be met by a continuous power P of

$$P = 350 \times 10^{12}/8760 = 40.0 \times 10^{9} = 40 \text{ GW}.$$

To provide 10%, i.e. 4 GW, wind farms with a capacity of ~12 GW would be required. This capacity would be provided by $(12 \times 10^{3}/5) = 2400$ 5-MW turbines.
A typical diameter D for a 5 MW turbine is 115 m. So, if each turbine occupied an area of $4D \times 7D$, the total area A required would be

$$A = 2400 \times 460 \times 805 = 8.9 \times 10^{8} \text{ m}^2 \approx 30 \text{ km by } 30 \text{ km}.$$

The area of the UK is ~200 000 sq km so 900 sq km is a small fraction of the land area and, if most of it were offshore, the impact of the wind farms would be minimal.

Noise from wind turbines has been a concern but improvements in blade design have reduced the noise from modern turbines. Although the magnitude of noise from a wind farm is relatively low (see Table 5.5), the perception of noise is partly subjective. When the wind is blowing strongly the noise from the turbines is masked by that from the wind itself. Only when close to built-up areas is noise generally a concern.

Table 5.5 Comparison of noise levels

Noise	Noise level (dB)*
Threshold of pain	140
Pneumatic drill at 7 m	95
Busy general office	60
Wind farm at 350 m	35–45
Rural night-time background	20–40
Threshold of hearing	0

*$I(\text{dB}) = 10 \log_{10}(I/I_0)$ where I_0 is the threshold of hearing (at 1000 Hz, $I_0 = 10^{-12}$ W m^{-2}).

5.14 Economics of wind power

The economics of wind power depend on the capital costs to build the wind turbines, the ongoing costs to run the equipment, called maintenance and operational costs (M & O), and the revenue from the sale of the electrical energy (kWh) produced over the lifetime of the turbines. The capital costs of future machines can be hard to estimate as the technology is still developing considerably but a guide is given by the costs of previous machines. A 3 MW turbine costs around 2 million euros (2005) with typically $\frac{2}{3}$ to $\frac{3}{4}$ of the cost on the turbine itself with the rest on electrical equipment and the installation of the turbine. The M & O annual costs have been estimated as ~2% of the capital. The increase in capital and M & O costs for offshore wind farms depends on distance and water depth. Generally, wind speeds offshore are higher than onshore. There are no fuel costs. The design lifetime of wind turbines is ~30 years and it is over that time that we need to consider their economic viability.

Wind power is a source of electrical energy and will be in competition with other supplies of energy, in particular from fossil fuels. With the reduction in the costs of turbines wind power is becoming competitive: onshore wind energy ~3.4 UK pence/kWh compared to 2.4–4.5 pence/kWh for a new coal-fired plant and 4.7 pence/kWh for a new nuclear plant. Currently offshore wind energy is estimated to be 30–100% more expensive, though as more farms are built this can be expected to drop.

Wind power does, however, have an additional value in saving a utility power company from using other fuels, and these direct savings are called **avoided costs**. Its use will also give some reduction in the requirement for conventional generating capacity. Wind energy has substantial environmental benefits, as discussed above, and these can be given a monetary value from the amount of CO_2 and other emissions saved. The external cost (mainly environmental) of coal-fired generation is estimated to be 2–15 pence/kWh compared to 0.2 pence/kWh for wind energy. The desire to reduce global warming can also result in requirements on utilities to use a certain percentage of renewable energy such as that from wind.

Using wind power alone would require alternative generating capacity for when the wind was not blowing. But if we are considering wind providing a relatively small fraction ($\lesssim$20%)

of the total power on a grid, then its variability can be accommodated in the same way as demand variations by turning on and off conventional generating capacity (spinning reserve, see p. 285). It has been estimated that, with 10% of the demand supplied by wind, only a small increase in the amount of back-up supply would be needed, ~1% of the total supply. Over a large region, such as Europe or the USA, the interconnectivity of the electricity grids helps smooth out variations in supply.

We will now consider the cost of wind energy and the competitiveness of wind power. A simple estimate is the time required to pay back the cost of manufacture and installation. A modern wind turbine generates enough electricity to recoup these costs in some 3–10 months. Furthermore, its decommissioning costs, a particular concern with nuclear power plants, are roughly covered by its scrap value.

But this payback estimate does not quantify what return on its investment a utility company can expect. To calculate this we must take into account that the value of revenue received in the future is worth less than if it were received today. For example, UK £100 invested today at 10% interest would be worth £110 in a year's time, so the value of £110 of revenue in a year's time would only have a present value of £100. This translation of future revenue to present value is called **discounting** and the interest rate used is called the discount rate R. Discounting is particularly important when revenue is expected over many years, as from a wind turbine. We show in Chapter 11 how to calculate the cost of producing energy and the rate of return on the capital used to build it.

The cost of wind power at the best sites is now competitive with that from fossil fuels. A kWh costs 3–10 US cents with wind power compared with 4–9 cents with coal and 3–5 cents with gas. This comparison is affected by the price of gas and coal and also by changes to the cost of wind power. As production increases, costs fall and a learning curve (see Chapter 11) can be drawn from which future cost estimates can be made. The average price for wind-generated electricity was ~3.5 pence/kWh in 2003 when the global cumulative production was ~100 TWh. From the learning curve, wind will become competitive, at a price of ~2.5 pence/kWh, with CCGT when the cumulative production has reached ~400 TWh. This amount of production corresponds to approximately 150 GW of installed capacity and would occur in about 2010, if the present annual increase in capacity of ~25% were sustained.

5.15 Outlook

The total world land-based theoretical potential from wind energy resources was estimated by Grubb and Meyer (1993) as ~500 000 TWh per year. Of this amount they estimated that ~10%, 53 000 TWh per year, could be exploited. How this technical wind potential is distributed over the world is shown in Fig. 5.14. World electricity demand by 2020 is predicted to be 26 000 TWh per year (equivalent to 3 TW continuous, compared with ~20 TW continuous estimated total energy demand), so there is a huge resource in wind energy.

A similar estimate was obtained in 1994 by the World Energy Council (WEC 1994) based on the following information and assumptions.

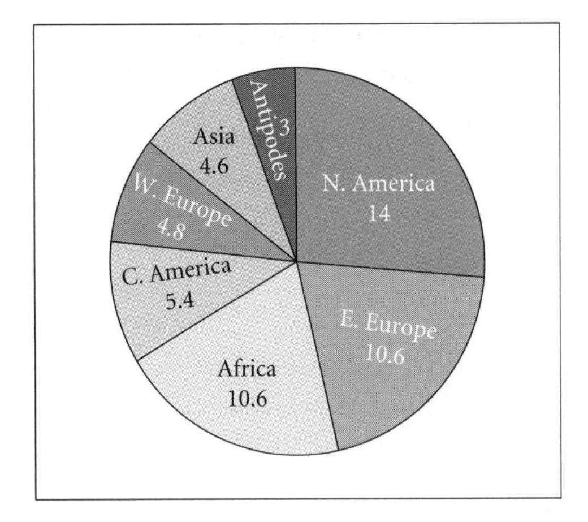

Fig. 5.14 Technical potential (TWh/y) wind resources by area.

1. The land area of the world is 107×10^6 sq km and the wind has an average speed of >5.1 m s^{-1} over 27% of this land area.
2. Wind turbines can be sited with a generating capacity of 8 MW per sq km.
3. The capacity factor of the wind turbines is 0.23 or 16 000 MWh per sq km per year.

These numbers give a gross potential of 480 000 TWh per year. A more recent estimate by Archer and Jacobson (2005) based on 80 m high turbines is 627 000 TWh per year for the gross potential. There are practical, environmental, and social constraints that limit the area of land from which wind energy can be obtained. Applying these constraints gives the **technical potential**. The WEC assumed that only 4% of the land area with suitable wind speeds could be used. This percentage was based on experience in the Netherlands and in the USA. The WEC technical potential was therefore 20 000 TWh a year, still a very large source of electricity.

Offshore there is a huge technical potential, which is estimated in Europe to be 3030 TWh a year. Placing wind turbines out to sea avoids planning constraints and helps public acceptability. Offshore wind speeds are similar to those found at good onshore sites. Costs offshore, though, are currently up to double those for good onshore sites because of additional transmission, installation, and M & O costs. The turbines need to be reliable and robust, but can have a higher tip speed as noise is not a problem, increasing the conversion efficiency slightly. It is expected that offshore wind will expand significantly over the next decade.

Areas with great potential are in Northern Europe along the North Sea, in South America near the southern tip of the continent, in Tasmania, in the Great Lakes region of North America, and along the north-east and north-west coasts of North America.

Two further estimates of energy supply are often quoted: the **practicable** and the **economic potentials**. (It should be noted that definitions of potential differ.) The practicable, or **accessible**, potential is that amount of the technical potential that can be utilized by a particular time. For the practicable potential various factors need to be taken into account,

Table 5.6 Potential energy supply and cost of electricity from wind—estimates for 2025

Wind region	Cost (UK pence/k Wh)	Potential energy supply (TWh/year)		
		Economic	Technical	Practicable
Offshore	2.5–3.0	100	~3500	100
Onshore	<3.5	58	317	8

such as the effects of competing land use, planning permission, grid limitations, and the rate of construction of new turbines. The economic potential is the amount of the technical potential that is economically competitive. This depends on the cost of alternative supplies and can be affected by policies such as a carbon tax that would favour renewable sources.

In the UK estimates made by the DTI for 2025 for the practicable, economic, and technical potentials and costs for on- and offshore wind are shown in Table 5.6. In 2002 the consumption of electricity in the UK was 349 TWh so the economic potential is a very significant fraction of the UK demand. That the practicable potential for onshore wind is estimated as less than the economic potential is a consequence of the estimated rate of build of new wind turbines, with only a capacity for ~8 TWh/year built by 2025.

The UK has the largest onshore technical potential in Europe, estimated as 317 TWh/year, which is close to the annual electricity consumption in 2002 of 349 TWh, but in 2002 the UK only produced 1.45 TWh by wind power, equivalent to 0.5%. The current UK policy, though, is to provide 10% of the electricity demand in 2010 through renewables, with about half coming from wind, the remainder from solar and biomass. By 2020 the UK hopes to increase the percentage from renewables to 20%. In Europe, the largest producers (in TWh/y, 2002) of wind-generated electricity were Germany (18.5), Spain (12.0), and Denmark (5.3). These amounts corresponded to 3.5, 5, and 15%, respectively of their annual consumption of electricity.

The installed capacity by the end of 2002 (2001) was in the EU 23 (17.5) GW, in North America 5 (4.5) GW and in the rest of the world 3 (2.5) GW, a global total of 31 (24.5) GW. The annual growth of 27% in 2002 continues a marked growth seen in installed capacity since the beginning of the 1990s (Fig. 5.15). The amount of electricity produced by 31 GW of

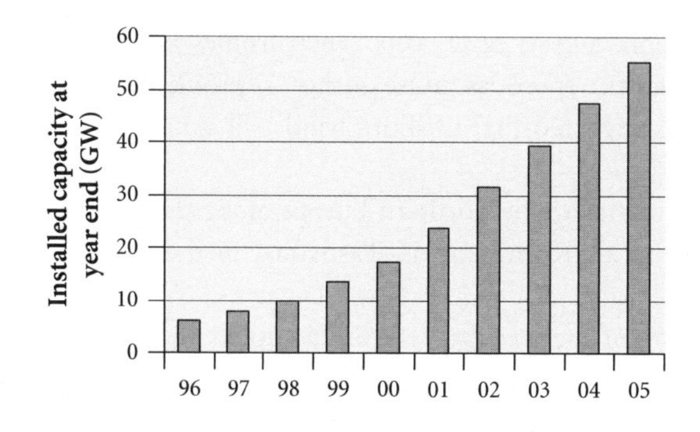

Fig. 5.15 Total world wind turbine installed capacity.

installed capacity is ~11 GW or 90 TWh/y, compared to a total global energy demand in 2002 of 14.3 TW or 125 000 TWh/y.

5.16 Conclusion

While wind power is only currently producing a relatively small amount of electricity worldwide, it has the potential to make a significant contribution to carbon-free electricity. Improvements in the storage of electricity would reduce any impact from the variability in the wind, as would strengthening the electrical grid.

In particular, conversion of electricity to hydrogen through the use of fuel cells would provide a valuable store of energy, if it could be developed to be cost-effective. However, the effect of the wind's variability has only a small impact at a contribution of less than 20%, because extra capacity has to be available to meet the variability in demand.

It is crucial to expand production and deployment of wind turbines for costs to decrease and market incentives are very important in making the technology competitive. But already (2005) wind power from the best sites is starting to be competitive with fossil fuel alternatives. The rapid growth in installed capacity is expected to continue and expand.

SUMMARY

- Global gross wind potential ~600 000 TWh, some 5 times the current global energy use and 40 times the electricity usage. Constraints on land give a potential of ~20 000 TWh.
- Power in the wind is proportional to the cube of the wind speed. Sites with wind speeds greater than 6 m s^{-1} at 10 m height are favoured. Power extracted by turbine is limited to 16/27 of the incident power (Betz limit) and is given by

  $$P = \frac{1}{2} C_P \rho u_0^3 A$$

  where C_P is the power coefficient, ρ is the density of air, u_0 is the wind speed, and A is the area swept out by the turbine blades. C_P is typically ~0.45 at the optimum tip-speed ratio $\lambda \equiv v_{tip}/u_0$, where v_{tip} is the blade tip speed. λ is typically between 6 and 10.
- Modern wind turbines are horizontal-axis wind turbines (HAWTs) and have maximum (rated) outputs of typically 1.5–5 MW with diameters $D = 70–115$ m. On wind farms turbines are spaced ~$4D \times 7D$. The annual output is typically a third of the rated output, i.e. a capacity factor of a third.
- Installed capacity is currently increasing at ~25% per annum and was 31 GW in 2003. At a capacity factor of 0.3 this is equivalent to providing ~0.5% of the global electricity demand of ~2000 GW.
- Wind is a potentially large source of carbon-free electricity. The cost per kWh for the best wind sites is starting to be competitive with fossil fuel alternatives. Onshore wind farms have an output power density of ~50 kW/ha and much of the land between turbines is available to animals for grazing or to crops. Offshore sites are likely to be more acceptable to the public.

FURTHER READING

Manwell, J.F., McGowan, J.G., and Rogers, A.L. (2004). *Wind energy explained.* Wiley, Chichester. Good discussion of modern wind turbines.

Spera, D.A. (ed.) (1994). *Wind turbine technology.* ASME Press, New York. Useful information on wind turbines.

European Wind Energy Association (2004). *Wind energy—the facts.* EWEA, Brussels. Good overview.

Twidell, J. and Weir, T. (2006). *Renewable energy resources*, 2nd edn. Taylor and Francis, London. Good discussion of wind characteristics.

Boyle, G. (ed.) (2004). *Renewable energy.* Oxford University Press, Oxford. Useful discussion of wind resources.

WEB LINKS

www.windpower.org	Good overview of wind power.
www.ewea.org	European Wind Energy Association.
www.awea.org	American Wind Energy Association.

LIST OF MAIN SYMBOLS

ρ	air density	α	angle of attack
A	cross-sectional area	P	power
u	wind speed	T	thrust
$\mathcal{L}$	lift	W	width of turbine blade
$\mathcal{D}$	drag	D	diameter of turbine swept area
λ	tip-speed ratio	α_s	wind shear coefficient
C_P	power coefficient	z_0	surface roughness parameter
a	induction factor		

EXERCISES

5.1 Calculate the power in a wind blowing with a speed of $12\,\mathrm{m\,s^{-1}}$ incident on a wind turbine whose blades sweep out an area of diameter 110 m.

5.2 A simple drag machine (Fig. 5.16) consists of two flaps attached to a rotating belt. The drag force F_D is given by $\frac{1}{2}C_D\rho A u_{rel}^2$, where A is the cross-sectional area of a flap, C_D is the drag coefficient for the flap, ρ is the density of air, and u_{rel} is the wind speed relative to the flap. Show that the power P_D is given by

$$P_D = \tfrac{1}{2}C_D A(u_0 - v)^2 v$$

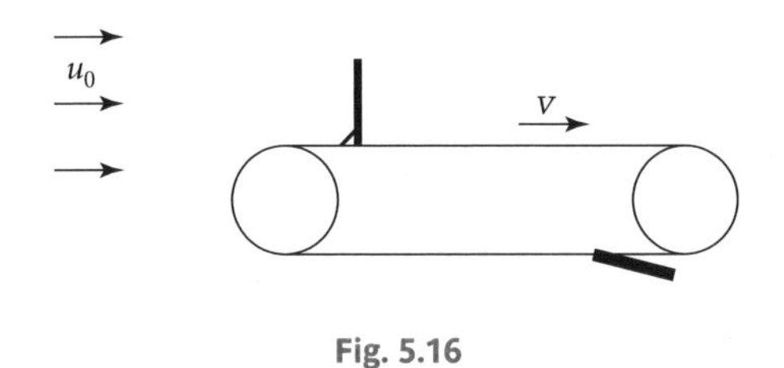

Fig. 5.16

and that the maximum power $P_D(\max)$ is given by

$$P_D(\max) = \tfrac{1}{2}(4/27)C_D\rho A u_0^3.$$

Deduce that the power coefficient C_P for such a drag machine is equal to $4C_D/27$. The maximum value of C_D is ~1.5 for a cup-shaped flap so the maximum efficiency of a drag machine is 22%.

5.3 Consider stream-tubes of air (Fig. 5.17) before and after a turbine, but not across the turbine because the flow is unsteady and not streamlined. Applying Bernoulli's principle and noting that $p_0 = p_2 =$ atmospheric pressure and that from the conservation of mass $u_1 = u_1^*$, show that

$$(p_1 - p_1^*)/\rho = \tfrac{1}{2}(u_0^2 - u_2^2),$$
$$F_{\text{thrust}} = \tfrac{1}{2}\rho(u_0^2 - u_2^2)A_1.$$

Hence show that the maximum thrust is given as

$$F_{\text{thrust}} = \tfrac{1}{2}\rho u_0^2 A_1 \times 8/9.$$

This is similar to a circular disc of area A_1 which has a drag force

$$F_D = \tfrac{1}{2}C_D\rho u_0^2 A_1 \text{ and } C_D \sim 1.$$

5.4 Using the result for F_{thrust} in Exercise 5.3, calculate the force in tonnes weight on a turbine in a wind speed of 10 m s^{-1} whose blades have a radius of 50 m.

5.5 Deduce the form of eqn (5.2), $P \propto A\rho u^3$, by using dimensional analysis.

5.6 Show that the maximum power coefficient for two identical turbines placed one behind the other in line with the wind direction is 16/25.

5.7 Take the extent of the disturbance of the wind in the direction of the incoming wind, i.e. parallel to the axis of the turbine, as πW, where W is the width of the turbine blade. Show

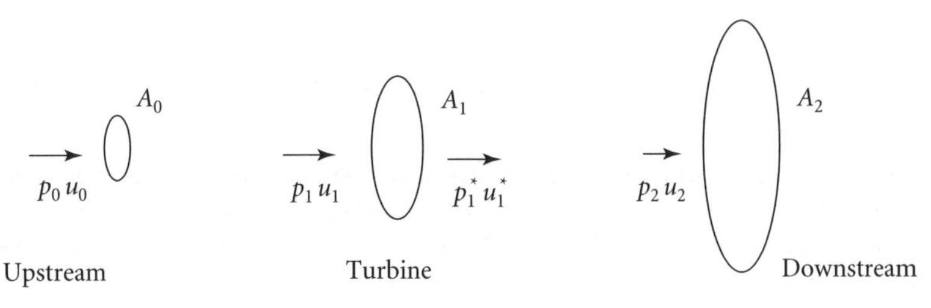

Fig. 5.17

that the ratio R_t of the time for the wind to travel a distance πW to the time for one blade to reach the position of the next blade is given by

$$R_t > 4\pi/(3\lambda).$$

Hence show that the reaction of the blades occurs over a significant fraction of the area swept out by the blades.

5.8 Calculate, using eqn (5.25), the width and twist of the optimal turbine blade of length $R = 48$ m for a tip-speed ratio $\lambda = 10$ at radii of 12, 24, 36, and 48 m. Take $\alpha_0 = 5^\circ$ and $C_L = 1$ and $n = 3$.

5.9 The two blades of a vertical-axis wind turbine (VAWT) are rotating at a speed v in a wind speed u (Fig. 5.18). Show from considering the direction of the wind over the blades that the turbine will rotate as illustrated no matter what direction the wind comes from.

5.10* Show that the value of a that minimizes $dP = \frac{1}{2}\rho(2\pi r\, dr)4a(1-a)^2u_0/(1+a')$ subject to the constraint $a'(1+a') = a(1-a)/\lambda_r^2$ is given by the equation $\lambda_r^2 = (4a-1)^2(1-a)/(1-3a)$, and that $a' = (1-3a)/(4a-1)$. Hence show that for $\lambda_r > 2, a \approx 1/3$ and $a' \approx 2/(9\lambda_r^2)$. (The method of Lagrange multipliers states that the extremum values of a function $f(a, a')$ subject to the constraint $g(a, a') = 0$ are given by the solutions to

$$\partial f/\partial a + \lambda \partial g/\partial a = 0, \quad \text{and} \quad \partial f/\partial a' + \lambda \partial g/\partial a' = 0.)$$

5.11 Calculate the number of cycles (N) a turbine blade would make in 30 years on a wind turbine with $D = 100$ m, $\lambda = 10$, and a mean wind speed of 12 m s^{-1}. The data for the fatigue strength of two possible turbine materials can both be represented by $\sigma/\sigma_0 = 1 + b\log_{10}(N)$. Material x has $\sigma_0 = 100$ MPa and $b = 0.10$, and material y has $\sigma_0 = 120$ MPa and $b = 0.11$. What material would you choose?

5.12 The pressure P_t acting on a wind turbine is given by $P_t = (4/9)\rho u_0^2$ (Exercise 5.3). Estimate the variation in P_t across the height of the circle swept out by the turbine blades arising at a site with a wind shear characterized by $z_0 = 0.1$, $u_s = 10$ m s^{-1} at $z_s = 10$ m, hub height $= 70$ m, and $D = 50$ m.

5.13 For a wind farm with a spacing of $5D \times 5D$, estimate the energy density (MW ha^{-1}) for a farm with 20 turbines each with a rated output of 5 MW, hub height $= 90$ m,

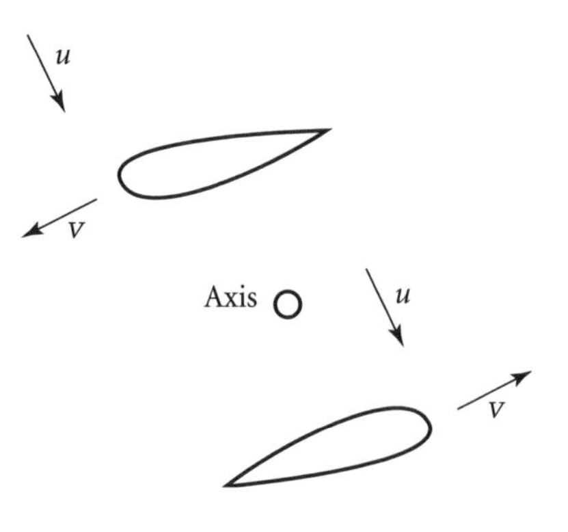

Fig. 5.18

and diameter $= 115$ m. Assume: (a) a capacity factor of 0.3; (b) use the formula $P_o \approx 0.2D^2 \langle u(z) \rangle^3$, with $u_s(10) = 9\ \mathrm{m\,s^{-1}}$ and $z_0 = 0.01$.

5.14 Using the data provided on the Cuillagh Mountain wind farm, estimate: (a) the variation of C_P as a function of wind speed; (b) the value of the optimum λ. (c) Assuming that the Rayleigh wind speed distribution given in Fig. 5.12 describes the variation in wind speed at the farm, calculate the mean power expected per turbine and compare with the observed output of ~235 kW. (d) What is the energy density (MW ha^{-1}) for this farm?

5.15* The angular speed ω of a fixed pitch wind turbine is controlled by setting the generator torque $\tau_{gen} = K\omega^2$, where $K = \frac{1}{2}\,\rho A R^3 C_P^{(max)} / \lambda_{opt}^3$. The angular acceleration of the turbine $d\omega/dt$ is proportional to $(\tau_{wind} - \tau_{gen})$, where $\omega\tau_{wind} = P_{wind} = \frac{1}{2}\rho A C_P u^3$ and $\omega = \lambda u/R$. Show that the turbine will alter speed so that $\lambda = \lambda_{opt}$, provided $C_P > C_P^{(max)} \lambda^3 / \lambda_{opt}^3$.

6 Solar energy

List of Topics

- ☐ Solar spectrum
- ☐ Semiconductors
- ☐ p-n junction
- ☐ Solar cells
- ☐ Efficiency of cells
- ☐ Silicon cells
- ☐ Thin film cells
- ☐ Light trapping
- ☐ Multilayer devices
- ☐ Developing technologies
- ☐ Solar panels
- ☐ Economics of photovoltaics
- ☐ Environmental impact
- ☐ Outlook for photovoltaics
- ☐ Solar thermal power plants

Introduction

The average solar power incident on the Earth is $\sim$1000 W m^{-2} ($\sim$100 mW per cm^2) or about 100 000 TW. This power is far larger than the current world power consumption of $\sim$15 TW. Currently, $\sim$11% of the world's power is supplied by biomass, while 85% is derived from fossil fuels. Both are the consequence of photosynthesis, in which plants use solar energy to convert water and carbon dioxide into carbohydrates. While biomass is not necessarily a net producer of CO_2, the burning of fossil fuels definitely is. However, biomass is not a good converter of solar energy as the efficiency of biomass production is low ($\sim$0.2 – 2%).

A more efficient conversion ($\sim$15%) of solar energy directly to electrical power is provided by photovoltaic (PV) cells. Currently (2004) these provide a peak power of $\sim$2.5 GW that is predicted to rise to $\sim$1000 GW by 2030. The current price of PV cells is too high to be competitive with fossil or nuclear power, for electricity supply to a national grid, but is expected to decrease as new systems are developed. However, PV cells are already very competitive for applications in areas far from a grid.

In this chapter we will evaluate PV cells and solar power plants, and discuss biomass in the next chapter. We will first look at the solar energy spectrum and at how a solar PV cell works. We will see what limits its efficiency and the potential for improvement. We will then look briefly at solar power plants, and at schemes to utilize the temperature difference that occurs in oceans between surface and deep water due to solar heating.

6.1 The solar spectrum

The smooth spectrum shown in Fig. 6.1 is that of a black body at 5800 K. This spectral shape is close to that incident from the Sun on the Earth's atmosphere. The effect of passing through the thickness of the atmosphere is to reduce the total intensity from 1.36 kW m^{-2} for sunlight incident on the atmosphere, called AM0, to 1.0 kW m^{-2} for that passing through a typical thickness of the Earth's atmosphere taken to be 1.5 times its height, called AM1.5. AM1.5 corresponds to sunlight incident at an angle of 48° to the vertical. The effect of absorption by water vapour, carbon dioxide, and methane is nearly all in the infrared region, corresponding to photon energies below ~1.7 eV. The energy of the photons in the visible part of the solar spectrum ranges from ~3 eV (0.4 μm) to ~1.7 eV (0.7 μm).

Radiation reaches the Earth's surface by direct radiation (focusable by mirrors) and diffuse radiation (unfocusable). The diffuse percentage is strongly dependent on how clear the sky is, and a typical yearly average is about 30%. The total amount of radiation varies considerably with cloudiness, season, and location, reducing from a yearly total on a horizontal surface of ~2300 kWh m^{-2} in the tropics to ~800 kWh m^{-2} by the Arctic circle (latitude 66.5°). The average intensity (watts per m^2) on a cloudy day is typically ~10% in the UK and ~50% in the tropics of the flux on a sunny day. Note that the sky would be black (except for stars and planets) in the absence of diffuse radiation and is blue because short wavelengths are more scattered than long ones.

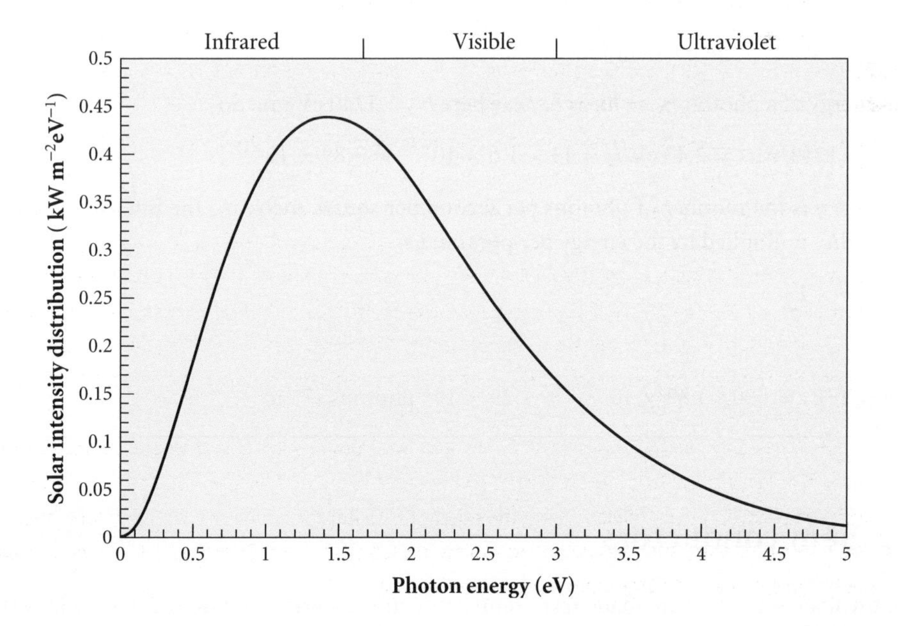

Fig. 6.1 Black-body spectrum at $T = 5800$ K. The total intensity is normalized to that of the AM1.5 solar spectrum of 1 kW m^{-2}.

EXAMPLE 6.1

Direct sunlight of average intensity 200 W m^{-2} is incident normally on a solar cell. The area of the cell is 0.1 m^{-2}. What is the total incident energy in one day in kWh and in MJ? How is this total energy altered if the sunlight falls at angle of 30° to the normal to the surface of the cell?

The incident power P is the intensity I multiplied by the area A of the cell. So

$$P = I \times A = 200(0.1) = 20 \text{ W} = 0.02 \text{ kW}.$$

The incident energy E is the power multiplied by the time t. So

$$E = P \times t = 0.02(\text{kW}) \times 24(\text{hours}) = 0.48 \text{ kWh}$$

or

$$E = P \times t = 20(\text{W}) \times 24 \times 60 \times 60(\text{seconds}) = 1.73 \text{ MJ}.$$

If the sunlight falls at 30° to the normal to the cell, then the incident power is reduced by $\cos(30^\circ)$, as $A\cos(30^\circ)$ is the projected area of the cell normal to the beam. The incident energy is then

$$E = 0.42 \text{ kWh or } 1.50 \text{ MJ}.$$

The variation of the inclination of sunlight with latitude and season is part of the variation in the Sun's energy m^{-2} d^{-1} or insolation on a horizontal surface with location.

EXAMPLE 6.2

A source of light with a wavelength of 510 nm (green) of intensity 500 W m^{-2} is incident on a solar cell. What is the incident flux of photons?

The energy of a photon $E_\gamma = h\nu = hc/\lambda$, where $hc = 1240$ eV nm. So

$$E_\gamma = 1240/510 = 2.43 \text{ eV} = 2.43 \times 1.6 \times 10^{-19} = 3.89 \times 10^{-19} \text{ J}.$$

The flux F is the number of photons per second per square metre, so the intensity I is given by the flux multiplied by the energy per photon, i.e.

$$I = F \times E_\gamma.$$

So

$$F = I/E_\gamma = 500/(3.89 \times 10^{-19}) = 1.29 \times 10^{21} \text{photons s}^{-1} \text{ m}^{-2}.$$

6.2 Semiconductors

Photovoltaic solar cells are made from semiconductor materials and to understand how they work, we first need to realize why some materials are conductors and some insulators, before looking at semiconductors.

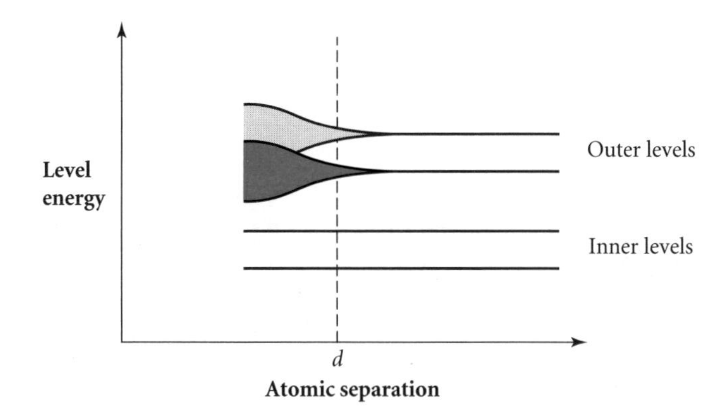

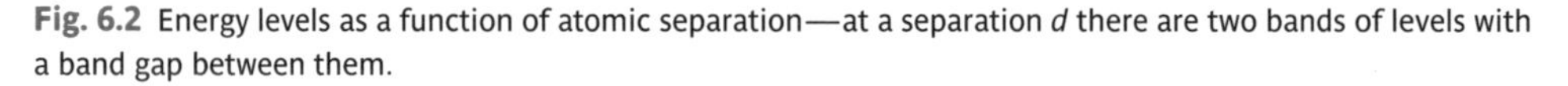
Fig. 6.2 Energy levels as a function of atomic separation—at a separation *d* there are two bands of levels with a band gap between them.

In a metal that conducts electricity easily, such as copper, the atoms are $\sim 10^{-10}$ m across and are separated by $\sim 2.5 \times 10^{-10}$ m. The electrons in the outer shells can move quite easily from one atom to another and so the metal conducts well. But this does not explain why some materials do not conduct and are insulators. To understand this we need to look at what states are available to the electrons in a solid. For atoms that are well separated the electron energy levels are also well defined as shown in Fig. 6.2. As the atoms come closer the levels spread and overlap. At a separation of d, there are still spaces, called **band gaps**, between them. The effect is larger for the outer electrons which overlap most.

A single level, which corresponds to a certain spatial distribution of an electron, will spread into $2N$ states, where N is the number of atoms in the crystal. The factor 2 comes about because the electron has two spin states (spin $\frac{1}{2}$ up or down). So, for example, for sodium with one electron in the outer 3s level only half of the states in the 3s band will be occupied. When an electric field is applied electrons start to accelerate and gain energy, and so occupy higher states. A steady drift velocity (current) is quickly attained when the number of electrons being scattered down into lower energy states equals the number being promoted up by the electric field. The electrons are scattered by defects in the crystal and by the thermal vibrations of the atoms. As these vibrations increase with temperature we expect the conductivity to decrease if the temperature rises, and this is what is measured.

For materials where there are two electrons in an outer s shell we would expect the band to be full. On applying an electric field, there are no empty energy levels available close by for electrons to be promoted into, so the material is an insulator. (Unless an empty band overlaps a full band, as in magnesium, when the material will be a conductor.) Conductivity can only occur through electrons that have been thermally excited to the next band. This will only occur to a significant extent when the gap between the filled (valence) band and empty (conduction) band is relatively small, ~ 1 eV. Such materials are called **semiconductors**. We expect their conductivity to increase with temperature as more electrons will be thermally excited into the conduction band, and this is what is found. Figure 6.3 illustrates these three cases.

A particularly important semiconductor is silicon, which as an isolated atom has four electrons in a half-filled shell. In a solid these electrons are shared (covalent bonds) between

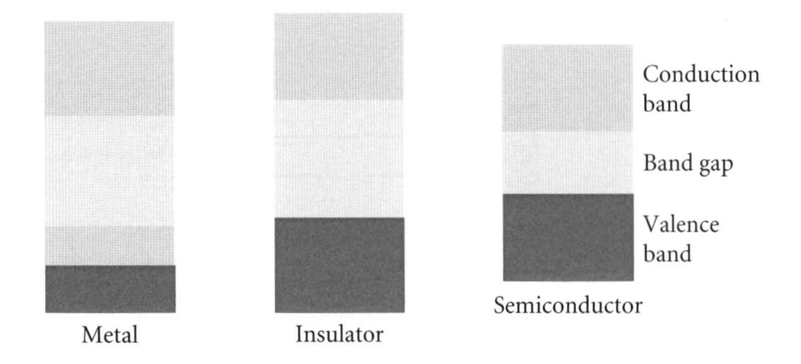

Fig. 6.3 Occupancy and band structure of a metal, an insulator, and a semiconductor.

four silicon atoms and a full valence band is formed with a 1.1 eV gap to the next band. Besides silicon, there are a number of other semiconductor materials that are increasingly important for use in photovoltaic cells: GaAs, CdTe, and $CuInGaSe_2$ (CIGS).

The presence of impurity atoms can give rise to electron states within the band gap and this contamination must be reduced before the material acts as a pure semiconductor, called an intrinsic semiconductor. The development of the technique of zone-refining, in which sections of a rod of semiconductor are heated sequentially so the molten region moves from one end to the other, thereby concentrating the impurities at one end of the bar, was a key part in the development of semiconductor devices. Silicon can be produced with less than 1 part in 10^{11} impurities.

The addition, called **doping**, of certain impurity atoms to intrinsic material can significantly alter the conduction properties of semiconductors. If we include atoms with five outer electrons (pentavalent atoms) within a silicon crystal, then each of these atoms will have one electron which is only weakly bound and can be easily excited into the conduction band. Such atoms are called **donors** and the doped silicon is called n-type. Atoms with only three outer electrons (trivalent atoms) can gain an electron quite easily through thermal excitation of an electron from the top of the valence band, which leaves a positively charged vacancy called a hole. Such atoms are called **acceptors** and the doped silicon is then called p-type. A particularly important device, and the basis of the photovoltaic solar cell, can be made by forming a junction between p- and n-type material called a **p-n junction**.

6.3 p-n junction

Figure 6.4 shows separate pieces of p- and n-type material and a single piece of semiconductor doped to form a p-n junction. We can see thermally excited electrons in the conduction band of the n-type and positive (i.e. absence of electrons) holes in the valence band of the p-type.

In the piece with both p- and n-type materials touching, Fig. 6.4(b), electrons have diffused across the junction from the conduction band on the n-side, because of the concentration gradient, and these have filled the vacancies (holes) in the top of the valence band on the p-side. An electric field is set up across the junction and causes a drift of electrons, which

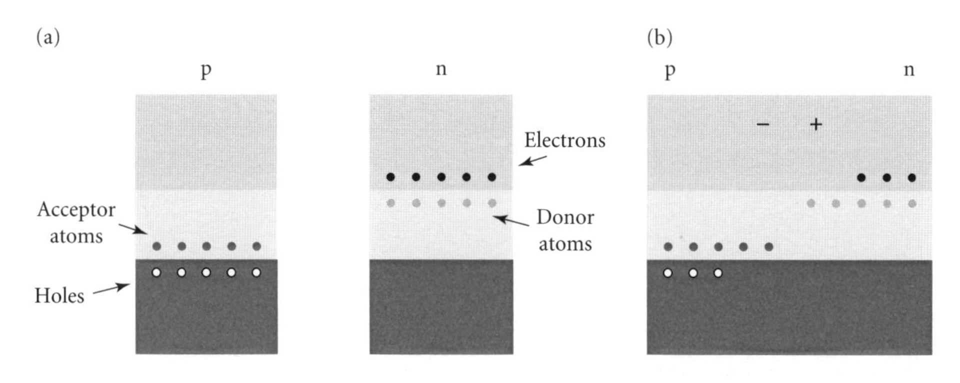

Fig. 6.4 (a) p- and n-type material. (b) p-n junction.

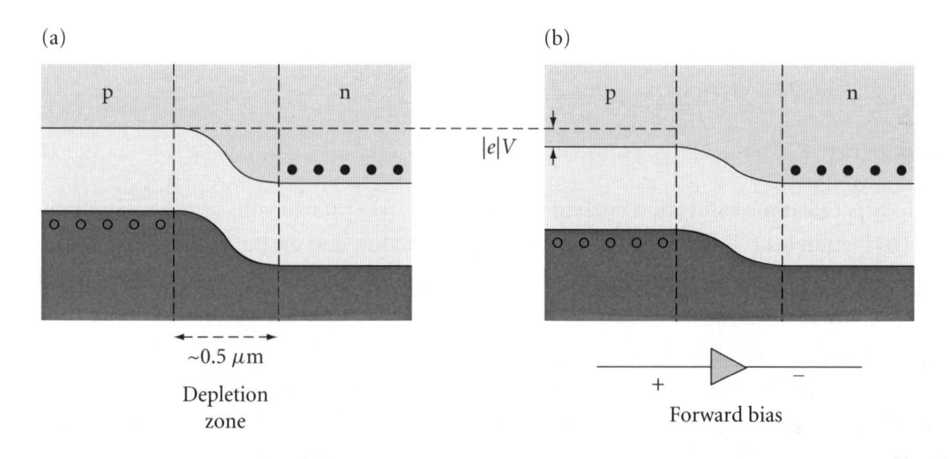

Fig. 6.5 (a) Energy levels and depletion zone in a p-n junction. (b) Effect of positive bias on a p-n junction with standard circuit symbol for a diode below.

balances the electron diffusion due to the concentration gradient. (There are corresponding currents of holes.) As a result the n-side becomes positively charged while the p-side becomes negatively charged. The energy of the electrons on the n-side is therefore lowered compared with electrons on the p-side and the junction region is normally drawn as shown in Fig. 6.5(a).

The region about the junction where there are no charges is called the depletion zone. The electrons (holes) in the conduction (valence) band on the p- (n-) side of the junction are called the minority carriers; on the n- (p-) side the majority carriers are electrons (holes). If the n- and p-sides are connected by a wire, then the contact potentials that arise at the junctions of the wire to the n-and p-regions cancel the internal potential across the p-n junction and no current will flow.

Figure 6.5(b) shows the effect of biasing the p-region positively by a potential V relative to the n-region. The junction is then **forward-biased**. (The circuit symbol for a diode is shown below the junction.) This lowers the conduction band on the p-side relative to that on the n-side of the junction by $|e|V$, where e is the charge, of the electron. This alters the balance between the diffusion and drift currents across the depletion zone and gives a net electron

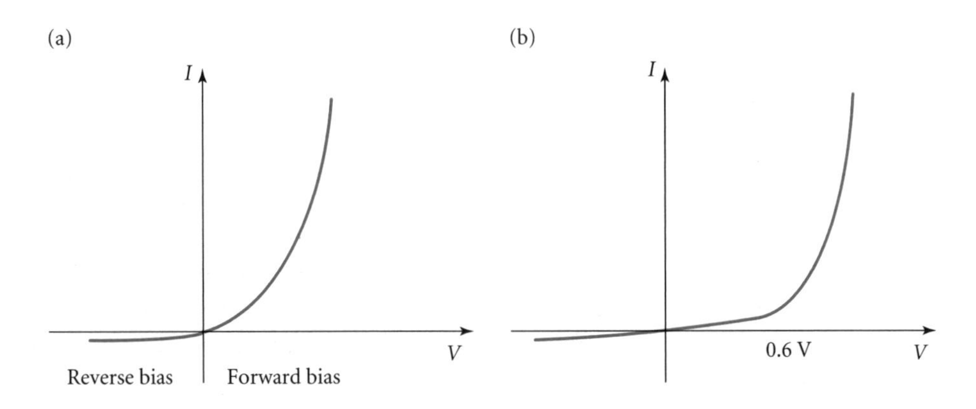

Fig. 6.6 (a) IV characteristics of ideal p-n junction. (b) Actual IV characteristic for silicon diode.

(hole) current from the n- (p-) to the p- (n-) side, called the forward current. The total forward current I is given by

$$I = I_S\{\exp(V/V_T) - 1\} \tag{6.1}$$

where I_S is called the **saturation current** and $V_T \equiv kT/|e| \approx 0.026$ volts at room temperature (see Derivation 6.1). I_S depends on the area of the junction and on the doping. The direction of positive charge flow is across the junction from the p- to the n-side and the ideal current–voltage characteristics given by eqn (6.1) are shown in Fig. 6.6(a).

We can see that the p-n junction acts as a diode allowing current to flow easily only when forward biased. Figure 6.6(b) shows the actual characteristic for a silicon p-n junction: until a forward bias of ~0.6 V conduction is less than for an ideal junction. This is caused by trapping and generation of electrons and holes within the depletion zone. But the idealized equation for the forward current is a very useful approximation and we will now see how such a device can act as a photocell.

Derivation 6.1 Current – voltage characteristic of a p-n junction

The effect of biasing the p-region positively by V relative to the n-region lowers the conduction band on the p-side relative to that on the n-side of the junction by $|e|V$, Fig. 6.5(b). This alters the balance between the diffusion and drift currents across the depletion zone and gives a net electron (hole) current from the n- (p-) to the p- (n-) side, called the forward current. We can calculate the size of this current by considering what happens at the edge of the depletion zone.

The energy shift of $|e|V$ has the effect of increasing the electron concentration at the edge of the depletion zone on the p-side by the Boltzmann factor $\exp(|e|V/kT) = \exp(V/V_T)$, where $V_T \equiv kT/|e| \approx 0.026$ volts at room temperature. The excess concentration $\Delta n_e{}^P$ at the edge is therefore equal to $n_e^P\{\exp(V/V_T) - 1\}$; n_e^P is the concentration of the minority carriers on the p-side when V is zero. This excess drops to zero over the distance that the electrons diffuse in the p-region before recombining with holes. This concentration gradient gives rise to a diffusion current proportional to Δn_e^p, i.e. to $n_e^p\{\exp(V/V_T) - 1\}$. Outside the depletion zone the electric field is very small and there is essentially no electron drift current. At the edge of the zone the total forward electron current is therefore equal to the diffusion current.

The current is constant throughout the device. Thus, as the diffusion current decreases, as electrons drop into the vacancies in the valence band, i.e. recombine with holes, the drift current of electrons in the valence band moving away from the junction into vacancies increases. This latter is normally described as a hole drift current towards the junction. (Minority carrier drift currents are much less than majority carrier drift currents.) There is a similar contribution from the minority hole diffusion and majority electron drift currents on the n-side giving the total forward current I as

$$I = I_S\{\exp(V/V_T) - 1\}$$

where I_S is called the saturation current (this is eqn (6.1)). I_S depends on the area of the junction and on the doping. The direction of positive charge flow is across the junction from the p- to the n-side and the ideal current–voltage characteristics given by eqn (6.1) are shown in Fig. 6.6 (a).

6.4 Solar photocells

When light falls on a silicon p-n junction some of the photons can create electron–hole pairs through the **photoelectric effect**, in which a photon is absorbed by an electron promoting it from the valence band to the conduction band. The minimum energy that the photon must have equals the band-gap E_{gap}. For silicon this is 1.1 eV and corresponds to a wavelength of $\sim$1.1 μm, using the Einstein relation $E_\gamma = h\nu = hc/\lambda = (1.24/\lambda)$eV, with λ in μm. Figure 6.7 shows photons incident on a junction producing electron-hole pairs.

The built-in field across the depletion region sweeps the electrons to the n-side and the holes to the p-side of the junction. This produces a reverse current I_L (as the electrons flow across the junction from the p- to the n-side), and the photocell current I_C is given by

$$I_C = I_L - I_S\{\exp(V/V_T) - 1\} = I_L - I_S\{\exp(I_C R/V_T) - 1\} \tag{6.2}$$

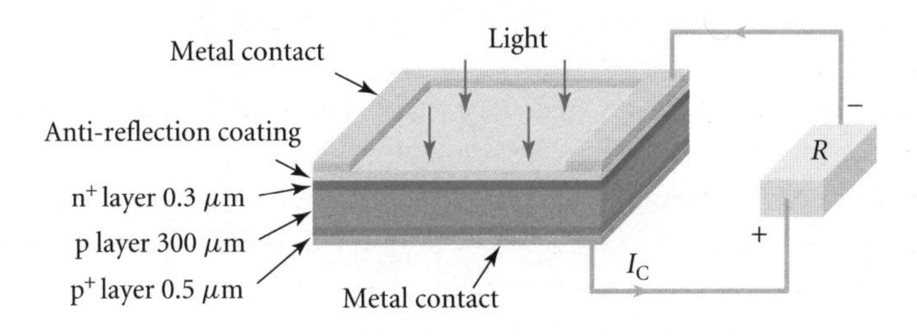

Fig. 6.7 Operation of a solar photocell.

where the forward bias V across the junction equals $I_C R$. This forward bias produces a current given by the second term (eqn (6.1)) flowing across the junction from the p-side to the n-side, which is in the opposite direction to the light-induced current I_L.

The top metal electrode is in narrow strips to let the light fall on the junction. An anti-reflection coating increases the transmission of light to within the junction. To reduce the series resistance, the top of the n-layer is more highly doped (labelled n^+). Minimizing surface recombination at the rear contact is achieved by placing a highly doped p-layer (p^+) just in front of the contact. This reduces the electron concentration in this region and hence the chance of recombination.

When R is infinite then I_C is zero and the open-circuit voltage V_{OC}, by which the p-side is more positive than the n-side of the junction, is given by

$$V_{OC} = V_T \ln(1 + I_L/I_S) \approx V_T \ln(I_L/I_S), \tag{6.3}$$

as $I_L \gg I_S$. (V_{OC} is less than E_{gap}, as the maximum voltage must be less than the band gap.) Note $V_T \equiv kT/|e| \approx 0.026$ volts at room temperature.

When R is zero then V is zero, and the short-circuit current I_{SC} equals I_L. For a finite resistance R the photocell current I_C generates power P_C given by

$$P_C = I_C V = I_C^2 R. \tag{6.4}$$

As V increases P_C increases until V is slightly less than V_{OC} as shown in Fig. 6.8.

For this solar cell $I_{SC} = 1$ mA, $I_S = 10^{-14}$ amperes. So from eqn (6.3), $V_{OC} = 0.66$ V. We can generate the curve of power P_C versus voltage by using eqns (6.2) and (6.4). First calculate I_C for a given V and then calculate P_C. Table 6.1 gives the results for a number of voltages between 0 and 0.66 V. Note $I_L = I_{SC}$.

At the maximum power point $P_C = P_m = 0.55$ mW, $V = V_m = 0.58$ V, and $I_C = I_m = 0.95$ mA. We notice that I_m and V_m are both close to I_{SC} and V_{OC}, respectively. The fill factor (FF) is defined as the ratio

$$FF = P_m/(I_{SC} V_{OC}). \tag{6.5}$$

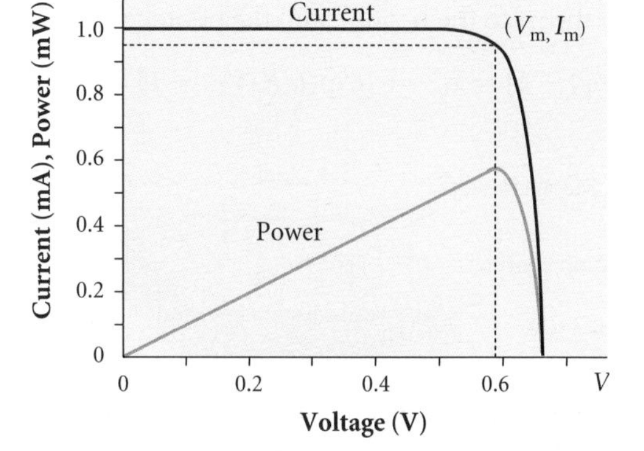

Fig. 6.8 Characteristics of a photocell.

Table 6.1 I_C, P_C and V for cell with $I_{SC} = 1$ mA, $I_S = 10^{-14}$ A

V (volts)	0.10	0.30	0.50	0.55	0.58	0.60	0.62	0.64	0.66
I_C (mA)	1.0	1.0	1.0	0.98	0.95	0.89	0.77	0.51	0.0
P_C (mW)	0.1	0.3	0.5	0.54	0.55	0.53	0.48	0.33	0.0

For the example in Table 6.1 the fill factor FF equals 0.83. FF is a measure of how close the IV characteristic is to a rectangle. It is useful for quality control with good solar cells having $FF > 0.7$: typically FF lies between 0.75 and 0.85.

EXAMPLE 6.3

A photocell has a saturation current $I_S = 2 \times 10^{-12}$ amps, a short circuit current $I_{SC} = 30$ mA, and an area of 1 cm^2. Find the maximum power output, the fill factor, and the conversion efficiency of the cell. What resistance across the cell is required to give the maximum output?

From eqn (6.3) $V_{OC} = 0.61$ V. Calculating I for a number of voltages V below V_{OC} using eqn (6.2), and then calculating the power P from eqn (6.4) gives the results in the following table

V (V)	0.50	0.52	0.53	0.54	0.56
I (mA)	29.6	29.0	28.6	27.9	25.5
P (mW)	14.8	15.1	15.2	15.1	14.3

Peak power P_m is 15.2 mW when $V_m = 0.53$ V and $I_m = 28.6$ mA. From eqn (6.5)

$$FF = P_m/V_{OC}I_{SC} = 0.0152/\{(0.61)(0.03)\} = 83\%.$$

Approximately 100 mW per cm^2 of solar radiation falls on the Earth, so this cell has an efficiency of $\sim$15%.

As voltage equals current times resistance, the resistance R required for maximum output is given by $V_m = I_m R$, so

$$R = V_m/I_m = 0.53/0.0286 = 18.5 \text{ ohms}.$$

We will now look at why solar cells have efficiencies of typically only $\sim$10% to $\sim$30%.

6.5 Efficiency of solar cells

The conversion efficiency is defined as the ratio of the maximum power output to the incident solar power, which for AM1.5 solar radiation (Fig. 6.1) is close to 100 mW cm^{-2}. This is not

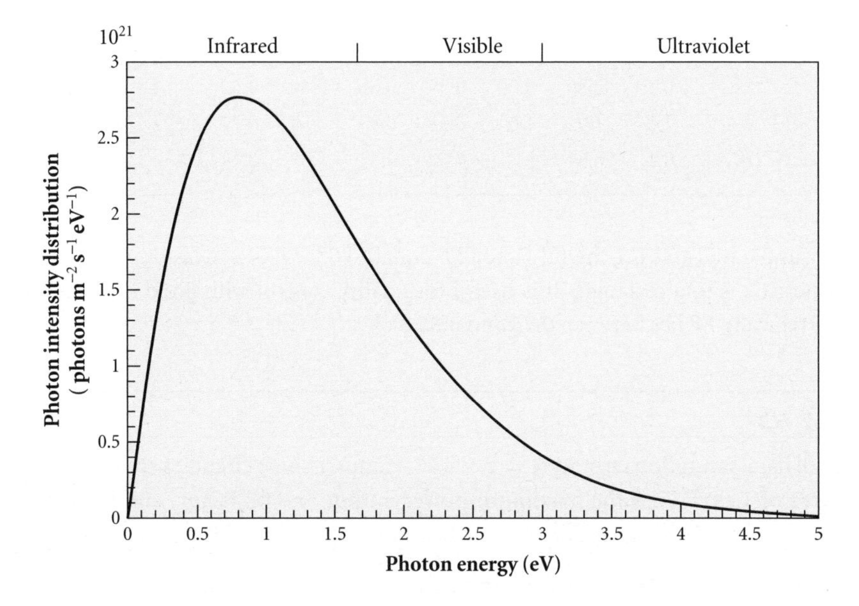

Fig. 6.9 Photon intensity for AM1.5 solar radiation (100 mW cm^{-2}), assuming a black-body distribution ($T = 5800$ K).

100% for several reasons: one is that not all the photons have sufficient energy (>1.1 eV for a silicon cell) to produce electron–hole pairs, as can be seen in Fig. 6.9.

The photons with energies less than 1.1 eV carry 23% of the incident solar energy. Only 1.1 eV of the energy of any higher energy photons is required to produce power. The result is that only 47% of the incident solar energy contributes to the power with 30% going to heat. So 47% is the limiting efficiency from the solar spectrum for a silicon cell. Another significant reduction is from the voltage factor, which is the ratio of $|e|V_m/E_{gap} \approx 0.7/1.1 = \sim 0.65$. This is the ratio of the energy given to an electron to the minimum energy required to produce an electron. The final significant loss comes about because not all of the electron–hole pairs that are produced are collected by the field across the junction; about 10% recombine.

A potentially large loss ($\sim$40%) from reflection from the front surface of the silicon can be reduced a large amount by using quarter-wavelength layers of material to act as an anti-reflection coating. The reflectance ρ between two media with refractive indexes n_1 and n_2 is

$$\rho = (n_1 - n_2)^2/(n_1 + n_2)^2. \tag{6.6}$$

Silicon has a complex refractive index, as it is partly conducting, which is frequency-dependent and averages about 3.5. Substituting this value into eqn (6.6) gives $\rho \approx 40\%$. If we add an odd number of quarter-wavelength thick layers to the silicon with a refractive index (n_1) that is intermediate between that of silicon (n_2) and air ($n_0 = 1$), then ρ can be reduced considerably. Figure 6.10 shows a ray of light incident almost normally on such an arrangement.

The odd number of quarter wavelengths thickness makes the reflected components a and b out of phase. The reflectance at each surface must be equal so $(n_1)^2 = n_0 n_2$. The effect over

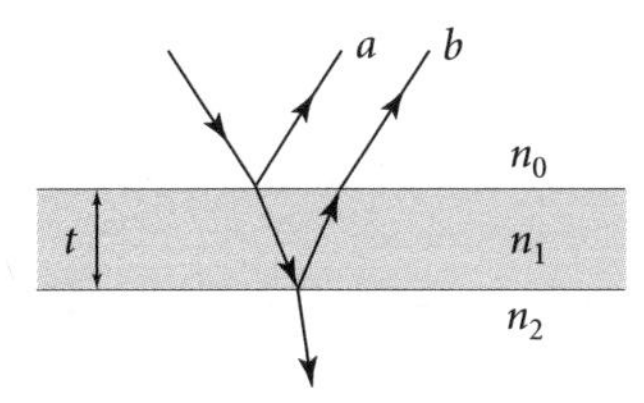

Fig. 6.10 Quarter wavelength anti-reflection coating (the angle of incidence is exaggerated for clarity).

the range of solar wavelengths is to reduce ρ to ~6%. Multiple reflection coatings can reduce it still further to ~1%.

There are small losses from contacts on the front surface and there is very little loss from photons not absorbed, due to optimizing the silicon thickness plus adding reflecting layers on the back of the cell (total ~3%). The overall efficiency η_C from multiplying all these factors together is

$$\eta_C \approx (0.47)(0.65)(0.9)(0.96) \approx 26\%. \tag{6.7}$$

We can see that the efficiency is dependent on the band gap E_{gap}: decreasing E_{gap} increases the photocurrent, as more light can produce electron–hole pairs, but decreases the maximum output voltage as $|e|V_{OC} < E_{gap}$. So there is an optimum E_{gap}^{opt} which is ~1.4 eV, and the semiconductors GaAs and CdTe have band gaps close to this optimum value (see Derivation 6.2).

We will first look at the construction of silicon crystalline cells before looking at thin film solar cells and at devices under development.

Derivation 6.2 Efficiency dependence on band gap

The maximum power of a cell from eqn (6.5) is

$$P_m = FF \times I_{SC} V_{OC}.$$

We can estimate how P_m depends on $E_{gap} \equiv E_g \equiv |e|V_g$ by taking FF as a constant and looking at how $I_{SC} \equiv I_L$ and V_{OC} are affected by changing E_g. As I_{SC} is proportional to the area of the cell, the result for the current density J_{SC} (mA cm^{-2}) is plotted in Fig. 6.11.

The photocurrent density J_L is given by the product $N_g e$, i.e.

$$J_L = N_g e \tag{6.8}$$

where N_g is the number of photons per unit area per second with $E_\gamma > E_g$; this assumes that all the electrons and holes contribute to the photocurrent, i.e. none are lost through recombination. The number N_g is

$$N_g = \int_{E_g}^{\infty} n(E_\gamma) dE_\gamma \tag{6.9}$$

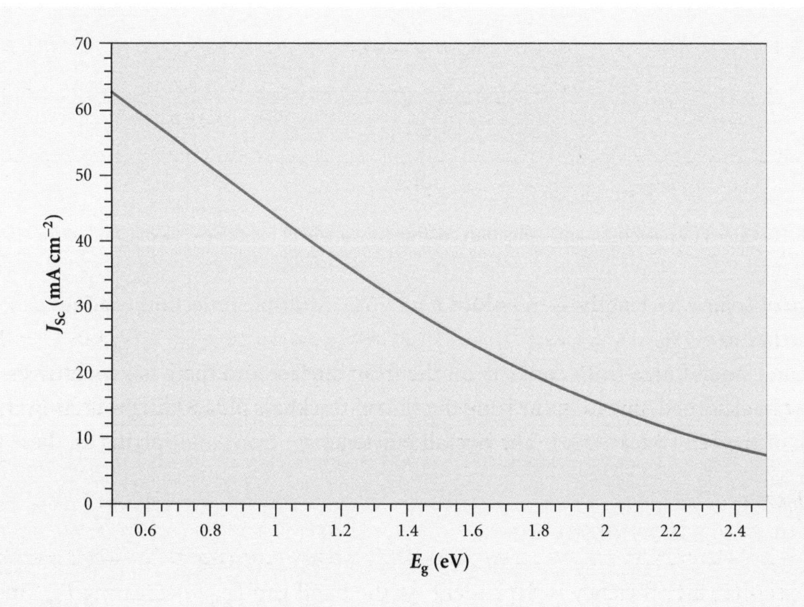

Fig. 6.11 J_{SC} dependence on E_{gap}.

where the number of photons per unit area per second per eV, $n(E_\gamma)$, is given by the Planck distribution:

$$n(E_\gamma) = \frac{D(kT)^2 y^2}{e^y - 1} \tag{6.10}$$

where $y \equiv E_\gamma/kT$, and D is a constant dependent on the solid angle subtended by the Sun (see Exercise 6.4). This expression for $n(E_\gamma)$ assumes the solar spectrum can be approximated by that of a black body at a temperature T. For the Sun, $T = 5800$ K is a good approximation.

The number N_g can therefore be expressed as

$$N_g = D(kT)^3 \int_x^\infty \frac{y^2\, dy}{e^y - 1} \tag{6.11}$$

where

$$x \equiv E_g/kT \equiv eV_g/kT. \tag{6.12}$$

The incident solar power density J_{inc} is given by

$$J_{inc} = \int_0^\infty n(E_\gamma) E_\gamma dE_\gamma = D(kT)^4 \int_0^\infty \frac{y^3 dy}{e^y - 1} = D(kT)^4 \frac{\pi^4}{15}. \tag{6.13}$$

Combining eqns (6.8) and (6.11)–(6.13) gives the relation

$$V_g J_L = J_{inc} x \frac{15}{\pi^4} \int_x^\infty \frac{y^2\, dy}{e^y - 1}. \tag{6.14}$$

The integral can be expressed in terms of a sum

$$J_L = J_{inc} \left(\frac{15}{V_g \pi^4} \right) \sum_{n=1}^{\infty} e^{-nx} \left(\frac{x^3}{n} + \frac{2x^2}{n^2} + \frac{2x}{n^3} \right). \tag{6.15}$$

For $x > 2$ only the first term with $n = 1$ is required to give a good approximation for $J_L (= J_{SC})$. The solar spectrum AM1.5 is taken as a standard and for this spectrum $J_{inc} = 1$ kW m^{-2}. Figure 6.11 shows a plot of eqn (6.15) with $J_{inc} = 1$ kW m^{-2} as a function of E_g.

Notice in Fig. 6.11 that, for $0.5 < E_g < 1.75$ eV, J_{SC} is approximately linearly dependent on E_g, and hence on V_g. This dependence can be written as

$$J_{SC} = J_L \approx A - BV_g \tag{6.16}$$

where $A = 80$ mA cm^{-2} and $B = 34$ mA cm^{-2} V^{-1}.

The open circuit voltage is given by eqn (6.3) as $V_{OC} = V_T \ln(J_L/J_S)$, as both I_L and I_S are proportional to the area of the cell. The saturation current density J_S is determined by the concentration of electrons n_e^p on the p-side of the junction and this depends on the Boltzmann factor $\exp(-E_g/kT)$. An approximate expression for J_S is

$$J_S = J_0 \exp(-E_g/kT) \tag{6.17}$$

with $J_0 = 2 \times 10^9$ mA cm^{-2}. This gives $J_S \sim 10^{-12}$ A cm^{-2} for silicon ($E_g = 1.1$ eV). Substituting eqn (6.17) into the expression for V_{OC} gives

$$V_{OC} = V_g - V_T \ln(J_0/J_L) \approx (V_g - 0.46), \tag{6.18}$$

over the range of J_L from the Sun corresponding to $0.5 < E_g < 1.75$ eV. The maximum power P_m is therefore

$$P_m \approx FF(A - BV_g)(V_g - 0.46) \text{ mW cm}^{-2}. \tag{6.19}$$

This has a maximum assuming the fill factor FF is constant when $V_g = \{(A/2B) + 0.23\} \approx 1.4$ V. The semiconductors GaAs and CdTe both have V_g close to this optimum value.

Table 6.2 Typical (2004) conversion efficiencies for different types of solar cell

		Efficiency (%) for cell of area	
Material*	**Band gap (eV)**	**~1 cm^2**	**~1 m^2**
GaAs	1.4	~24	—
GaAs (multi)	1.8–0.7	~34	—
Si(sc)	1.1	~24	—
Si(mc)	1.1	~20	~13
a-Si(single)	~1.7	~12	~7
a-Si(multi)	~1.7–1.3	~13	~10
CdTe	1.5	~16	~8
$CuIn_{1-x}Ga_xSe_2$	~1.2	~19	~12

*sc, single crystal; mc, multicrystalline; single, single junction; multi, multiple junctions.

For a silicon cell that has $E_g = 1.1$ eV and $FF = 0.8$, this expression (6.11) gives $P_m \approx$ 25 mA cm^{-2}. Hence the conversion efficiency is ~25%, which is close to the maximum that has been obtained (Table 6.2).

6.6 Commercial solar cells

There are two main types of solar cell in production today: crystalline silicon and thin-film cells. The desire to reduce the cost of cells led to the research and development of thin-film cells. Considerable progress has been made over the last 25 years, although crystalline silicon still has the largest share of the market (~93% in 2004). We will first describe crystalline silicon-based cells. Then we will discuss the newer thin-film technologies—first single layer devices and then multiple layer cells. At the end of this section we will consider some of the technologies that are under development.

6.6.1 Crystalline silicon cells

Crystalline silicon cells are produced either as single crystal or polycrystalline (multicrystalline) cells. For the single crystal cells, thin wafers, ~200–400 μm thick, are sawn from silicon crystal ingots. The technology for making these ingots has been developed for the semiconductor industry. Much of the material used at present for silicon solar cells is surplus microelectronics material, which has helped keep costs down. But the wafers are still expensive and techniques have been explored to reduce these costs. One way is to use ingots of polycrystalline silicon made from casting silicon, which are cheaper but less efficient. Another way, developed to avoid the losses caused by sawing, is to make polycrystalline silicon in the form of a ribbon.

The construction of a silicon cell is illustrated in Fig. 6.7. The thickness of the silicon needs to be ~ 200–300 μm in order to capture most of the sunlight, as the light absorptivity of silicon is low. This is a consequence of the nature of the band gap, being indirect rather than direct. To use a thinner silicon layer (~50 μm) requires light trapping techniques. These scatter or reflect the light causing it to pass through the silicon layer several times. The use of highly light-absorbing semiconductor material would reduce the amount of material required considerably, which could help reduce the cost of the cell. This has been the motivation to develop thin-film solar cells.

6.6.2 Thin-film cells

There are a number of materials that have good solar light absorption; in particular, GaAs, CdTe, $CuInGaSe_2$ (CIGS), and amorphous hydrogenated silicon (a-Si:H). All of these films only need to be about (~1 μm) thick, so much less material is required (<~1%) than for silicon cells. Materials are only part of the cost and the challenge in thin film technology is to develop techniques for fast deposition of films while maintaining film quality. Cells with a large area (~1 m^2) generally have more imperfections and so have less efficiency than small area ($\sim$1 cm^2) devices. Typical percentage conversion efficiencies that have been obtained (2004) are shown in Table 6.2.

GaAs has a band gap of 1.4 eV, close to optimal. It can withstand high temperatures as the band gap is sufficiently large to keep thermal excitation small. This enables concentrators to be used with GaAs cells. These focus the solar radiation on to the active cell area and can increase the flux by over a 100 times. The active region is grown epitaxially on a very thin GaAs single crystal layer. This layer is, however, relatively expensive so only small area GaAs cells have been made. They find use where high performance is required, e.g. in space applications.

CdTe, CIGS, and a-Si:H can all be made into large area ($\sim$1 m^2) solar cells and so are all candidates for solar power plants. One promising semiconductor is CdTe, which also has a good band gap of 1.4 eV.

EXAMPLE 6.4

A 4 cm^2 GaAs solar cell has a saturation current of 4×10^{-15} mA. Under normal illumination of AM1.5 solar radiation the short-circuit current is 127 mA. What is the conversion efficiency under normal and under $\times 100$ illumination?

The open circuit voltage V_{OC} under normal illumination is given by eqn (6.3) as

$$V_{OC} = V_T \ln (I_{SC}/I_S) = 0.026 \ln \{127/(4 \times 10^{-15})\} = 0.988 \text{ V}.$$

We could find the maximum output power P_m by finding I and P for different V, as in Example 6.3. However, good approximations (see Exercise 6.7) for V_m, I_m, and P_m are given by

$$V_m = V_{OC}(1 + x_{OC} \ln x_{OC}), \text{ and } I_m = I_{SC}(1 - x_{OC}). \tag{6.20}$$

where $x_{OC} = kT/\{|e|V_{OC}\} = V_T/V_{OC}$, $V_T \equiv kT/|e| \approx 0.026$ volts at room temperature. So the fill factor FF is

$$FF = (1 - x_{OC})(1 + x_{OC} \ln x_{OC}). \tag{6.21}$$

For normal illumination $x_{OC} = 0.0263$. Substituting in eqn (6.20) gives

$$V_m = 0.988\{1 + 0.0263 \ln (0.0263)\} = 0.89 \text{ V},$$
$$I_m = 127(1 - 0.0263) = 124 \text{ mA}.$$

Therefore,

$$P_m = 0.89(124) = 110 \text{ mW}.$$

Under 100 times illumination the open circuit voltage increases as the short circuit current is 100 times larger, i.e.

$$V_{OC} = 0.026 \ln \{12700/(4 \times 10^{-15})\} = 1.108 \text{ V}.$$

The value of $x_{OC} = 0.0235$ and the maximum output power occurs when $V_m = 1.01$ V and $I_m = 12.4$ A corresponding to

$$P_m = 1.01(12.4) = 12.5 \text{ W}.$$

Under AM1.5 illumination the solar intensity is 100 mW cm^{-2}, so as the area of the cell is 4 cm^{-2} a solar power of 400 mW falls on the cell. The conversion efficiency is the output power P_m over the incident solar power, i.e. 110/400 = 28%.

Under 100 times illumination the incident solar power is 40 W and the output power P_m is then 12.5 W, giving a conversion efficiency of 31%. Increasing the illumination gives an improvement of ~10% in conversion efficiency. The fill factor remains approximately constant, changing from 0.88 to 0.89.

CdTe solar cell

A schematic diagram of the construction of a CdTe cell is shown in Fig. 6.12. The cell is fabricated on a thin sheet of glass (2–4 mm thick). This is normally coated with an anti-reflection film. The first layer on the glass is a transparent conducting oxide, usually tin or indium tin oxide, which provides a good electrical contact to the thin CdS layer. The polycrystalline CdS layer is n-doped. Its band gap is ~2.4 eV and is transparent down to wavelengths of ~500 nm. Below this wavelength light is attenuated but some is still transmitted to the CdTe, as the CdS is only ~100 nm thick.

The CdTe is polycrystalline and p-doped. Its energy gap is ~1.5 eV and is well matched to the solar spectrum. The CdTe is less doped than the the CdS so most of the depletion zone lies in the CdTe layer, which is typically ~10 μm thick. An aluminium or gold contact is generally used to make contact with the CdTe layer.

This is an example of a **heterojunction** device since the p- and n-regions are in different semiconductors. While these cells have been made with good performance, CdTe cells have shown instabilities, which are assumed to be associated with the difficulty of making good contact with the CdTe region. Two other semiconductors that may have even more promise are a-Si and CIGS. We will look at CIGS first.

CIGS solar cells

The structure of a CIGS cell is shown in Fig. 6.13. The molybdenum contact layer is sputtered on to the glass substrate, followed by the CIGS p-type layer (1.2 μm). This can be formed by first depositing an indium gallium selenium compound, $(InGa)_2\ Se_3$. Then this layer is reacted with Cu and Se, followed by In and Ga evaporation in the presence of Se. This technique can give a material in which the band gap varies with depth (see multilayer cells below). Thin

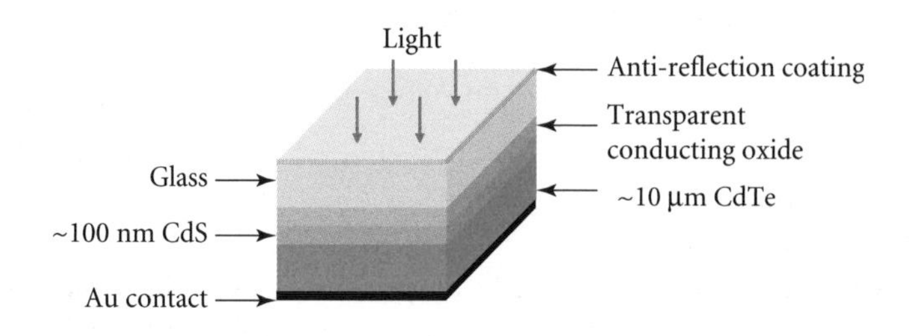

Fig. 6.12 Construction of a CdTe cell.

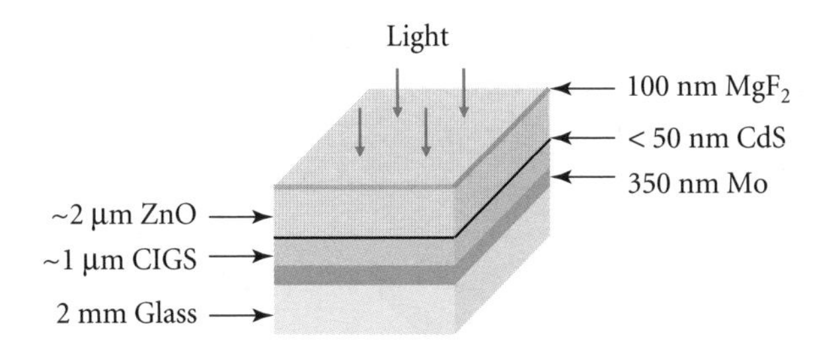

Fig. 6.13 Layer structure of a CIGS cell.

CIGS layers with a uniform band gap can also be made by simultaneous deposition of all the elements. The n-type CdS layer is then put down to form the heterojunction, and the ZnO layer is deposited to provide the transparent contact. Finally a 100 nm MgF_2 layer provides an anti-reflection coating.

By altering the ratio of In to Ga the band gap can be varied between 1.1 and 1.2 eV. For a small area device with $E_g = 1.2$ eV illuminated under AM1.5 ($1\ \mathrm{kW\,m^{-2}}$) radiation, $J_{SC} = 34.6\ \mathrm{mA\,cm^{-2}}$, $J_S = 5 \times 10^{-8}\ \mathrm{mA\,cm^{-2}}$, $FF = 79.7\%$, and a conversion efficiency of 19.3% has been obtained. Large area modules have been made with efficiencies of ~12%. The technology is less developed than for a-Si cells but it has great potential.

Amorphous silicon solar cells

The first amorphous silicon (a-Si) solar cell was made in 1976. a-Si is produced by electrically decomposing silane, SiH_4, together with a small amount of boron dopant. The hydrogen provides additional electrons that combine with dangling bonds in the a-Si producing an intrinsic semiconductor a-Si:H. The material is a silicon hydrogen alloy with 5–20 (atomic)% of H and the resultant band gap is ~1.7 eV. The disordered arrangement of atoms, together with the hydrogen, also gives a high optical absorption, which allows the device to be only ~1 μm thick. Figure 6.14(a) shows the layout of an a-Si solar cell and Fig. 6.14(b) the current–voltage characteristic of a small area device.

The cell is built upon a glass substrate with the transparent indium tin oxide providing the electrical contact to the p-doped region of a hydrogenated amorphous SiC layer. The SiC has a larger bandgap than a-Si and so allows a large fraction of the solar radiation through to the intrinsic hydrogenated a-Si layer (a-Si:H). Below the a-Si:H there is an n-doped region with a contact layer.

In a crystalline silicon cell, electrons and holes produced by light within a few diffusion lengths of the junction contribute to the photocurrent. But in the p- and n-regions of the a-Si the diffusion lengths are quite small and most of the contribution is from electron-holes produced in the a-Si:H region. The field between the doped regions causes the holes to flow to the p-region and the electrons to the n-region. The layers form a single p-i-n junction, where i stands for the intrinsic a-Si:H layer.

A problem with such cells is that their efficiency decreases under illumination. This effect, first seen by Staebler and Wronski, results from the creation of metastable defects. The effect has been reduced to a ~20% decrease by diluting the silane with hydrogen and by optimizing

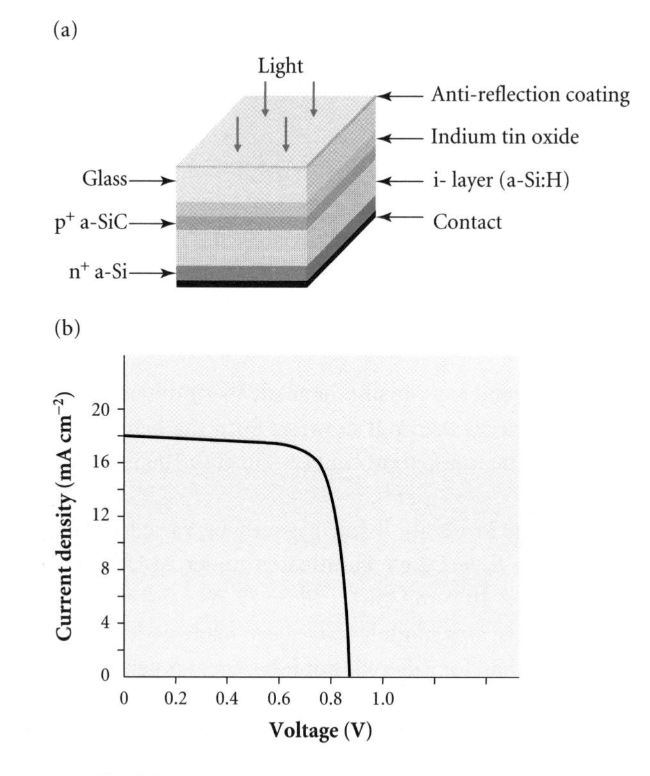

Fig. 6.14 (a) Layout of a-Si cell. (b) a-Si IV characteristic.

the growth conditions. The result is that large area single junction modules have been produced with a long term efficiency of ~7%.

By changing the deposition conditions it has proved possible to make good quality multicrystalline (mc) Si films as well as a-Si:H layers. But for these to be useful the probability of absorbing the light in the thin mc-Si layer must be improved. This is possible by using light trapping. This method is described in Box 6.1 and can lead to an increase in path length of $\sim 4n^2$, where n is the refractive index of the material For silicon n is ~3.5, so this increase of ~50 can enable relatively thin layers (~10 μm) of mc-Si to be used.

The efficiency of a thin-film cell can be increased by making a multilayer device consisting of several p-i-n junctions on top of one another. We will now consider these multilayer thin film cells.

Box 6.1 Light trapping in thin films

Silicon has a low absorption, so light trapping is required if thin layers are to be used. This allows layers to be made by deposition, which uses less material. There is also less recombination in these thin layers, which increases the open circuit voltage and hence the efficiency. One way of enhancing the light absorption is to texture the top surface of the silicon and make the back surface reflective, as illustrated in Fig. 6.15.

The texturing of the top surface can cause the light to make two reflections. This reduces the overall amount reflected, but more important is the increase in path length for light within the silicon. Silicon has a high refractive index n of $\sim$3.5 so if silicon is in a medium with a refractive index of one then only light incident within a cone of half-angle $\theta_c = \sin^{-1}(1/n)$ will escape. Otherwise the light is totally internally reflected, as shown in Fig. 6.15(a).

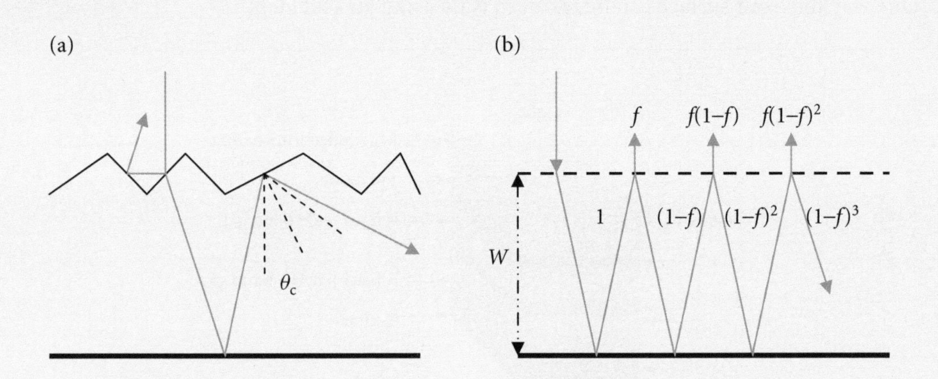

Fig. 6.15 (a) Textured top and reflective back. (b) Schematic of multiple reflections.

Light passing through the top surface of the silicon undergoes refraction and as the surface is irregular it acts as a diffuser. For an ideal diffusing surface the intensity per unit solid angle $B(\theta)$ is related to the incident intensity I_0 by

$$B(\theta) = (I_0/\pi)\cos\theta.\text{(Lambert's law.)} \tag{6.22}$$

After each reflection the fraction f of light that escapes is given by

$$f = \int_0^{\theta_c} (1/\pi)\cos\theta\, 2\pi \sin\theta\, d\theta = \sin^2(\theta_c) = 1/n^2, \tag{6.23}$$

which is the fraction of light that falls within the critical angle θ_c. Figure 6.15(b) shows the amounts lost and intensities remaining after each reflection. (The actual light is scattered in all directions.) The mean distance D that light travels between each reflection off the back surface is given by

$$D = \int_0^{\pi/2} (1/\pi)\cos\theta\,(2W/\cos\theta)2\pi\sin\theta\, d\theta = 4W, \tag{6.24}$$

where W is the depth of the silicon layer and $(2W/\cos\theta)$ is the distance travelled by light reflected at an angle θ.

The mean path P travelled is given by summing up the fraction of light undergoing 1, 2, 3... reflections, each multiplied by $4W$, which from Fig. 6.15(b) is

$$P = 4W + (1-f)4W + (1-f)^2 4W + \cdots = 4W/f = 4n^2 W \tag{6.25}$$

where the identity $(1-x)^{-1} \equiv 1 + x + x^2 + \cdots$ has been used to give the sum of the geometric series. As n for silicon is ~3.5 then this path length is ~50 times longer than with no texturing of the front surface and no back reflective surface. In practice this technique allows mc-Si films of thickness of the order of 10 μm to have good absorptivity. One way the front surface can be textured is by using an acid etch.

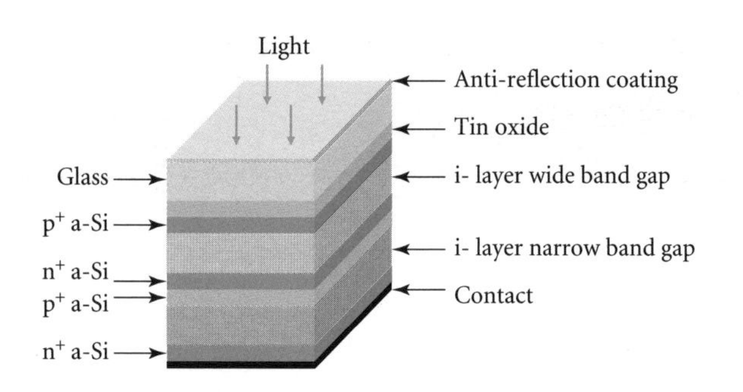

Fig. 6.16 Two-layer a-Si solar cell.

Multilayer thin film cells

A multilayer device can utilize different regions of the solar spectrum. For example, we will consider a two layer cell with a wide band gap material as the upper layer and a narrow band gap as the lower. High energy photons will be absorbed in the upper layer while lower energy light will be transmitted by the upper and be absorbed in the bottom layer. The construction of such a two layer a-Si solar cell is shown in Fig. 6.16.

The layers are in series so the current through both layers is the same and is limited by the layer producing the smallest current. So the layers must be matched and one possible combination would be a-Si:H (E_g~1.7 eV) with a-SiGe:H ((E_g~1.1 eV). The latter is made by co-depositing Ge and Si using a mixture of SiH_4 and GeH_4. An alternative for the wide band gap material is a-SiC made with CH_4 and SiH_4. Another possibility for the lower layer is microcrystalline Si with light trapping.

A four layer device with band gaps of 1.8, 1.4, 1.0, and 0.7 eV has a theoretical efficiency greater than 50%.

EXAMPLE 6.5

A two layer a-Si solar cell has band gaps of 1.7 and 1.1 eV. The photocurrent density from photons with energies between 1.1 and 1.7 eV is 19 mA cm^{-2}, while that from photons with energies greater than 1.7 eV is 21.3 mA cm^{-2}. Compare the output power for this two layer solar cell with that of a simple silicon solar cell ($E_g = 1.1$ eV). Assume the open circuit voltage of each layer, and of the silicon cell, is given by $V_g - 0.4$ V. Take the fill factor to be 0.8.

The two layers are in series so the voltage across the cell will be the sum of the voltages across each layer. The short circuit current density through the layers is that from the lower layer, 19 mA cm^{-2}, as it is the smaller. The open circuit voltage is the sum of V_{OC} from each layer, so

$$V_{OC} = (1.1 - 0.4) + (1.7 - 0.4) = 2.0 \text{ V}.$$

The maximum power P_m is given by eqn (6.5)

$$P_m = FF \times V_{OC}I_{SC} = 0.8(2.0)19 = 30.4 \text{ mW cm}^{-2}.$$

The simple silicon solar cell has a short circuit current given by

$$I_{SC} = 19 + 21.3 = 40.3 \text{ mA cm}^{-2}.$$

The open circuit voltage is $(1.1 - 0.4) = 0.7$ V. The output power is therefore

$$P_m = 0.8(0.7)40.3 = 22.6 \text{ mW cm}^{-2}.$$

We can see that the efficiency is significantly higher in the two layer cell.

6.7 Developing technologies

6.7.1 Electrochemical cells

The production costs of crystalline silicon and of thin-film cells have been steadily decreasing (see Section 6.9), but the cost per watt is still too high to be competitive with conventional power production. Thus cells using a lower cost technology than those requiring vacuum evaporation or crystal-growing techniques are required. In 1991 dye-sensitized solar cells (DSSC) were invented by M. Gratzel and B. O'Regan and these have attractive features.

In a DSSC a stack of titanium dioxide TiO_2 nanoparticles ~20 nm in diameter are coated with dye molecules and the whole is immersed in an electrolyte (Fig. 6.17(a)). Light is absorbed by the dye and electron–hole pairs are produced. The electrons go into the conduction band of the n-type semiconductor TiO_2 and diffuse to a transparent conductive substrate. The holes in the dye molecule are filled by electrons from negatively charged ions in the electrolyte, which come from the opposite electrode. The electron to ion transfer at this electrode is catalysed by a layer of platinum or carbon.

The energy levels involved are shown in Fig. 6.17(b) and the maximum potential difference, the open circuit voltage, is given by the difference in the potential of the conduction band in TiO_2 and the potential of electrons bound in the ions in the electrolyte (the redox potential). The recombination of electrons and holes is much slower than the injection of the electrons into the conduction band of TiO_2. Therefore diffusion of the electrons to the conducting substrate and the refilling of the holes by electrons from the ions in the electrolyte occurs. By

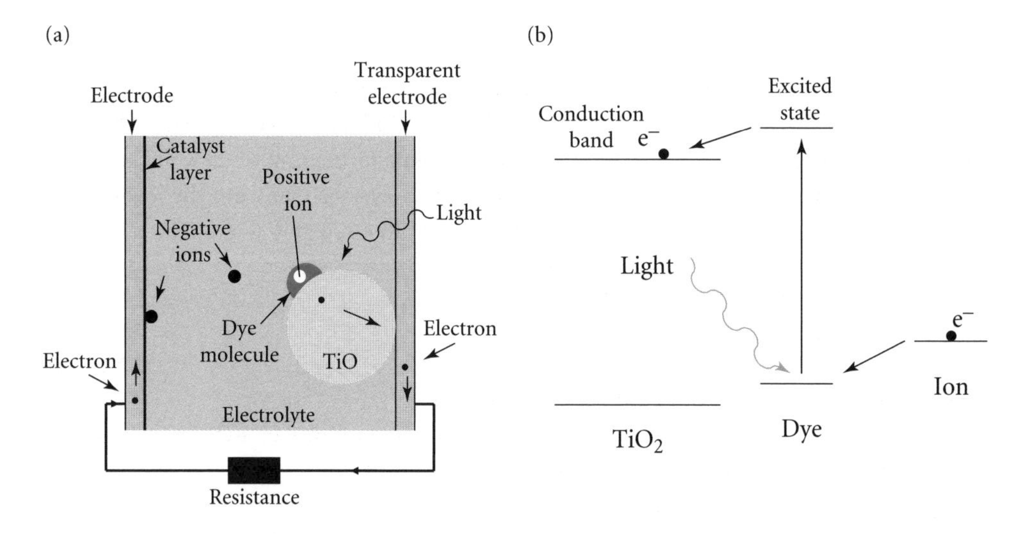

Fig. 6.17 (a) DSSC. (b) Level structure in DSSC.

this means the electrons and holes produced by the solar radiation are separated. Unlike in a p-n junction, the separation is not brought about by an internal electric field.

The percentage of light absorbed by the dye molecules attached to the TiO_2 nanospheres is very small, but the large surface area of all the nanospheres increases the absorption considerably. TiO_2 is cheap and the technology shows promise. Currently conversion efficiencies of laboratory cells exceed 11%, though long-term stability needs to be established.

Recent research on nanocrystals of lead sulfide and lead selenide, called quantum dot absorbers, has indicated that such crystals might enhance the performance of the TiO_2 cells if used in place of the dye molecules. When illuminated with light these nanocrystals produced three electrons on average per photon absorbed, rather than just the single electron in a p-n junction cell, and would give a larger solar photocurrent and greater efficiency. Such crystals are called quantum dots when their size is comparable with that of the orbital radius of electron–hole excitations (excitons) within the crystal.

6.7.2 **Concentrators**

We can avoid the need to make a large PV array by using reflectors or lenses to concentrate the light on to a single smaller area device. Silicon based cells with 27% conversion efficiency at an intensity up to 400 times that of normal sunlight (400 Suns) and GaAs with 28% efficiency at 1000 Suns have been developed. The concentrators need to track the movement of the Sun, which increases their cost. Fresnel lenses and parabolic reflectors have been used but their market impact is small at present.

We anticipate an improvement in conversion efficiency with illumination since the open circuit voltage is expected to increase from eqn (6.3) above. This relation predicts at 1000 Suns that V_{OC} would be 0.18 V larger, which for silicon would correspond to a 25% increase in both V_{OC} and in efficiency (see Example 6.4).

6.7.3 **Organic semiconductor solar cells**

In the search for cheap solar cells there has been considerable research and development (R & D) on organic thin film solar cells, which might be produced at low cost by being fabricated on a flexible plastic substrate. But conversion efficiencies have so far been very low at ~1%.

A recent development that has given improved efficiency has been a p-i-n type organic film solar cell made from the p-type organic semiconductor ZnPc, a phthalocyanine derivative, and a n-type semiconductor fullerene (C_{60}), separated by a layer of mixed ZnPc and C_{60} molecules in nearly equal numbers. This mixed layer acts macroscopically as an intrinsic layer. The structure is similar to that of a p-i-n a-Si solar cell.

The organic solar cell consists of a transparent electrode, an organic buffer layer to provide good electrical contact, 5 nm of ZnPc, 15 nm of mixed C_{60}–ZnPc, and 30 nm of C_{60}. This has given 4% conversion efficiency under AM1.5 illumination, which could be increased if recombination in the intrinsic region can be reduced.

6.7.4 **Thermo-photovoltaic cells (TPC)**

Solar radiation can be focused on to an intermediate absorber rather than directly on to a photocell. The absorber re-emits the solar energy as thermal radiation, which is detected by photocells surrounding the absorber. The thermal radiation is from a surface at a lower temperature than the sun, but the radiation can in principle be utilized more efficiently. Thermal radiation can also be generated by burning gases. The application of TPCs is discussed in Box 6.2.

Box 6.2 Thermo-photovoltaic cells (TPC)

Solar radiation can be focused on to an intermediate absorber rather than directly on to a photocell. The absorber re-emits the solar energy as thermal radiation, which is detected by photocells surrounding the absorber. The temperature of the emitter is typically 1500 K, rather than ~6000 K for the Sun, and the mean photon energy is ~0.45 eV compared with ~1.8 eV for solar radiation. The photocurrent density J_L (amps m^{-2}) produced by a photocell with band gap, $E_g \equiv eV_g$, is proportional to the incident radiation intensity J_{inc} (W m^{-2}). Assuming the thermal radiation is that of a blackbody then J_L also depends on the ratio $x \equiv E_g/kT$. As shown in Derivation 6.2 , for $x > 2$, J_L is given to a good approximation by

$$J_L = J_{inc}\left(\frac{15}{V_g \pi^4}\right) e^{-x}(x^3 + 2x^2 + 2x). \tag{6.26}$$

The open circuit voltage V_{OC} depends on J_L slightly (see Derivation 6.2) and can be estimated by the relation (6.18)

$$V_{OC} = V_g - V_T \ln (J_0/J_L),$$

where the empirical constant $J_0 \approx 2 \times 10^{10}$ A m^{-2}.

The amount of energy that is emitted per second per unit area J_E by a black-body radiator is given by the **Stefan–Boltzmann law** (p. 21)

$$J_E = \sigma T^4. \tag{6.27}$$

The value of the Stefan–Boltzmann constant σ is 5.67×10^{-8} W m^{-2} K^{-4}. The incident radiation intensity J_{inc} is proportional to J_E, with the proportionality constant dependent on the geometry of the device.

While the fraction of photons with $E_\gamma > E_g$ is lower than for a solar spectrum, the use of an intermediate absorber allows the energy of the photons with $E_\gamma < E_g$ that are transmitted through the cell not to be lost. These lower energy photons are returned to the emitter by a reflective layer on the back of the photocell. At the emitter their energy is absorbed and then re-emitted. The efficiency of conversion can be further improved by using selective emitters that radiate in a fairly narrow band, rather than with a broad thermal spectrum. An example is the rare earth oxide Er_2O_3, which emits in a band around 1500 nm. This is matched to photocells with an $E_g < 0.8$ eV.

So as to reduce the amount of thermal radiation with $E_\gamma < E_g$ falling on the photocell and heating it, a filter can be added. The filter can absorb long wavelength radiation or, by using dielectric quarter wavelength coatings, reflect it back on the absorber. A schematic of a TPC system is shown in Fig. 6.18; only one photocell of the several that would surround the emitter and filter is shown. As the mean photon energy is only 0.45 eV, a low band gap photocell is used, such as GaSb with $E_g = 0.72$ eV, or Ge with $E_g = 0.66$ eV. The predicted efficiency of such a system using GaSb cells is ~25%.

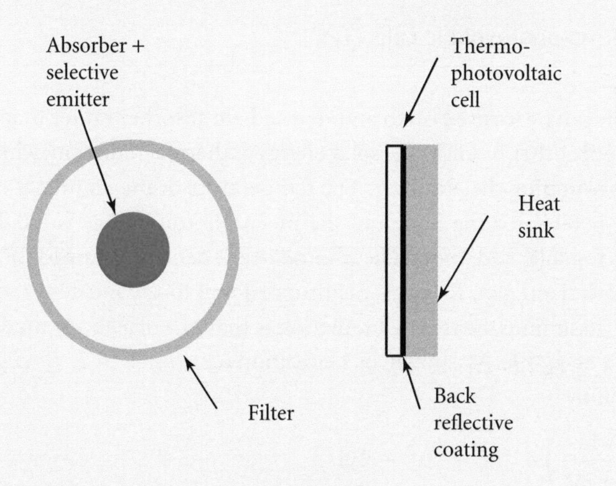

Fig. 6.18 Schematic of a TPC system—only one photocell is shown.

A further advantage of using a TPC is that the emitter can be heated by burning gas after dark. Alternatively, the TPV system can be run entirely with gas heating. A ceramic radiator burning methane can operate at ~1500 K and the total system can provide both heat and electricity. One possibility for producing methane is using an anaerobic digester (see Chapter 7). A TPC system would then provide heat and power in regions well away from a grid, such as some rural parts of developing countries. The lack of moving parts makes TPC attractive as maintenance should be low.

EXAMPLE 6.6

A methane heated radiator operates at a temperature of 1500 K. The cylindrical heater is 0.25 m long and 0.02 m in diameter. The TPCs surrounding the radiator form a cylinder 0.08 m in diameter. The photocells are GaSb ($E_g = 0.72$ eV) with a fill factor of 0.7. Estimate the electrical output of the system. Assume that the emitted radiation is that from a black body.

From eqn (6.27) the emitted intensity from the radiator

$$J_E = 2.87 \times 10^5 \text{ W m}^{-2}.$$

The geometry is cylindrical so

$$J_{inc} = J_E \times (r_{radiator}/r_{photocell}) = J_E/4 = 7.18 \times 10^4 \text{ W m}^{-2}.$$

The ratio

$$x = E_g/kT = 0.72 \times 1.6 \times 10^{-19}/(1.38 \times 10^{-23} \times 1500) = 5.57,$$

so, using eqn (6.26),

$$J_L = 1.44 \times 10^4 \text{ A m}^{-2}.$$

From eqn (6.18)

$$V_{OC} = 0.72 - 0.368 = 0.352 \text{ V}.$$

The current produced by the photocells is the current density multiplied by the area, i.e.

$$I_L = J_L 2\pi r_{photocell} L_{photocell} = 1.44 \times 10^4 (2\pi \times 0.08) 0.25 = 1810 \text{ A}.$$

Finally, eqn (6.5) gives the power output

$$P_m = FF V_{OC} I_L = 0.7(0.352)(1810) = 0.45 \text{ kW}.$$

The radiant power P_E is

$$P_E = J_E 2\pi r_{radiator} L = 2.87 \times 10^5 (2\pi \times 0.02) 0.25 = 9.0 \text{ kW}.$$

The conversion efficiency will be higher than $P_m/P_E = 5\%$, as no allowance has been made for photons returned to the radiator and for the use of selective emitters. This TPV system could provide both heating and power to a home.

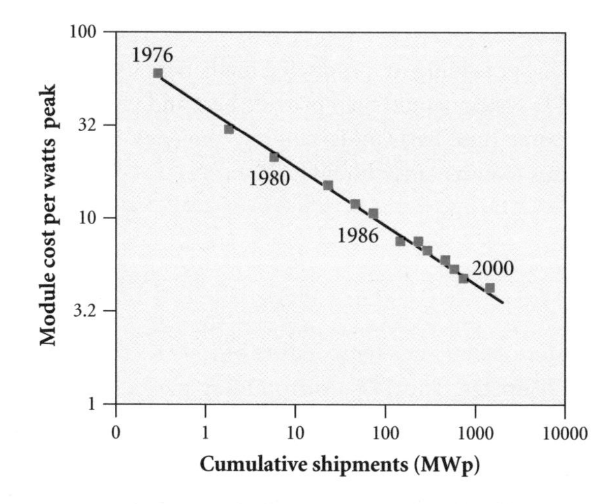

Fig. 6.19 Module cost per watts peak versus cumulative shipments.

6.8 Solar panels

Solar panels are built up from single units. For example, a silicon cell made from a thin textured multicrystalline wafer ($\sim$125 μm) might be 125 cm^2 in area with a $V_{OC} \sim 0.6$ V, and a panel might contain about 40 such cells connected in series with an output voltage of 24 V. The 0.5 m^2 area panel when illuminated with solar radiation of 1000 W m^{-2} would provide $\sim$75 W, corresponding to 15% conversion efficiency. This output under 1000 W m^{-2} illumination is termed the watts peak (Wp) output and the cost of panels is often given in terms of US $/Wp or €/Wp. Figure 6.19 shows how the module (panel) cost per Wp has decreased and how the volume has increased since 1976.

The amount of power that a solar panel will produce over a year depends on how much sun the location receives. A 1 kWp array of panels gives an output of 1 kW when the solar intensity is 1 kW m^{-2}. The yearly output from a 1 kWp array is therefore numerically the same as the annual amount of solar energy per square metre. A 1 kWp array will provide $\sim$1800 kWh/y in southern California, 850 kWh/y in northern Germany, and 1600–2000 kWh/y in India and Australia. As 1 kW continuous for a year is equivalent to 8760 kWh, we see that we obtain 10–20% of the peak output of a panel, dependent on location.

An average British household uses $\sim$3000 kWh of electricity per year and the available roof area per house is about 10 square metres. On this area we could install $\sim$1 kWp which would provide in the UK around 750 kWh per year, which is about a quarter of the total consumption. The rooftop area available in the EC could provide some 1000 GWp, but at present subsidies would be required for this to be economic.

EXAMPLE 6.7

A solar panel is made up of 40 silicon cells in series each of area 0.01 m^2, open circuit voltage 0.6 V, and fill factor 0.7. The short circuit current density of a panel under AM1.5 illumination is 400 $A\,m^{-2}$. In the UK there is about 750 $kWh\,m^{-2}\,y^{-1}$ of solar radiation. If an area of 8 m^2 is available on a house, estimate the amount of energy per year that could be provided by solar panels.

AM1.5 illumination gives 1 $kW\,m^{-2}$. An annual amount E_S of 750 $kWh\,m^{-2}$ is equivalent to a continuous illumination P_{inc} given by

$$P_{inc}t = E_S$$

where t is one year, i.e. 8760 hours. Substituting gives $P_{inc} = 85.6\,W\,m^{-2}$. The short circuit current density J_{SC} is proportional to the intensity of solar radiation so

$$J_{SC} = (85.6/1000)400 = 34.2\,A\,m^{-2}.$$

A solar panel has an area of 0.4 m^2 and $V_{OC} = 40 \times 0.6 = 24$ V. The short circuit current $I_{SC} = J_{SC} \times 0.01 = 0.342$ A.

With 8 m^2 of solar panels the total short circuit current $I_{SC} = 0.342(8/0.4) = 6.84$ A. The average output power P_{out} is given by eqn (6.5) as

$$P_{out} = FF(I_{SC}V_{OC}) = 0.7(6.84)(24) = 115\,W = 0.115\,kW.$$

The amount of energy E produced in a year is given by the product of the power by the time, so

$$E = 0.115(8760) = 1007\,kWh \approx 1\,MWh.$$

6.8.1 Applications of solar panels

Solar panels are often used with battery storage. This allows operation of equipment at night, but also the battery can provide a load that is quite close to optimal. The voltage across a battery remains reasonably constant while being charged. The power provided by the panel is the product of the battery voltage and the current produced by the solar panel. This power can be quite close to the maximum power point over a wide range of current and hence solar intensity. For example, a 12 V battery requires a charging voltage between 12 and 15 V. This voltage can be provided by a 30 cell silicon module with an insolation varying between 0.2 and 1 $kW\,m^{-2}$ (see Exercise 6.9).

Solar panels can provide power in remote locations, for example, for telecommunications equipment and lighting, and also for small electronic devices. Where AC power is required, an inverter is used that converts the DC output to AC. Resistive loads give a voltage proportional to the current and so are not well matched to a solar panel supply. For such loads a DC–DC convertor is used, where the input DC voltage is close to the optimal voltage for the panel.

6.9 Economics of photovoltaics (PV)

We can see from Fig. 6.19 that the price per Wp has dropped significantly over the last 25 years to ~\$4/W in 2005. Whether this makes solar power economic depends very much on location.

Table 6.3 Comparative costs of energy sources

	€ cents/kWh	Capital costs (€/kW)
coal	4–9	1200
Gas	3–5	550
Wind	3–10	750–1000
Hydro	3–14	900
Biomass	7–20	1100
Solar PV	25–30	5000–9000

Besides depending on how sunny a place is, much more important is whether there is an electricity grid available. For locations that are far from a grid then the cost of solar power in industrial applications such as rural telecommunication or water pumping can be ~0.25 that of alternative sources. These sources may require transport, and in domestic applications, such as village lighting and TV, PV costs can be ~0.5 that of the alternatives. As a result there have been considerable sales of solar PV systems in remote areas. The cost of connection to a grid can be ~$5000 per kilometre. This cost can make PV competitive if a house is more than only a couple of kilometres from the grid.

The market shares for remote industrial (22%) and remote domestic (17%) over the last 5 years are still less than that for grid-connected applications (59%). (The remaining 2% is for items such as calculators.) In grid-connected systems any excess supply in the day can be exported to the grid, while any shortfall during the night is imported. Battery storage systems can be included to give security in the case of power cuts. The module (panel) cost is about 40–50% of the cost of a solar power unit, the remaining amount being taken up with the cost of installation, which will depend on the system–grid-connected or standalone, with or without battery storage.

At about 8000 euros per kWp the cost of generating electricity by PV systems is about 30 cents per kWh. The cost will depend on the discount rate, as discussed in Section 11.2.1 in Chapter 11 (see Example 6.8). This is some ten times the cost of generation by a combined cycle gas turbine (CCGT) thermal plant of ~3–5 euro cents/kWh, while wind and biomass generation costs are about two and three times larger than CCGT (see Table 6.3).

Electricity charges (tariffs) to consumers vary significantly globally with tariffs in cents per kWh (1999 prices) varying from 8 (US), 12 (UK), 14 (Argentina), 15 (Germany), to 21 in Japan. Government subsidies and policies also affect the market. For example in 1999 Germany initiated a 100 000 Roofs Solar Programme together with a high price (initially ~43 euro cents per kWh) that utilities must pay solar power producers, which enables the cost of the PV installation to be paid off after a few years. The relatively high cost of electricity in Japan has favoured solar power, and Japan together with Germany has the largest share of the present PV market (2004, 69%) with Japan being the largest producer.

EXAMPLE 6.8

The capital cost of manufacturing and installing an array of solar panels that will produce 1 kWp is 8000 euros. The annual solar energy density in the location where the panels will be installed is 2000 kWh m^{-2}. Calculate the cost of electricity per kWh. Take the lifetime of the solar panels as 30 years and assume the discount rate is 6%.

As the area of solar panels produces 1 kWp, the annual amount E_{elec} of electricity produced will equal (numerically) the annual solar energy density, i.e. $E_{elec} = 2000$ kWh. The capital cost $C_{capital}$ is 8000 euros, the discount rate R is 6%, and the lifetime N is 30 years, so using eqns (11.4) and (11.5) (see Section 11.2 in Chapter 11), we find the annual cost A_{cost} that repays the capital from

$$C_{capital} = A_{cost}\{1 - (1 + R)^{-N}\}/R.$$

Thus

$$A_{cost} = R \times C_{capital}/\{1 - (1 + R)^{-N}\} = 0.06(8000)/\{1 - (1.06)^{-30}\} = 581 \text{ euros.}$$

The cost of electricity C_{elec} is given by

$$C_{elec} = A_{cost}/E_{elec} = 581/2000 = 0.29 = 29 \text{ cents per kWh.}$$

This is in the range given in Table 6.3.

6.10 Environmental impact of photovoltaics

Solar PV power in operation produces no pollutants and in particular no greenhouse gases. It is visually unobtrusive and there are no moving parts, which reduces maintenance and also results in no noise pollution. This means that planning permission is generally straightforward. A large area, but this can be on rooftops, is required to produce MWs of power. One square kilometre will produce an average annual generation of ~10–40 MW, but PV is ideal for distributed power generation not requiring a grid.

In production some hazardous materials such as Cd and As are used but the quantities are small. With effective safeguards and regulations the risks can be kept very small and acceptable. Energy is required to manufacture the modules and if it is from fossil fuels then there will be an associated CO_2 emission. For the energy to manufacture not to be significant then the time required to produce this amount of energy should be small compared to the lifetime of the solar cell, which is about 30 years. This time, called the energy payback time (EPBT), has been estimated (ZSW 2002) to be ~7 y for mono-Si, ~5 y for multi-Si, ~3 y for a-Si, ~2 y for CIS, and ~1.5 y for CdTe cells.

For example, in one study in Australia in 2000 it was calculated that ~1000 kWh/m^2 were required to build a silicon PV panel and that in Sydney the electricity production would be ~150 kWh/y/m^2, giving a 7 year payback time. Assuming a 30 year lifetime such a PV system would produce the equivalent of ~7/30 the amount of CO_2 that would be produced by a fossil fuel system. The amount generated is dependent on the type of fuel: black coal ~1.0 kg

CO_2/kWh; gas ~0.6 kg CO_2/kWh; CCGT ~0.4 kg CO_2/kWh. So we see that, particularly for the thin-film technologies, the equivalent CO_2 emissions are $\lesssim$ 15% of fossil fuels.

6.11 Outlook for photovoltaics

The production of PV panels increased by 64% in 2004 relative to 2003 to 927 MWp, and it is estimated to rise to 3.26 GWp by 2010. The market is dominated by silicon-based panels and this is expected to be so for several years. Growth in 2005 is predicted to be limited by silicon production, which is presently largely a byproduct of the electronics semiconductor production. Silicon consumption for solar cells will start to overtake that of the semiconductor industry and will lead to silicon shortages and price rises. This increase in consumption will put pressure on developing dedicated silicon production for the solar industry. It will also be an incentive to find techniques for reducing the wafer thickness and waste. If such a dedicated production is developed (only 99.999% purity is required rather than the 99.999999% for the electronics industry), then it is predicted that the cost per watt would be US$2/Wp by 2010.

The development of thin-film, both silicon and non-silicon, technologies may lead to a way of producing panels at less than $1/Wp, when PV power production will start being quite competitive with alternatives, but this may well take another decade or two based on the slow but steady progress achieved over the last 30 years.

Photovoltaics provide low-carbon electricity that is already very competitive in off-grid applications and has great potential for providing a significant contribution to the world's energy requirements. Shell has made a prediction (scenario) for future energy supplies, in which they anticipate solar playing a significant role by the year 2060, producing between 15% and 20% of the energy supplies.

6.12 Solar thermal power plants

Innumerable methods have been devised for using solar radiation to provide heating in houses and offices but as yet there have been only a few large-scale solar power plants. An early example was at Barstow in California where in the 1980s and 1990s a large array of mirrors was used to direct the Sun's rays on to a tank on top of a central tower. Oil and then molten salt (at over 500°C) were used to transfer heat to a boiler for a conventional thermal power plant that produced 10 MW. The molten salt was also used as a store of heat. Another large plant in the Mojave Desert in California used parabolic trough collectors to provide heat for an 80 MW plant from ~500 000 m^2 of collector area. A tank containing oil was heated to ~390°C and an average conversion efficiency of 18% was achieved.

6.12.1 Ocean thermal energy conversion (OTEC)

The oceans collect a huge amount of solar radiation. As a result there is a vast amount of heat energy stored within the top hundred metres of the oceans, where the temperature is ~20–25°C higher than deep below the surface. Ideas on how to exploit this resource, called ocean thermal energy conversion (OTEC) have been developed (see Box 6.3). The small temperature differences, however, result in low thermal efficiencies. To date there are significant technical challenges to be met and the process is not economically competitive.

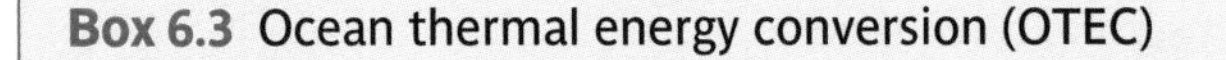

Box 6.3 Ocean thermal energy conversion (OTEC)

The oceans cover a huge area of the Earth and absorb an enormous amount of energy per day. Approximately 1000 MW falls on a square kilometre of sea. The result is a small temperature difference of ~20–25°C between the surface and water below ~1000 m. Using this temperature difference to drive a heat engine is the basis of OTEC. The temperature of the top ~100 m of the sea is roughly constant. It then decreases until it is ~5°C at ~1000 m, and below that depth it remains approximately the same.

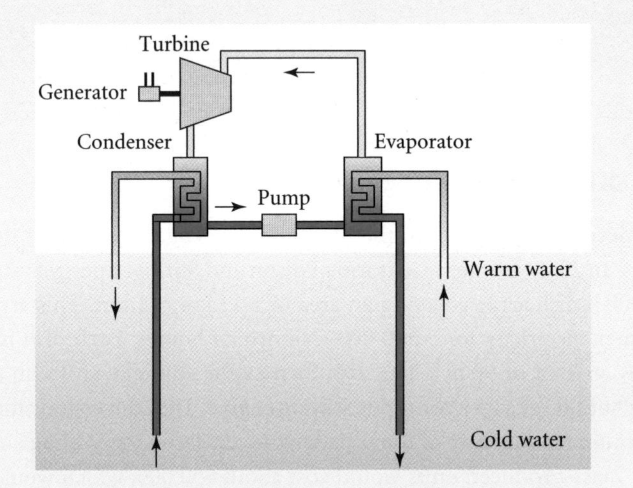

Fig. 6.20 Schematic of a possible OTEC system.

A schematic of a possible OTEC system is shown in Fig. 6.20. From the second law of thermodynamics there is a limit on the output power P_{out} given by

$$P_{\text{out}} < \eta_{\text{Carnot}} P_{\text{flow}} \tag{6.28}$$

where η_{Carnot} is the Carnot efficiency,

$$\eta_{\text{Carnot}} = (T_{\text{h}} - T_{\text{c}})/T_{\text{h}} = \Delta T/T_{\text{h}}, \tag{6.29}$$

and P_{flow} is the heat flow. If the pump draws a volume of sea water per second Q, and the specific heat capacity and density of the sea water are C and ρ, respectively, then

$$P_{\text{flow}} = \rho C Q \Delta T. \tag{6.30}$$

So the output power is limited to

$$P_{\text{out}} < (\rho C Q / T_{\text{h}})(\Delta T)^2. \tag{6.31}$$

Note the quadratic dependence on ΔT, which favours regions with large ΔT. Sea water has $\rho \approx 1000\ \text{kg m}^{-3}$, $C \approx 4200\ \text{J kg}^{-1}{}^{\circ}\text{C}^{-1}$ so, for $\Delta T = 20^{\circ}\text{C}$ and $P_{\text{out}} = 1$ MW, $Q > 0.18\ \text{m}^3\,\text{s}^{-1} \equiv 180$ litres per second is required.

Such a system would need large and expensive equipment: big pumps and in particular very efficient heat exchangers. Encrustation by marine organisms, bio-fouling, is a serious problem. The manufacture of long large diameter cold water pipes is also difficult. Dredging up the cold water from the deep will release some CO_2 dissolved in the sea water. But, the amount is less than 1% of that generated burning coal to produce the same amount of energy. An OTEC system, however, may affect the climate and the fauna and marine life of the region. OTEC is not yet economically competitive but, with global warming and technical developments, it is closer to becoming a viable option in some tropical regions.

6.12.2 Solar driven Stirling engines

A more recent development has been the use of Stirling engines placed at the focus of solar collecting dishes. In 2005 Southern California Edison and Stirling Energy Systems, Inc. agreed to set up a 20 000+ dish array covering an area of ~11 km×11 km. This array will generate 500 MW, sufficient electricity for ~500 000 Californian homes. Each dish is made up of 82 mirrors and has an area of 90 m^2. The dish focuses the sunlight on to an area of ~20 cm diameter on the hot side of a 25 kW output Stirling engine. The conversion efficiency is close to 30%. The maximum temperature of the sealed hydrogen gas in the Stirling engine is ~720°C. It is hoped that mass-produced units would cost about $50 000, which would correspond to $2 per watt, competitive with current solar panels.

The Stirling engine was conceived in 1816 by the Rev Dr Robert Stirling, who thought it a safer alternative to a steam engine whose boiler if poorly constructed could explode. In his engine a gas is sealed in a cylinder and alternately heated and cooled. In the process it drives a piston, which is connected to a crankshaft that drives a generator. The heat supply is external and the cold side can be an air-cooled heat exchanger. The internal pressures are lower than in a steam engine. But it is slow to warm up, less compact than a steam engine, and requires precise machining, and as a result was never competitive.

However, it has several attractive features: it is very quiet; it can be made to run very reliably; it has high thermal efficiency; it has a completely external heat supply; and it has no emissions. As a result it has already found important applications in submarines and in space. And it is now part of a major solar power plant. (More details on the Stirling cycle are given in Box 6.4.)

Box 6.4 Stirling engine

An ideal Stirling cycle is represented on a $T-V$ diagram in Fig. 6.21(a). In Fig. 6.21(b) a schematic is shown of how the sealed gas is moved from the heated to the cooled cylinder through a porous matrix, called a regenerator that acts as temporary heat store or supply.

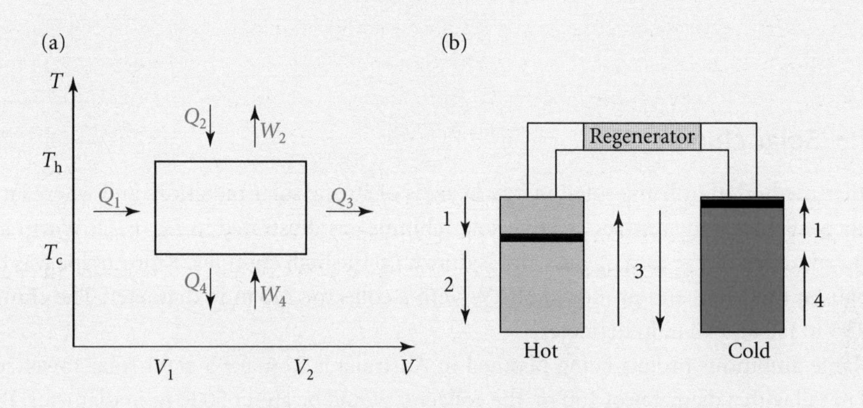

Fig. 6.21 (a) Stirling cycle. (b) Schematic of Stirling engine operation.

The cycle starts with the left hand piston at the top and the right-hand piston about two thirds of the way up the cold cylinder. At the end of the step 1 (as shown) heat Q_1 will have been absorbed from the regenerator in the isochoric (constant volume) process. In step 2 heat Q_2 is absorbed, the gas expands and the piston moves fully down, doing work W_2 in the isothermal process at the temperature T_h of the hot cylinder. Step 3 is isochoric and heat Q_3 is rejected to the regenerator. Finally, in step 4 work W_4 is done compressing the gas and rejecting heat Q_4 to the cold cylinder in the isothermal process at T_c.

The values for the heat flows Q_1 and Q_2 are given by

$$Q_1 = \alpha Nk(T_h - T_c), \qquad Q_2 = W_2 = NkT_h \ln(V_2/V_1) \tag{6.32}$$

where $\alpha(\alpha = C_V/R)$ for a monatomic gas is 1.5 and for a diatomic gas is ~2.5. $\Delta U = Q_2 - W_2 = 0$ as we are assuming an ideal gas, so $U = U(T)$. Likewise,

$$Q_3 = \alpha Nk(T_h - T_c), \qquad Q_4 = W_4 = NkT_c \ln(V_2/V_1). \tag{6.33}$$

Neglecting the regenerator, i.e. assuming Q_3 is lost from the system and Q_1 has to be provided, then the efficiency ε is given by

$$\varepsilon = (W_2 - W_4)/(Q_1 + Q_2) = \varepsilon_C/\{1 + \alpha\varepsilon_C/\ln(V_2/V_1)\} \tag{6.34}$$

where $\varepsilon_C = (T_h - T_c)/T_h$ is the efficiency of a Carnot cycle. (As $Q_3 = Q_1$, a perfect regenerator would result in the Stirling engine having an efficiency equal to that of a Carnot cycle.) High ε requires a monatomic gas and a large volume ratio (V_2/V_1). Actual Stirling engines are more complicated thermodynamically, but very good thermal efficiencies can be achieved.

6.12.3 Solar chimneys

Another method of utilizing solar energy in areas of strong solar radiation, and where land is readily available, is to construct a large solar chimney as illustrated in Fig. 6.22. Warm air is produced under a large area of glass and is drawn up the high chimney. A prototype was built in Spain in the 1980s and produced 50 kW with a collector 240 m in diameter. The chimney was 195 m tall and 10 m in diameter.

A large ambitious project being planned in Australia is to make a giant solar tower some 1000 m tall with a diameter of 150 m; the collector would be about 5000 m in diameter. Black plastic pipes full of water placed on the ground under the glass will heat up during the day and will give out their heat to the air during the night, providing output throughout 24 hours. The output power would be 200 MW, sufficient to provide power for ~200 000 homes. The conversion efficiency is ~2% but, where land is cheap and there is good sunshine, then the cost per kWh is estimated to be quite competitive at Australian $0.15/kWh. The principle of the solar chimney is explained in Box 6.5.

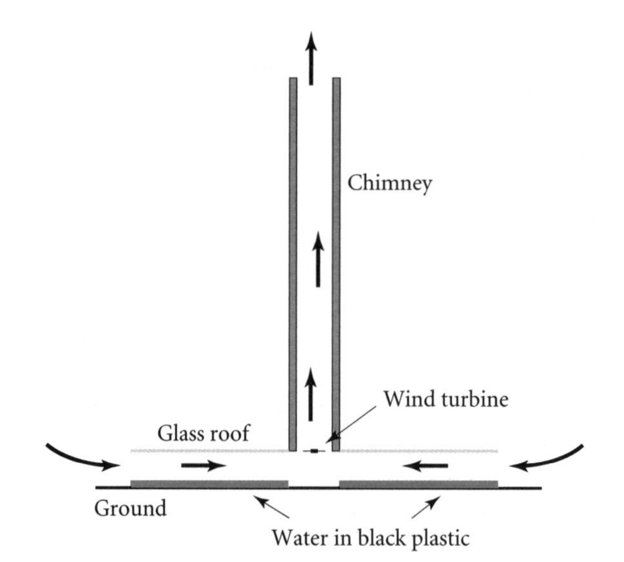

Fig. 6.22 Solar chimney.

Box 6.5 Solar chimney

The updraught depends on the height of the chimney. A simple model of the chimney shows how a pressure difference arises through the difference in the density of the heated air within the chimney at temperature T_i and that outside at temperature T_o. Figure 6.23 is a schematic of the chimney and the air pressure variation inside and outside.

The rate of change in air pressure with height is given by the equation for hydrostatic equilibrium (see eqn (3.3))

$$dp/dz = -\rho g, \tag{6.35}$$

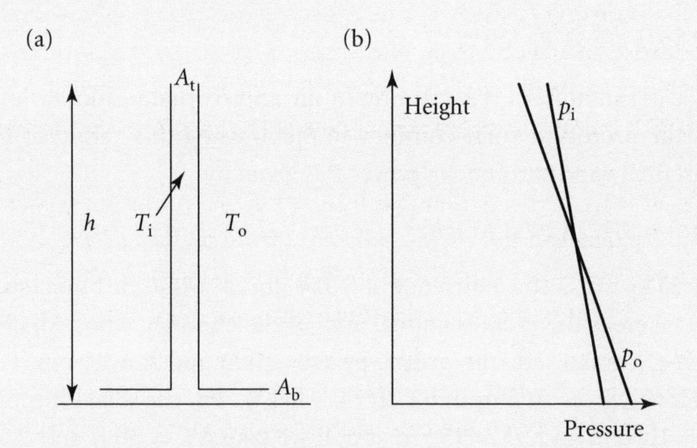

Fig. 6.23 (a) Solar chimney schematic. (b) Pressure variation in solar chimney.

which equates the change in pressure dp in a distance dz to the pressure exerted by a thickness dz of air of density ρ.

Making the assumption that the air inside the chimney is all at T_i and that outside at T_o, then

$$\Delta p_o = \rho_o gh \text{ and } \Delta p_i = \rho_i gh, \tag{6.36}$$

where the small change in ρ with Δp is neglected in comparison with the much larger relative change due to the difference in temperature between inside (T_i) and outside (T_o) the chimney.

As ρ_i is less than ρ_o there is less pressure drop inside the chimney than outside, which gives rise to a pressure that is lower at the bottom and higher at the top than the air pressure outside halfway up the chimney (Fig. 6.24(b)).

These differences in pressure drive air in at the bottom and out at the top, and this process, called the stack or chimney effect, is used to provide natural ventilation in buildings. The overall pressure difference Δp driving the air flow is given by

$$\Delta p = (\rho_i - \rho_o)gh = \rho(\Delta T/T)gh. \tag{6.37}$$

We can estimate the flow of gas through an opening (orifice) using Bernoulli's principle. Consider an air stream of cross-sectional area A_1 flowing slowly with velocity u_1, which then passes through a smaller opening A_2. By conservation of mass (eqn 3.1) its velocity through the opening will be u_2 where

$$u_2 = u_1 A_1/A_2.$$

The pressure difference $p_1 - p_2 = \Delta p$ is given by Bernoulli's theorem (eqn (3.2)),

$$\Delta p/\rho = \tfrac{1}{2}(u_2^2 - u_1^2) \approx \tfrac{1}{2}u_2^2, \tag{6.38}$$

if $u_1 \ll u_2$, i.e. $A_1 \gg A_2$. There will be energy lost due to turbulence as the air flows through the opening and empirically the velocity u_2 will be given by

$$u_2 = C_T(2\Delta p/\rho)^{1/2} = C_T\{2(\Delta T/T)gh\}^{1/2}, \tag{6.39}$$

where C_T is a constant <1. We can obtain an approximate estimate of the power P generated by the prototype solar chimney in Spain using this value for u_2 as the wind speed incident on a wind turbine. Its power P is given by

$$P = \tfrac{1}{2}\varepsilon A\rho u^3 \approx 0.1D^2\{2(\Delta T/T)gh\}^{3/2}. \tag{6.40}$$

where $\rho = 1.25\ \mathrm{kg\,m^{-3}}$, the efficiency $\varepsilon \sim 0.4$ for a wind turbine (see Chapter 5), $A = \pi D^2/4$ is taken as the cross-sectional area of the chimney whose diameter is D, and we have taken $C_T \sim 0.8$. For the prototype $D = 10$ m and $h = 195$ m. For $\Delta T = 20$ K and $T = 300$ K this gives an output power of $\sim$40 kW. For the solar chimney planned for Australia, $h \sim 1000$ m and, $D \sim 150$ m and, if we take $\Delta T = 30$ K and $T = 300$ K, then $P \sim 200$ MW.

These estimates are rather rough as the model that the speed of airflow up the chimney is given by that through an opening is very approximate. But they do show how a considerable amount of power can be generated by a very tall solar chimney.

SUMMARY

- There is a huge resource in solar radiation with the average incident solar power more than 5000 times current world power consumption. The solar intensity on a clear sunny day is $\sim$1000 W m^{-2}.
- Solar cells are currently mainly silicon semiconductor p-n junction devices. Their output current density J_C is given by

  $$J_C = J_{SC} - J_S\{\exp(V/V_T) - 1\},$$

  where J_{SC}, the short circuit current density, is proportional to the light intensity, J_S is the saturation current density ($\sim 10^{-8}\ \mathrm{A\,m^{-2}}$), and $V_T = kT|e| \approx 0.026$ volts.
- The maximum output power density P_m of a photocell is given by

  $$P_m = FF \times J_{SC}V_{OC}$$

where V_{OC} is the open circuit voltage and FF is the fill factor, typically ~0.8.

- Silicon solar cells under AM1.5 illumination have $V_{OC} \sim 0.6$ V, $J_{SC} \sim 400$ A m^{-2}, and $FF \sim 0.7$. Output power density $P_m \sim 150$ W m^{-2}, equivalent to a conversion efficiency of ~15%.
- Solar PV power density is ~100–400 kW ha^{-1}.
- Large area thin film cells, in particular amorphous silicon, $CuInGaSe_2$ (CIGS), and CdTe, are under development. There is R & D on new technologies, such as electrochemical, organic photocells, and thermophotovoltaic systems. Large power plants using concentrators and Stirling engines are also being developed.
- Cost of generating electricity by photocells has dropped significantly over last 20 years but is still ~10 times that from fossil fuels (CCGT). But cost to consumer is starting to be competitive in regions where fossil fuels are scarce.
- Solar power as a source of low-carbon electricity has great potential and is expected to be a significant source of energy in the world by 2060.

FURTHER READING

Twidell, J. and Weir, T. (2006). *Renewable energy resources.* Taylor and Francis, London. Provides useful information about photovoltaics.

Boyle, G. (ed.)(2004). *Renewable energy.* Oxford University Press, Oxford. Good qualitative discussion of solar photovoltaic technologies.

Wurfel, P.(2005).*Physics of solar cells.* Wiley, New York. More advanced textbook on physical principles of solar cells.

Sze, S.M.(1985). *Semiconductor devices.* Wiley, New York. Good textbook on the physics of photocells.

Rosenberg, H.M.(1988). *The solid state.* Oxford University Press, Oxford. Good introductory textbook for condensed matter physics.

WEB LINKS

www.eere.energy.gov/RE/ Good overview of solar cell technologies.

www.wikipedia.org/wiki/Solar_cell Good summary of solar cells.

LIST OF MAIN SYMBOLS

V	voltage	I_S	saturation current
I	current	I_C	photocell current
P	power	p^+	highly doped p material
R	resistance	n^+	highly doped n material
V_T	thermal motion voltage ($\equiv kT/|e|$)	FF	fill factor

V_{OC}	open circuit voltage	I_L	light induced current
I_{SC}	short circuit current	E_g	band gap energy
J	current density	V_g	band gap voltage
I_S	saturation current	Wp	Watts peak

EXERCISES

6.1 Blue light of wavelength 475 nm falls on a silicon photocell whose band gap is 1.1 eV. What is the maximum fraction of the light's energy that can be converted into electrical power?

6.2 Sunlight of intensity 600 W m^{-2} is incident on a building at 60 degrees to the vertical. What is the solar intensity on: (a) a horizontal surface; (b) a vertical surface?

6.3 When heated to temperatures of about 1700 K, approximately 20% of the radiation emitted by the rare earth oxide, ytterbium oxide Yb_2O_3, is in a narrow band around 1000 nm. Would this source of radiation be suitable for a Si or a GaAs photocell? (Band gaps: Si 1.1 eV, GaAs 1.4 eV.)

6.4 (a) The intensity I_{AM0} of solar radiation incident on the Earth's atmosphere is given to a good approximation by

$$I_{AM0} = \sigma T^4 \Omega_S / \pi \text{ W m}^{-2},$$

where Ω_S is the solid angle subtended by the Sun and σ is the Stefan–Boltzmann's constant. Find T given that the intensity I_{AM0} is 1367 W m^{-2} and $\Omega_S = 6.8 \times 10^{-5}$ sr. (b) Using eqn (6.13), find the constant D in terms of σ, k and Ω_S.

6.5 Explain why some materials are insulators, some conductors, and some semiconductors.

6.6 Plot the current through an ideal p-n junction, which has a saturation current of 10^{-11} A, for bias voltages -0.4 to $+1.0$ V in 0.2 V steps.

6.7* The current–voltage relation for an ideal diode is given by

$$I = I_L - I_S\{\exp(V/V_T) - 1\},$$

where $V_T \equiv kT/|e| \approx 0.026$ volts at room temperature. By differentiating the power P, given by $P = VI$, with respect to V and equating the derivative to zero, show that the maximum power P_m occurs when

$$(1 + V/V_T)\exp(V/V_T) = I_L/I_S + 1.$$

By approximating $(1 + V/V_T)$ by V_{OC}/V_T, where $V_{OC} = V_T \ln(I_L/I_S)$, and $(I_L/I_S + 1)$ by I_L/I_S, show that V_m, I_m, are given by

$$V_m = V_{OC}(1 + x_{OC} \ln x_{OC}), \text{ and } I_m = I_{SC}(1 - x_{OC})$$

where $x_{OC} = V_T/V_{OC}$. Note $I_L = I_{SC}$.

6.8 (a) A silicon photocell has an area of 4 cm^2 and is illuminated normally with AM1.5 solar radiation. The short circuit current is 160 mA and the saturation current is 4×10^{-9} mA. Calculate the maximum power output and the corresponding load resistor. (b)* What is the output power when the load resistor is 10% higher than the optimum value?

6.9 A household uses 4000 kWh of electricity in a year. Estimate what area of solar panels would be required to produce 1000 kWh of electricity. The insolation is 800 kWh m^{-2} y^{-1}.

6.10 A 30 cell silicon solar panel has a saturation current density $J_S = 10^{-7}$ A m^{-2}. Show that this panel could be used to charge a 12 V battery by calculating the peak power voltages V_m for insolation values of 0.2, 0.4, 0.6, 0.8, and 1.0 kW m^{-2}. An insolation of 1 kW m^{-2} gives a short circuit current density of 400 A m^{-2}.

6.11 In a region where the solar insolation is 1800 kWh m^{-2} y^{-1}, estimate the area of solar panels that would be required to produce 100 MW of electricity.

6.12 A reasonable approximation to the dependence of the short circuit current density J_{SC} on the band gap E_g for $0.5 < E_g < 1.8$ eV under AM1.5 illumination is

$$J_{SC} = (80 - 34E_g) \text{ mA cm}^{-2}.$$

A three layer multi-junction solar cell has an upper layer with $E_g = 1.8$ eV.(a) Determine the optimal band gaps for the lower two layers. (b) Calculate the output power under AM1.5 solar illumination, assuming a fill factor of 0.8 and an open circuit voltage for each layer given by $V_{OC} = V_g - 0.4$ V. (c) What is the conversion efficiency?

6.13* A thin film silicon solar cell has a thickness W. The upper surface is polished flat and has an anti-reflection coating. On the back surface there is a perfectly diffusing reflective coating. Show that light will have an effective path length within the silicon of $\approx (4n^2 - 1)W$, where n is the refractive index for silicon.

6.14 A CIGS photocell of area 10 cm^2 and band gap 1.5 eV is illuminated with laser light of wavelength 800 nm. The photocell has a saturation current of 10^{-7} mA. The light power is 150 W. Use eqns 6.3, 6.20, and 6.21 for V_{OC}, I_m, V_m and FF to estimate the conversion efficiency of the photocell.

6.15 What are the advantages and disadvantages of single crystal compared with thin film solar cells?

6.16* A solar cell with an open circuit voltage of 0.4 V utilizes quantum dot photon absorbers. These absorbers emit two electrons when the energy of the photon $E_\gamma > 1.6$ eV and only one when $0.8 < E_\gamma < 1.6$ eV. Compare the conversion efficiency under AM1.5 illumination with that of a p-n junction solar cell with a band gap of 0.8 eV and an open circuit voltage of 0.4 V. Assume that both cells have the same fill factor.

6.17* In a TPV system, the central cylindrical emitter is surrounded by two concentric quartz cylinders. The quartz cylinders transmit radiation with wavelength $\lambda < \lambda_{max}$. Wavelengths with $\lambda > \lambda_{max}$ are absorbed and re-emitted in all directions. Surrounding the quartz cylinders are the photocells.

Show that only a third of the radiant energy with $\lambda > \lambda_{max}$ from the central emitter is transmitted by the quartz cylinders to the photocells.

6.18 In a Stirling engine the regenerator has an efficiency of ε_R, i.e. the heat input Q_1 required in eqn (6.32) equals $(1 - \varepsilon_R)Q_3$. Show that the efficiency ε_S of the Stirling engine operating between temperatures T_h and T_c is given by

$$\varepsilon_S = \varepsilon_C / \{1 + (1 - \varepsilon_R)\alpha\varepsilon_C / \ln(V_2/V_1)\}$$

where ε_C is the Carnot efficiency $(1 - T_c/T_h)$ and $\alpha = 1.5$.

Calculate the efficiency for $T_h = 325°$C, $T_c = 75°$C, $\varepsilon_R = 0.5$ and $V_2/V_1 = 5$.

6.19 What would the capital cost and installation charges need to be for 1 kWp of solar panels for the cost of electricity to be 20 US cents/kWh? The insolation is 1500 kWh m^{-2} y^{-1}. Take the lifetime of the panels to be 25 years and the discount rate to be 7%. Neglect any maintenance charges.

6.20 The capital cost of manufacturing and installing a 1 kWp array of solar panels is 7000 euros. The annual solar energy density in the location where the panels will be installed is 1800 kWh m^{-2}. (a) Calculate the cost of electricity per kWh. Take the lifetime of the solar panels as 35 years and assume the discount rate is 5%. (b) What will be the cost of electricity if there is an annual maintenance charge of 100 euros?

6.21 An OTEC system is proposed for a region where the temperature difference between the surface and the deep water is 25°C. The pumps have a capacity of 100 litres per second. Estimate the power output if the overall efficiency is 50% of the theoretical maximum.

6.22 A solar chimney is proposed to provide 100 MW of power. The temperature difference of the air inside and outside the chimney is predicted to be 35°C. Estimate the height of the solar chimney required if the diameter of the chimney is 100 m.

7 Biomass

List of Topics

- Photosynthesis and crop yields
- Energy storage in plants
- Biomass potential
- Anaerobic digestion
- Combustion and gasification
- Municipal solid waste
- Bioethanol
- Biodiesel
- Liquid biofuel yields
- Environmental impact
- Economics and potential
- Outlook

Introduction

Plants derive their energy to grow from the Sun's radiation. The primary process is **photosynthesis** in which carbon dioxide and water are converted to carbohydrate and oxygen. Animals eat plants and other animals and the whole of animal and plant material is called biomass. The burning of biomass or of materials derived from biomass is an important source of energy in the world, providing in 2002 about 12% of the world's requirements.

In an energy context biomass refers to plant- and animal-derived material such as straw, logs, dung, and crop residues that are used either directly or indirectly as fuels. These fuels are often called biofuels. The attraction of biomass as a source of energy is that it is carbon-neutral, as the amount of CO_2 released in its combustion has been previously removed from the atmosphere when CO_2 was converted by photosynthesis into making the plant material. We therefore have a sustainable source with zero net production of CO_2, provided we renew the biomass consumed.

We will first look at photosynthesis and estimate crop yields and will see that the overall efficiency of conversion of solar radiation is rather low at around 0.5%. Quite large areas of land are therefore required to produce significant amounts of energy. We will then look at how biomass is currently used as a source of energy. Finally, we will discuss its future potential, both in providing energy and also in producing liquid biofuels has alternatives to the fossil fuels petrol (gasoline in the US) and diesel.

7.1 Photosynthesis and crop yields

In photosynthesis carbon dioxide and water are converted to oxygen and carbohydrate (sugar).

$$CO_2 + H_2O + h\nu \rightarrow O_2 + [CH_2O]$$

where $h\nu$ represents light quanta (photons) and $[CH_2O]$ stands for carbohydrate. The products are ~5 eV per carbon atom higher in energy, which corresponds to ~16 MJ kg^{-1} for pure carbohydrate. The amount of energy per kilogram depends on the degree of oxidation of the carbon with zero available when fully oxidized as CO_2, ~16 MJ kg^{-1} as carbohydrate, and the maximum available when fully reduced as CH_4 at 55 MJ kg^{-1}.

In photosynthesis a minimum number of eight photons each with ~1.8 eV, so 14.4 eV in all, are needed to produce one O_2 molecule and one C atom fixed in carbohydrate that stores ~4.8 eV of energy. In sunlight, photons, at the red and blue ends of the visible spectrum, are absorbed by the chlorophyll pigment in the leaves of the plant. More green light is therefore reflected off the leaves and that is why leaves appear green. In a leaf the pigment molecules are close together and, when an absorbed photon excites an electron, this electron can transfer to an adjacent molecule rather than de-exciting and emitting a photon. In this way the energy of the electron can be used to form molecules in a series of complex chemical reactions, together called the electron transport pathway. The maximum efficiency of this process is ~33% (4.8/14.4). But only ~50% of the energy of the solar photons can be absorbed in photosynthesis because of the spread in wavelengths of solar radiation. Together with losses (~25%) from the leaves, mainly through reflection and transmission, the maximum efficiency is ~12%, and ~10% has been achieved under optimal conditions in a laboratory.

But in the field the conversion efficiency is much less, as only about one-third of the solar radiation falls in the growing period. Only approximately one-fifth of the radiation lands on the leaves. Approximately 60% of that is converted to biomass with the remaining ~40% used up in sustaining the plant through respiration. Multiplying these factors together gives us a rough estimate for the overall average annual efficiency of ~0.5%.

$$\text{Efficiency} \approx (0.12)(0.33)(0.2)(0.6) \approx 0.5\%.$$

In Europe the annual amount of solar energy is ~1000 kWh m^{-2}. This amount is equivalent to 10 GWh ha^{-1} or 36 TJ ha^{-1} (ha $\equiv 10^4$ m^2). The available energy for biomass production is therefore ~180 GJ ha^{-1} so the yield of carbohydrate biomass, which requires ~16 GJ t^{-1}, is approximately

$$\text{Biomass yield} \approx 10\ \text{t ha}^{-1}\ \text{y}^{-1} \cong 150\ \text{GJ ha}^{-1}\ \text{y}^{-1} \cong 5\ \text{kW ha}^{-1}. \tag{7.1}$$

The yield and energy content depend on the plant and growing conditions and should be only taken as a rather approximate estimate for the dry biomass yield. But we will find these values useful in estimating what land area is required for biomass supply.

Most plants have three-carbon based (C_3) compounds produced in photosynthesis as part of the Calvin cycle. Some tropical plants, e.g. sugar cane and maize, initially produce a four-carbon (C_4) based compound. The C_4 compound raises the CO_2 concentration in the leaf, which is then fixed in a Calvin cycle. These C_4 plants in high light levels and high temperatures (tropics) have greater conversion efficiency, while at lower temperatures and lower light levels (temperate regions) C_3 plants have a greater efficiency. For example, the yield of sugar cane, a C_4 plant, can be ~100 t ha^{-1} y^{-1}, though ~75% of that mass is water and the yield of sugar is typically ~10 t ha^{-1} y^{-1}. Algae, which are C_3 plants, can also have a high conversion efficiency, though the plant mass produced has a high water content.

Respiration is the reverse reaction to that occurring in photosynthesis and proceeds by enzyme-catalysed reactions that convert carbohydrate plus oxygen to carbon dioxide and water with the release of about 5 eV per carbon atom. Respiration occurs all the time in humans and is the source of energy for human activity, such as walking. In plants respiration occurs mainly at night but can occur during the day if there is little available water. Under these conditions forests can be a source rather than a sink of carbon dioxide.

Further conversion of carbohydrates to oils occurs when certain plants ripen and these oils provide a more compact form of energy storage; typically $\sim$38 MJ kg^{-1}, the details of which are described in Box 7.1.

Box 7.1 Energy storage in plants

Plants synthesize carbohydrates from CO_2 and H_2O using the energy from sunlight (photosynthesis, see above). The carbohydrates are part of a plant's structure and provide a store of energy. The simplest carbohydrates are sugars or monosaccharides, which have the composition $(CH_2O)_n$. Glucose, $C_6H_{12}O_6$, is the most common plant sugar and is called a 6-sugar as it contains six carbon atoms. The glucose molecule can exist in several forms, in which the atoms have different bonding and orientations, called structural isomers; in particular, as α-glucose and β-glucose. These two hexagonal ring forms are illustrated in Fig. 7.1. The upper illustrations indicate that the hexagonal rings are not actually flat in nature. Also the carbon and hydrogen symbols are left off the upper diagrams for clarity.

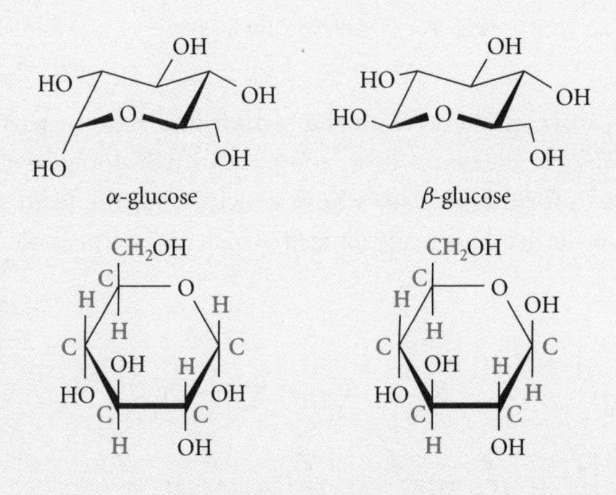

Fig. 7.1 α- and β-glucose.

Glucose forms a disaccharide by a condensation reaction

$$C_6H_{12}O_6 + C_6H_{12}O_6 \rightarrow C_{12}H_{22}O_{11} + H_2O \tag{7.2}$$

that is the reverse of a hydrolysis reaction. For example, maltose is formed by linking one α-glucose and one β-glucose and cellubiose by linking two β-glucose molecules. Further condensation reactions convert glucose to polysaccharides, in particular to starch and cellulose. These biopolymers store energy and provide bulk and structure in a plant.

The differences in the bonding in starch and cellulose significantly affect their structure, with starch much more amorphous than cellulose, which forms fibrous bundles. Amylose, which is a component of starch, is a polymer of α-glucose molecules linked by α-glycosidic bonds, while cellulose is a polymer of β-glucose molecules linked by β-glycosidic bonds. The linkages in amylose and cellulose are illustrated in Fig. 7.2. The hydrogen bonding gives increased stability and leads to long straight chains. These chains can hydrogen bond with each other giving rise to strong micofibrils.

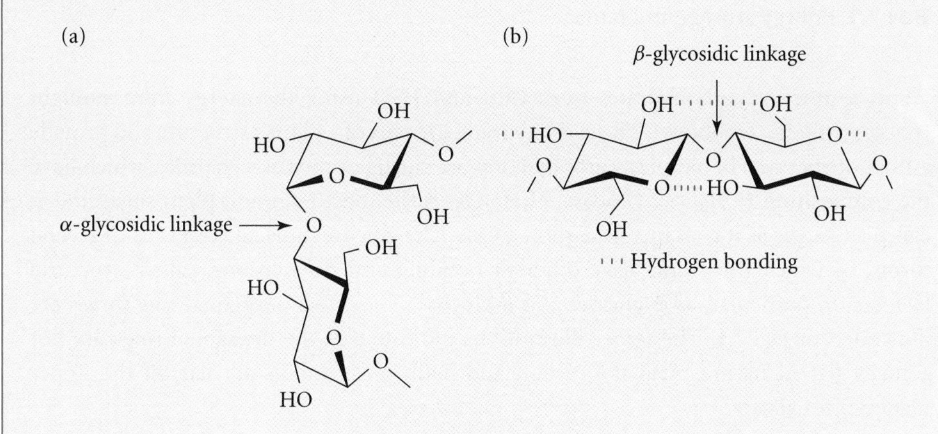

Fig. 7.2 (a) Amylose. (b) Cellulose.

The sugars are a store of energy with the carbon in a state of partial reduction; on combustion ~16 MJ/kg is released. A more compact form of storage is afforded by further reducing the sugars to form fatty acids whose structure has the form shown in Fig. 7.3. (The H of the carboxylic group is easily ionized, which makes the molecule an acid.)

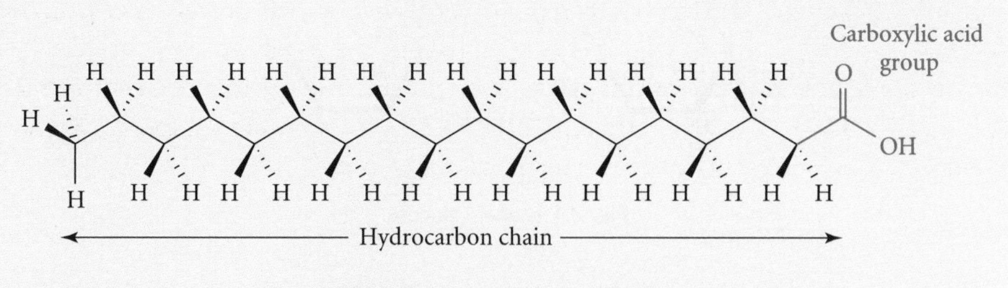

Fig. 7.3 Structure of a fatty acid.

The most common length of the hydrocarbon chain lies within 12 to 24 carbon atoms. These molecules have a very low solubility in water and they and their derivatives are called lipids. The carbon is almost fully reduced and has the highest ratio of H:C when there are no double bonds between the carbon atoms—such fatty acids are called saturated. The heat of combustion is therefore much higher than for carbohydrates with typically 38 MJ/kg released.

A common storage molecule is a triglyceride, which is a fatty acid ester: three fatty acids joined to a glycerol molecule with the removal of three water molecules. This transformation of carbohydrate to triglyceride occurs when certain plants, e.g. olives, ripen. The reverse process occurs when a seed starts to grow (germination) with the hydrolysis (uptake of water) of 1 gram of oil producing ~2.7 grams of carbohydrate. Shown in Fig. 7.4 is an unsaturated triglyceride—trilinolein.

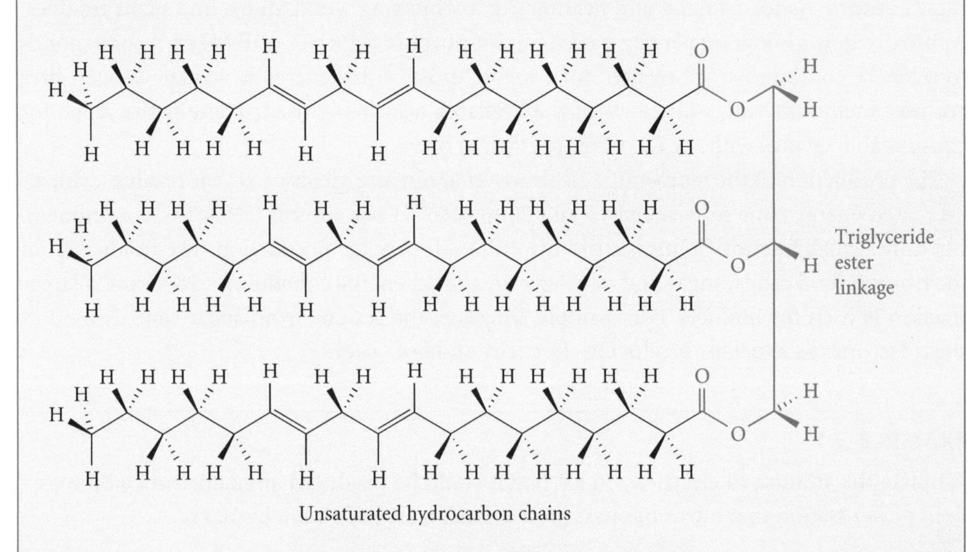

Fig. 7.4 An unsaturated triglyceride—trilinolein.

In this fat the double bonds cause the chain to kink at the indicated positions with the result that the intermolecular bonding is reduced as the molecules cannot pack together so closely as in a saturated triglyceride. As a result, such naturally occurring unsaturated fats tend to be liquids at room temperature while the saturated fats tend to be solids. For example, olive oil is composed of triglycerides made up mainly from the isomers oleic acid (55–85%), a monounsaturated acid (only one double bond), and linoleic acid (~9%), a polyunsaturated acid (more than one double bond).

7.2 Biomass potential and use

The mass of plants and animals on land produced each year is about 4×10^{11} t, and contains $\sim 1.5 \times 10^{11}$ t of carbon. An average of ~ 7.5 MJ kg^{-1} is stored in biomass making the annual amount of terrestrial bioenergy stored equal to $\sim 3 \times 10^{21}$ J. This is the amount of energy that would be produced in a year by a power output of ~100 TW, which is many times the global power usage of ~14 TW (2002). About half the stored bioenergy is used in sustaining the plant material, so the global biomass potential is ~50 TW. Of this we use $\sim 5 \times 10^{19}$ J y^{-1} ($\equiv$ 1.6 TW) as a source of energy, called biofuels, with $\sim 1.6 \times 10^{19}$ J y^{-1} ($\equiv$ 0.5 TW) for food. Biomass therefore supplies ~12% (2002) of the global power usage. An additional ~50% of bioenergy is stored in the oceans, but virtually none of this is used as a source of energy owing to its inaccessibility.

The current use of biomass for energy is mainly (~70%) for residential cooking and heating in the developing countries. In the developing countries biomass provides $\sim\frac{1}{3}$ of the energy consumption with, for example ~20% in China and ~40% in India. Its largest use in these countries is for cooking and heating, e.g. by burning wood, dung, and plant residues. Approximately 1 kilogram per day is used for this purpose, which at ~10 MJ kg^{-1} corresponds to ~120 W continuous. When used for cooking most of this energy is wasted, as open fires are very inefficient with only ~5% of the available heat being used. There is the scope for considerable savings with the use of more efficient ovens.

The production of the temperate cereals wheat and maize yields each year residues with an estimated energy content between 15 and 20 EJ (10^{18} J) per annum ($\cong$0.6 TW continuous). But only a small fraction of this is currently utilized for energy production. The residues from the tropical food crops, sugar and rice, have an annual energy content of ~18 EJ and a larger fraction is used for biofuels. For example, biogasse, the residue from sugar cane, is used in sugar factories as a fuel for producing electricity and hot water.

EXAMPLE 7.1

Estimate the amount of electricity in kWh that could be produced annually from a biomass-fired power station that burns biomass grown over an area of 10 km by 10 km.

Equation (7.1) gives an estimate of 5 kW ha^{-1}. This is equivalent to 0.5 MW per square kilometre, so the thermal power from $10 \times 10 = 100$ square kilometres is 50 MW. Taking the efficiency of the thermal power station as 40% would give an output of 20 MW_e (megawatts electrical power). In one year the amount of electricity would be $(2 \times 10^4)8760 \approx 1.75 \times 10^8$ kWh.

We will now concentrate on the commercial use of biomass as a source of energy, either directly in heating or indirectly as feedstock for the production of liquid and gaseous biofuels. We will then discuss the environmental impact and end with a discussion of the economics and potential of biomass.

7.3 Biomass energy production

Energy is produced from two main sources of biofuels: agricultural and municipal wastes and energy crops. These have rather low energy content per kilogram compared with fossil fuels and relatively low density, making them bulky and expensive to transport. Economic use for energy production therefore generally requires the biomass source to be readily available, e.g. waste dump or factory residue. For this reason bioenergy production is currently often combined with crop production or as a useful way of disposing of organic waste, both municipal and agricultural. Biomass supply in the future, though, lies in dedicated energy crops such as willow or switch grass. It is only by planting large areas with such crops that sufficient biomass can be produced.

The utilization of biomass is varied. We will concentrate on a few of the major ways in which biomass can contribute to energy production, all of which have significant savings in greenhouse gas emissions compared to fossil fuels. One of the most widespread uses is to take advantage of natural decomposition of organic waste by anaerobic bacteria. Anaerobic digestion is used in the developing countries for heating and cooking and in industrialized countries to provide gas for small power units.

On a larger scale there is the combustion and gasification of biomass to produce electricity and heat. There is also the conversion of biomass to liquid biofuels to replace oil-based fuels. In North and South America there is large-scale production of ethanol through fermentation and more recently, particularly in Europe, a significant increase in the manufacture of biodiesel from the extraction of oil from plants. We will first look at anaerobic digestion, then look at combustion and gasification before discussing the production of liquid biofuels.

7.3.1 Anaerobic digestion

Anaerobic digestion is the decomposition of organic matter in the absence of air by bacteria. Bacteria break down the organic matter and produce a gas consisting of methane (~65%) and carbon dioxide (~35%), with traces of other gases. The gas has a calorific value of ~17–25 MJ m^{-3} (STP, standard temperature and pressure, which is 0°C and 100 kPa ≡ 1 bar) and the conversion efficiency is typically 40–60%. Anaerobic digestion occurs naturally, e.g. in compost heaps, and is the source of marsh gas. It takes place in landfill sites over a period of years, with, typically, peak methane production occurring after 10 years, and in purpose-built digestors, where the process occurs at higher temperatures (30–60°C) after only a few weeks. In the latter the residue can have use as a fertilizer. In comparison with aerobic digestion in which organic matter plus oxygen is converted to a residue plus carbon dioxide and water (as in combustion but at much lower temperatures), anaerobic digestion produces considerably less residue (~5–10 times less) as well as a gas that can be used to produce power. Anaerobic digestion also occurs in cows and is a source of a significant amount of methane in the atmosphere—CH_4 production from cows and other ruminants is estimated to account for about 3% of the global warming associated with greenhouse gas emissions (FAO). The biochemical processes involved are explained in Box 7.2.

Anaerobic digestion is widely used in Asian villages where the biogas is used for heating and cooking. A Chinese fixed dome digestor is illustrated in Fig. 7.5. The technology for larger-scale anaerobic digestion of sewage and industrial sludges and of waste water is well developed. The potential within the EU from sludges and waste water is ~2.5 GWy, with a total potential including agricultural and municipal solid waste (MSW) of ~12.5 GWy. In municipal waste digestors the organic fraction of the waste needs to be separated out, which can be costly.

Box 7.2 Biochemical processes in anaerobic digestion

Anaerobic digestion consists of several processes in which organic matter is transformed by anaerobic bacteria to a gas consisting mainly of CH_4 and CO_2 and a residue called sludge or biosolids. Anaerobic digestion results in a significant fraction of the original biomass energy being stored in the methane, as the reactions that occur only give out a small amount of heat. These reactions are catalysed by enzymes, which are typically complex proteins whose action depends on their molecular shape. These enzymes increase the reaction rate by lowering the activation energy (the energy required to start a chemical reaction).

In anaerobic digestion there are three stages: hydrolysis; acidification; and methane production. The large molecules in the organic matter are initially broken down by hydrolysis, which is a reaction that causes molecules to break apart when water is added, i.e.

$$AB + HOH \rightarrow A\text{–}OH + H\text{–}B. \qquad (7.3)$$

The hydrolysis of cellulose and protein produces fatty acids, amino acids, and glucose. These are then converted to organic acids, such as ethanoic (acetic) and butanoic acid, by acidogenic and acetogenic (acid- and acetate-forming) bacteria, and H_2 and CO_2 are produced, e.g.

$$C_6H_{12}O_6(\text{glucose}) + 2H_2O \rightarrow 4H_2 + 2CH_3COOH(\text{ethanoic acid}) + 2CO_2. \qquad (7.4)$$

In the final stage (methanogenesis) bacteria digest the products of the acidification stage and produce methane via reactions such as

$$CH_3COOH \rightarrow CH_4 + CO_2, \qquad 4H_2 + CO_2 \rightarrow CH_4 + 2H_2O. \qquad (7.5)$$

The acidification and methane production stages are symbiotic as the H_2 and acetate inhibit the bacteria producing them, so their consumption by the methane-producing bacteria is beneficial. If the production were complete then the overall process would amount to

$$C_6H_{12}O_6 \rightarrow 3CH_4 + 3CO_2. \qquad (7.6)$$

The efficiency of the conversion is very high with ~90% of the stored energy in glucose being stored in the produced methane.

An anaerobic digestor consists of a feedstock holder, digestion tank with mixing system, biogas and residue recovery, and where necessary, e.g. northern Europe, fitted with heat exchangers to maintain the optimum temperature for the bacteria to produce the biogas (~30–60°C). Small digestors are mainly used for heat production while the larger units are used for electricity generation with power outputs of ~2 MW.

In the EU the dumping of biodegradable waste in landfill sites is being reduced. Although the methane can be utilized to provide energy, recovery is variable with typically only 50%

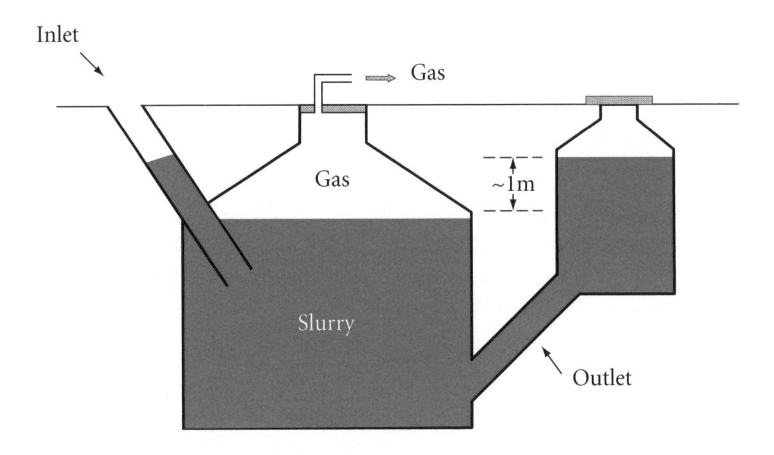

Fig. 7.5 A Chinese fixed dome anaerobic digester.

used. As methane is a very potent greenhouse gas, some 23 times more than carbon dioxide, reducing its emission is important. Moreover, there can be pollution from leachates from biodegradable waste in the landfill sites.

EXAMPLE 7.2

Calculate the energy efficiency of conversion of carbohydrate to methane in anaerobic digestion.

The molecular weights of glucose $C_6H_{12}O_6$ and methane CH_4 are 180 and 16, respectively, so 180 kg of glucose are converted in anaerobic digestion to 48 kg of methane. The heats of combustion of glucose and methane are $\sim$16 MJ kg^{-1} and $\sim$55 MJ kg^{-1}, respectively. The stored energy in 180 kg of glucose is therefore $\sim$2880 MJ, while in 48 kg of methane it is $\sim$2640 MJ, so the conversion efficiency ε_{AD} is given by

$$\varepsilon_{AD} \approx 2640/2880 \approx 92\%.$$

7.3.2 Combustion and gasification

There is considerable potential energy from the combustion and gasification of biomass, estimated as up to $\sim$10–15% of the energy required in UK and Europe. Gasification can be brought about by anaerobic digestion, as described above, and by burning the biomass in a reduced supply of air. One of the earliest ways that biomass was converted to produce a more useful fuel was the burning of wood with a reduced air supply to yield charcoal. Charcoal burns at a much higher temperature than wood and the higher temperatures that could be achieved using charcoal significantly advanced the extraction of metals from ores. Modern methods of biomass conversion by thermochemical processes concentrate on the gaseous and liquid products and, in particular, on gasification.

The first industrial use of gasification was in the production of coal gas or producer gas from heating coal in the presence of steam. With the advent of natural gas this production ceased but a similar process is involved in the gasification of biomass. The different thermochemical processes that occur are described in Box 7.3. The gas that results from burning biomass in a reduced air supply consists of CO, H_2, CO_2, CH_4, and nitrogen (from the air) and is called **producer gas.**

Producer gas has a low calorific value of 1000–1200 kCal m^{-3} at STP, which is equivalent to 4.2–5.0 MJ m^{-3}(STP). But producer gas burns cleanly with low emissions. The total

Box 7.3 Thermochemical processes in gasification

Various thermochemical processes occur at the same time in a gasifier. We will consider wood as an illustration. Drying of the wood occurs at ~150°C, which drives off the water as steam. The dried wood is decomposed by heat in the absence of air, a process called **pyrolysis.** This occurs between 150°C and 700°C and produces gases, in particular CO and H_2, liquids (oils), and a solid residue (charcoal). These products are oxidized by reacting with the oxygen in air supplied to produce CO_2 and H_2O by the reactions

$$C + O_2 \rightleftharpoons CO_2 + 394\ \text{MJ/kmol}, \tag{7.7}$$

$$H_2 + \tfrac{1}{2}O_2 \rightleftharpoons H_2O + 242\ \text{MJ/kmol}, \tag{7.8}$$

which occur between 700°C and 2000°C (a kmol of carbon is 12 kg.). These gases are reduced by the following reactions

$$CO_2 + C \rightleftharpoons 2CO - 173\ \text{MJ/kmol}, \tag{7.9}$$

$$C + H_2O \rightleftharpoons CO + H_2 - 131\ \text{MJ/kmol}, \tag{7.10}$$

$$C + 2H_2 \rightleftharpoons CH_4 + 75\ \text{MJ/kmol}, \tag{7.11}$$

$$CO_2 + H_2 \rightleftharpoons CO + H_2O + 41\ \text{MJ/kmol}, \tag{7.12}$$

which are predominantly endothermic, i.e. requiring heat. These reducing reactions therefore occur in a lower temperature range between 800°C and 1100°C.

Of particular importance in determining the yield of these reactions is the reaction temperature, with an increase in temperature at constant pressure favouring the products of an endothermic reaction. This is an example of an important rule (**Le Chatelier's principle**), which follows from the second law of thermodynamics, that states:

> When a reaction in equilibrium is perturbed, the equilibrium alters in the direction that reduces the perturbation.

This means that it is important that the temperature should not fall too far in the reduction reactions; otherwise the yield of CO and H_2 will be reduced. By altering the conditions hydrogen production can be enhanced and this can be used in fuel cells to provide electricity generation with high efficiency (see Chapter 10).

conversion efficiency can be ~60–70%. Each kilogram of air-dried biomass (10% water content) yields about 2.5 m^3 of producer gas equivalent to ~12 MJ kg^{-1}, agreeing with our rough estimate of 15 MJ kg^{-1} of biomass. (Biomass has a higher output if it is low in water content as water requires 2.4 MJ kg^{-1} to evaporate.) Producer gas can be used to provide heat, or in internal combustion engines and in gas turbines for electricity generation.

Liquid fuels in the form of oils can also be derived from heating biomass in the absence of air, a process called pyrolysis, but this has been so far less economic, producing fuels with too low a calorific value to substitute for petrol or diesel. The producer gas can also be purified to produce synthesis gas or syngas, which is carbon monoxide and hydrogen. Syngas can be used to synthesize chemicals and fuels such as methanol using catalysts.

We will illustrate the potential for combustion and gasification of biomass to contribute to the power generation capacity of an industrialized society by considering the potential in the UK.

Combustion and gasification in the UK

In a UK study by the Department of Trade and Industry (DTI) two main biomass energy chains have been identified: (1) the combustion and gasification of short rotation coppice willow, miscanthus, and straw, and (2) the gasification, anaerobic digestion, and combustion of municipal solid waste and sewage sludge (see Fig. 7.6). The conversion technologies involved in combustion and gasification require significant investment and have risks for the more advanced technologies. So a secure biomass supply and a market for the energy is required, which will require economic incentives.

The study concludes that short rotation coppice willow has the potential to provide ~9% of the UK electricity by 2020 using large-scale power plants, but plant construction rates will limit this to ~1%. The amount is very dependent on the area of land used with 9% assuming one million hectares planted. This would agree roughly with our estimate of ~5 kW per ha, which would give ~5 GW total output, ~10% of the UK electricity supply.

One million hectares is a substantial land area of 100 km by 100 km. For comparison there are 18 million ha of agricultural land in the UK with 0.5 million ha set-aside. On this

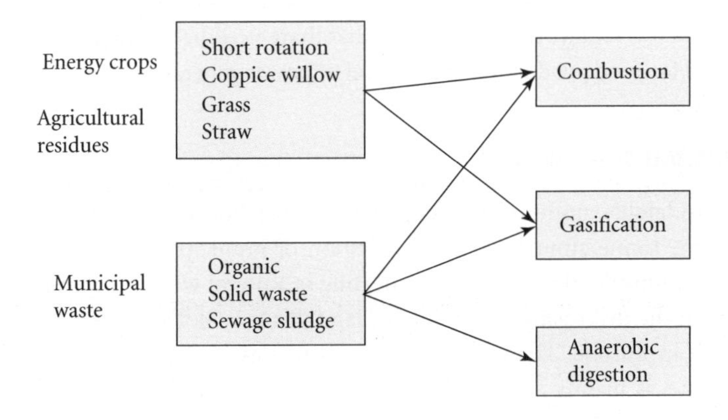

Fig. 7.6 Biomass sources in the UK for 2020.

scale gasification and combustion are quite competitive by 2020 with electricity generation at ~3 UK pence/kWh. There is also potential for ~5% contribution over this timescale using smaller scale technologies. Miscanthus and straw could provide up to 12% by 2020 but are slightly more expensive than short rotation coppice willow. In the long term gasification is cheaper than combustion and has the higher efficiency

Organic solid waste (OSW) could provide a small fraction, ~2% with the potential of 4% if there is no limitation from the build rate of plant. OSW requires efficient waste separation with gasification, anaerobic digestion, and combustion all being possible routes to power generation. Anaerobic digestion produces bio-fertilizer as a byproduct lowering the cost of generation, and combustion has the possibility of higher pollutant emissions. All are currently rather expensive in comparison with fossil fuel generation. For sewage sludge there is only ~0.5% potential contribution, but the reduction in greenhouse gas emissions from avoided energy use in waste disposal and the decrease in CH_4 emissions could be greater than that from avoided fossil fuel generation.

The total UK biomass potential is ~13% by 2020, which would be a significant source of continuous renewable electricity. Compared with generation using gas (CCGT) the annual CO_2 savings would be up to 19 million tonnes of CO_2. The CO_2 emissions associated with biomass generation are less than 50 g/kWh compared with 387 g/kWh for CCGT, the lowest emission fossil fuel generation. Transport of imported feedstocks, e.g. wood chips from the Baltics, increases CO_2 emissions up to ~138 g/kWh, but this is still only a little over a third that of CCGT.

Combustion and gasification in the OECD

A similar contribution of 15% to the electricity and heat production within the OECD countries has been suggested based on woody biomass, which is biomass from forestry and farming. The plan assumes that power demand in the industrialized countries in the OECD will double by 2020. It requires the use of a quarter of the agricultural residues and the putting aside of 5% of crop, farm, and woodland area to the growing of woody biomass. The total area of such land in the OECD is over 1500 million ha, so 5% is over 75 million ha. This area has been estimated to provide sufficient fuel to generate 200 GW of power, enough for over 100 million homes. This is in agreement with our rough estimate of 5 kW/ha, which would give a power of 375 GW. The study concludes that there need be no conflict between land use for biomass and for food production at this level of power generation.

7.3.3 **Municipal solid waste**

There is a considerable amount of waste generated per household in industrialized countries—about one tonne annually. The combustion of municipal waste with the use of the heat for electricity production or for space heating is a useful way of disposing of this waste, particularly with the decreasing availability of landfill sites. Globally, there is 3 GW installed with about half in Europe. The waste often needs processing before it can be used as fuel. Although the fuel is mainly organic the combustion is not carbon-neutral as some of the material is derived from fossil fuels, typically 20–40%. This analysis, called a life cycle analysis, works out the amount of CO_2 (and other gaseous emissions) per kWh of energy produced. While the burning of agricultural wastes gives less than 30 g per kWh, municipal solid waste, MSW, also called energy from waste, EfW, gives ~360 g per kWh compared with ~970 g per kWh from coal and ~450 g per kWh from a natural gas CCGT power plant. The typical energy content of MSW is ~10 MJ/kg.

Besides providing heat and power, biomass can provide fuels that reduce our dependence on oil-based fuels such as petrol (gasoline) and diesel. We will now look at how biomass has been used for transport fuel production and its future potential for such fuels.

7.3.4 Liquid biofuels

While sustainable electric power can be provided by other renewables, biomass is the only renewable source of carbon-based fuels and chemicals. In 2004, $\sim 40 \times 10^9$ litres of biofuel (mainly ethanol) were produced annually in the world. Besides the environmental benefit from less CO_2 emission, biofuels produced within a country provide energy security and economic development. Less petroleum needs to be imported and there is more immunity against international political disturbances affecting oil supplies. Energy and agricultural policies can be coupled with biocrops, aiding rural development as well as energy resources. Biofuels typically burn cleanly and life cycle analyses (LCAs) show that biofuels with modern biorefineries give net savings on CO_2 emissions compared to using petroleum-based fuels, including the energy used for farming and production of the biomass feedstock and conversion. But there is need for considerable improvement both in economic competitiveness and in CO_2 savings.

The use of about 1.6 Gtoe (gigatonnes oil equivalent) of fuel for transport in 2001 (see Fig. 7.7) was close to a quarter of the total world energy consumption for that year, with nearly all of it fossil-fuel based. The growth in car usage in developing countries is significant so non-fossil based transportation fuels would help reduce a huge source of CO_2.

There are two important liquid biofuels—bioethanol and biodiesel, the first derived from sugar-containing plants, the second from oil-containing plants. We will first look at how the biological fermentation of sugar-containing plants is used to produce ethanol.

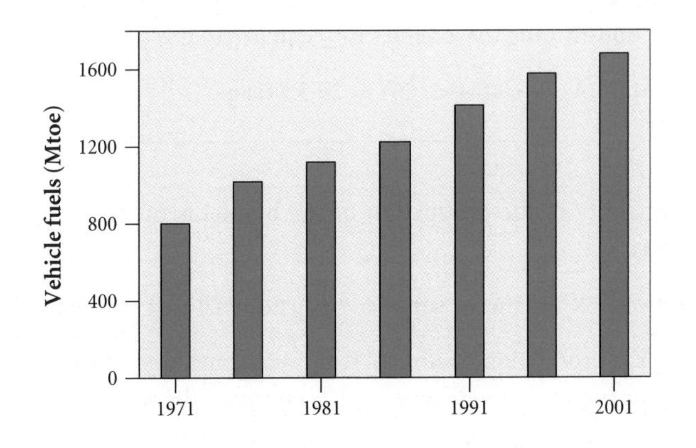

Fig. 7.7 Vehicle fuel consumption in Mtoe (megatonnes of oil equivalent).

Bioethanol from fermentation of biomass

Currently ethanol production is limited to using sources of soluble sugar or starch—in the US primarily from corn and in Brazil, where there is a long established programme, from sugar cane. Vehicles can run using petroleum blended with up to ~20% ethanol and an increasing number of cars are available that can use higher percentages. The amount of ethanol produced in each country is $\sim 1.5 \times 10^{10}$ litres per year and it is used as a substitute for petrol (gasoline) or to dilute petrol. While this is a large volume per year it corresponds to only ~2.5% of the petrol use in the US. The usage is ~4 litres per person per day, so bioethanol is a long way from being a significant alternative source of motor car fuel in the US. The next largest producers are China, $\sim 0.4 \times 10^9$ litres per year, from wheat and corn, and India $\sim 0.2 \times 10^9$ litres per year from sugarcane.

Sugar from sugar-containing plants can be directly fermented by yeast or bacteria, which reduce the carbohydrate to ethanol and produce CO_2

$$C_6H_{12}O_6 \rightarrow 2C_2H_5OH + 2CO_2 + 0.4\ \mathrm{MJ\,kg^{-1}}. \tag{7.13}$$

As the heat released is so small, nearly all the energy stored in the sugar is stored in the alcohol. The ethanol has a much higher heat of combustion 30.5 MJ kg^{-1} than that of glucose, cf 15.6 MJ kg^{-1}. It is sufficiently high that it can be used as a substitute for gasoline, which has a heat of combustion of ~45 MJ kg^{-1}. Almost half the weight of glucose (molecular weight, MW = 180) is converted to carbon dioxide (MW = 44) and the maximum conversion efficiency by weight to ethanol is 51%. The estimation of the heat of combustion of a biofuel is shown in Box 7.4.

EXAMPLE 7.3

Estimate the heat of combustion H_c of ethanol.

Ethanol has the chemical composition C_2H_5OH. Its molecular weight MW is therefore

$$MW = 2 \times 12 + 6 \times 1 + 16 = 46.$$

The fraction by weight of carbon C is therefore 24/46, that of hydrogen $H = 6/46$, and of oxygen $O = 16/46$. Substituting these values into eqn (7.16) gives

$$H_c = 32.8(24/46) + 142.9(6/46 - 2/46) = 29.5\ \mathrm{MJ\,kg^{-1}}.$$

An important quantity in the production of the biofuel is the fossil energy replacement ratio (FER), which is

$$\text{Fossil energy ratio (FER)} = \text{energy supplied to customer/fossil energy used.} \tag{7.14}$$

The energy ratio in the production of ethanol from sugarcane is good with an FER of ~8. The waste, bagasse, can be used to provide heat, which improves the ratio.

In the production of ethanol from starch, which is contained in corn, the glucose is in the form of a biopolymer. The glucose molecules are linked together by α-glycosidic bonds. These

Box 7.4 Heat of combustion of biofuels

We can estimate the heat of combustion of a biofuel by assuming all the carbon and hydrogen are fully oxidized when the fuel is burnt. The two reactions are

$$C + O_2 \rightarrow CO_2 + 32.8\ \mathrm{MJ\,kg^{-1}C}, \tag{7.15}$$

$$2H + \tfrac{1}{2}O_2 \rightarrow H_2O + 142.9\ \mathrm{MJ\,kg^{-1}H}. \tag{7.16}$$

The heat of combustion H_c of a fuel composed of C, H, and O is therefore given by

$$H_c = 32.8(C) + 142.9(H - O/8)\ \mathrm{MJ\,kg^{-1}} \tag{7.17}$$

where C, H, and O are the fractions by weight of carbon, hydrogen, and oxygen. The oxygen atoms in the fuel are assumed to combust with hydrogen atoms and form water. Using this formula we would estimate H_c in MJ $\mathrm{kg^{-1}}$ for carbohydrates ($[CH_2O]$) as 13.1(~15.5), for ethanol (C_2H_5OH) as 29.5 (29.7), and for octane (C_8H_{18}) as 50.2 (47.9), where the measured values are in brackets. So except for carbohydrates this simple estimate (Dulong's formula) is accurate to ~5%.

Olive oil is composed of triglycerides: mainly 55–85% oleic acid, a monounsaturated acid (only one double bond), and ~9% linoleic acid, a polyunsaturated acid (more than one double bond). These two acids are structural isomers and the composition of the triglcerides is $C_{57}H_{96}O_6$. Using the formula above, the heat of combustion is estimated to be 39.3 MJ $\mathrm{kg^{-1}}$, compared with 40 MJ $\mathrm{kg^{-1}}$ measured. This is only ~20% less than that released (47.9 MJ $\mathrm{kg^{-1}}$) when octane, the principal component of petrol (American, gasoline), is burnt

$$C_8H_{18} + 12.5O_2 \rightarrow 8CO_2 + 9H_2O. \tag{7.18}$$

In this process 114 kg of octane produce 352 kg of carbon dioxide, just over three times as much by weight.

The heat of combustion is also called the higher heating value (HHV) and assumes that any water vapour is condensed. If the water vapour produced in combustion of a fuel is not condensed, as in a car engine, then the available energy, called the lower heating value (LHV), is less than the HHV by about 5–10%. Values of LHVs are given for a number of fuels with the conversion factors at the end of the book.

bonds are easily broken apart using human and animal enzymes. The enzymes catalyse the decomposition by hydrolysis of starch to glucose, which can then be fermented to produce ethanol. Energy, though, is required in the conversion of the starch to ethanol and in the production of the corn.

The corn produced in the US is mainly used to provide animal feed, with some for ethanol production; in Iowa about a quarter of the corn is used to make ethanol. Energy is required in producing the ethanol and on the farm, for planting, harvesting, and for making fertilizers. Some critics of corn-produced ethanol have concluded that the resulting FER is less than one.

However, using the energy requirements of modern processing and production, it is now generally agreed that the FER for corn-ethanol is positive but small with an FER of ~1.2–1.4. As the FER is small the corn-ethanol does not reduce CO_2 emissions significantly, but it does reduce the amount of gasoline and hence oil required.

The cost of producing a gallon of ethanol from corn was ~\$1.20 per gallon in the US in 2005. The wholesale cost of petrol (gasoline) was \$1.50 per gallon in 2005 when the price of oil was \$55 a barrel. The sharp rise in the cost of oil in the last few years from a level of ~\$20 boe (barrel of oil equivalent) has made ethanol quite competitive. Emissions are lower with ethanol but the energy per gallon is less. About 50% more ethanol is required for the same stored energy. While corn-based ethanol can displace oil, we ideally need a glucose-containing feedstock that is both cheap to produce and process and has a good FER.

Moreover, the amount of ethanol produced in the US ($\sim 1.5 \times 10^{10}$ litres/y) is only equivalent to ~2.5% of the petrol (gasoline) that is consumed. To produce another 2.5%, which would be $\sim \frac{1}{3}$ EJ y^{-1} of ethanol, would require $\sim 5 \times 10^6$ ha. This area is ~3% of the US cropland. So we would like a plant that can grow in marginal land unsuitable for food crops. Cellulose-based plants such as switch grass could well provide such a feedstock.

Cellulosic feedstocks, such as wood and grasses, contain mainly cellulose, hemicellulose, and lignin, and are also called lignocellulosic-based feedstocks. Lignin is a biopolymer rich in phenolic components that confers stiffness, and makes up about a tenth to a quarter of biomass. It is the part of a plant that fossilizes and becomes coal. These plants can grow on marginal areas unsuitable for food crops. Switch grass, for example, is a deep-rooted perennial that prevents soil erosion and can restore degraded land. Switch needs only a small amount of fertilizer or pesticide and uses water efficiently. As a result the costs of production can be low.

Cellulose, which is the largest component of biomass (40–60%), is a biopolymer (polysaccharide) of glucose, as is starch, and consists of bundles of long chains of glucose molecules bonded together by β-glycosidic linkages. The fibre bundles are strong because of a high level of hydrogen bonding between the glucose chains and are resistant to cleavage. The other component of biomass, hemicellulose (20–40%), interlinks the cellulose and is mainly made up of the 5-sugar xylose. The hemicellulose and lignin enclose the cellulose bundles and protect them from microbial attack (Fig. 7.8).

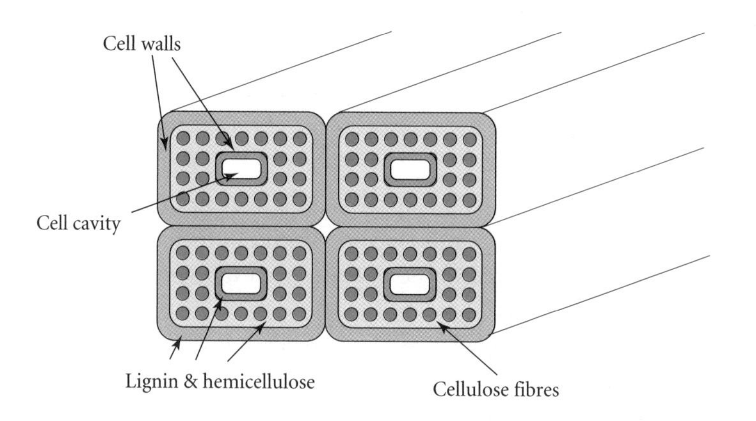

Fig. 7.8 Cellulose, lignin, and hemicellulose in a plant (schematic).

The hydrolysis of cellulose to glucose is less straightforward than in starch-containing plants, because of the hemicellulose and lignin that encase the cellulose. Pre-treatment with dilute acid combined with heat and pressure is used to separate the hemicellulose and lignin and expose the cellulose for hydrolysis. The hydrolysis can be acid-catalysed but this process is expensive as it can require pressure vessels and significant amounts of energy.

Enzyme hydrolysis is currently being actively pursued. This process was first noticed in the Second World War when a fungus was found that attacked cotton clothes and tents. The fungus was providing cellulase enzymes. The process requires low temperatures and can be carried out at atmospheric pressure. Energy requirements are low and reaction vessels are cheap to construct. The conversion efficiency is quite high. But currently enzymes are too expensive and act too slowly for good economy. Recently, a cocktail of three cellulase enzymes—endoglucanase, exoglucanase, and betaglucosidase—was discovered that showed good performance that should reduce costs. Enzyme hydrolysis (saccharification) and fermentation can also be combined to reduce the inhibition of enzyme action by sugar build up.

Yeasts that ferment 6-sugars have been known since antiquity but only recently have yeasts been developed that can ferment 5-sugars such as xylose, the main constituent of hemicellulose. A genetically engineered *Escherichia coli* micro-organism has been developed by the University of Florida that can ferment both 5- and 6-sugars. Bacteria are also under investigation as their speed of fermentation can be minutes rather than the hours required by yeast.

The successful development of ethanol production from lignocellulosic feedstocks, such as wood or grass, which are less expensive than corn, will significantly reduce the cost of ethanol, provide a good FER, and could make it competitive with fossil-fuel-derived petrol (gasoline). Typical values for the FER and alcohol yields from corn, switch grass, and sugarcane are shown in Table 7.1.

The lignin is a byproduct that cannot be fermented, but can be used as a fuel, which helps improve the FER value.

Biodiesel from plant oils

The production of biodiesel is growing rapidly, particularly in Europe. Essentially zero in 1995 the amount produced globally by 2003 was 1.5×10^9 litres per year. Use of 100% biodiesel

Table 7.1 Typical values for the FER and yields for corn, switch grass, and sugarcane. Values are dependent on conditions and vary considerably

			Ethanol	
Feedstock	**FER**	**Plant (t/ha)**	**(t/ha)**	**(gallons/acre)**
Corn	1.34	9	2.4	320
Switch	3–5	12	3	400
Sugarcane	8	10*	5	650

*Sugar yield.

occurs in some European countries such as Germany, but also as a blend (5–25%) throughout N. America and Europe. Biodiesel is made by the chemical transesterification of vegetable oils from oilseed crops such as rapeseed and sunflower, or from other sources such as waste cooking oil.

Diesel demonstrated his engine in Paris in 1900 using peanut oil, but the availability and cheapness of diesel fuel derived from fossil fuel oil meant vegetable oils were not used. While diesel engines can run on pure vegetable oils, their viscosity is rather high and transesterification produces a lower viscosity fuel that starts more easily. Fatty acid methyl esters (FAME) are the product and biodiesel is also called FAME. The process involves a relatively simple reaction of the oil with either methanol or ethanol using sodium or potassium hydroxide as a catalyst; the chemical processes involved are described in Box 7.5. The efficiency of the process is high (>97%) and requires ~10% by weight of alcohol; the resultant FER is quite good at about 3.2.

The diesel engine, like the petrol engine, normally has a four-stroke cycle: intake, compression, expansion, and exhaust. But in a diesel engine it is just air that is taken in and compressed, while in a petrol engine it is a fuel–air mixture. This means that the compression ratio of intake to compressed volume, corresponding to the piston being at the bottom and top of the cylinder, respectively, can be much higher in a diesel than in a petrol engine, typically 20:1 as compared to 9:1. This is because the compression must be limited in a petrol engine to avoid

Box 7.5 Transesterification of plant oils

Triglycerides can be used neat in diesel engines but better starting is obtained in cold weather by lowering the viscosity. This can be done by mixing the triglyceride with a solution of methanol and sodium hydroxide, the sodium hydroxide acting as a catalyst:

$$\begin{array}{lcccccc} CH_2OOR_1 & & & \text{catalyst} & & & CH_2OH \\ | & & & \downarrow & & & | \\ CHOOR_2 & + & 3CH_3OH & \Leftrightarrow & 3CH_3OOR_x & + & CHOH \\ | & & & & & & | \\ CH_2OOR_3 & & & & & & CH_2OH \\ \textit{Vegetable oil} & & \textit{Methanol} & & \textit{Biodiesel} & & \textit{Glycerin} \end{array} \tag{7.19}$$

The methanol and sodium hydroxide form sodium methoxide plus water and the sodium methoxide then successively converts the triglyceride to methyl esters plus glycerin. The first step can be represented by

$$NaOH + CH_3OH \rightarrow NaOCH_3 + HOH, \tag{7.20}$$

$$HOH + NaOCH_3 + (\text{–}CH_2OOR_1) \rightarrow (\text{–}CH_2OH) + CH_3OOR_1 + NaOH. \tag{7.21}$$

This process continues until all three methyl esters have been formed. The catalyst, sodium hydroxide, is not used up in the reaction, and the triglycerine molecules (glycerin esters) have been converted to methyl esters, hence the description of the process as a **transesterification.**

pre-ignition (knocking) as the temperature of the fuel–air mixture rises as it is compressed. In a diesel engine fuel is injected into the hot (~900°C) compressed air whereupon it ignites and rapidly expands. (It is a very fast burn, not an explosion.) The maximum temperature (and pressure) is higher in a diesel than in a petrol engine and as a result the efficiency is higher, as would be expected from the maximum thermodynamic efficiency being given by $(1 - T_{min}/T_{max})$.

In a modern diesel engine the air that is taken into the cylinder is pre-compressed so that the final maximum pressure is higher. This increase in air (oxygen) allows more fuel to be injected and more power to be obtained. This process, called turbocharging, makes the performance of diesel engines comparable to that of petrol engines. Their efficiency is greater at ~40% as compared to ~30% for car engines, while for large diesel engines their efficiency can be ~50%. Better control of the fuel burning has also reduced emissions from diesel engines.

A summary of the main processes involved in producing liquid biofuels from biomass is given in Fig. 7.9. Of particular importance is the yield of biofuel per hectare and we will see that this limits how much can be produced easily without impacting on normal agricultural use of the land.

Liquid biofuel yields

We expect roughly 10 t ha^{-1} y^{-1} of dry carbohydrate and, as the maximum conversion efficiency by mass to alcohol is ~50%, then the annual yield for bioethanol is about ~5 t ha^{-1}. In Europe the annual yield from sugar beet is ~4.5 t ha^{-1} and from wheat 2.1 t ha^{-1}, while in the US from switch grass it is ~2.8 t ha^{-1}.

In Europe most biodiesel comes from rapeseed and the annual yield is ~1.3 t ha^{-1}. Higher yields are obtained from the jatropha plant whose seeds have an oil content of 37%. It grows in tropical regions, e.g. India, and can withstand arid conditions. Typical biodiesel annual yields from jatropha are 2.3–3 t ha^{-1}. Palm trees give an even higher annual yield of ~4–6 t ha^{-1} and production of palm oil is currently concentrated in Malaysia and Indonesia. Certain algae produce quite a high yield of oil but with a high percentage of water that can make extraction

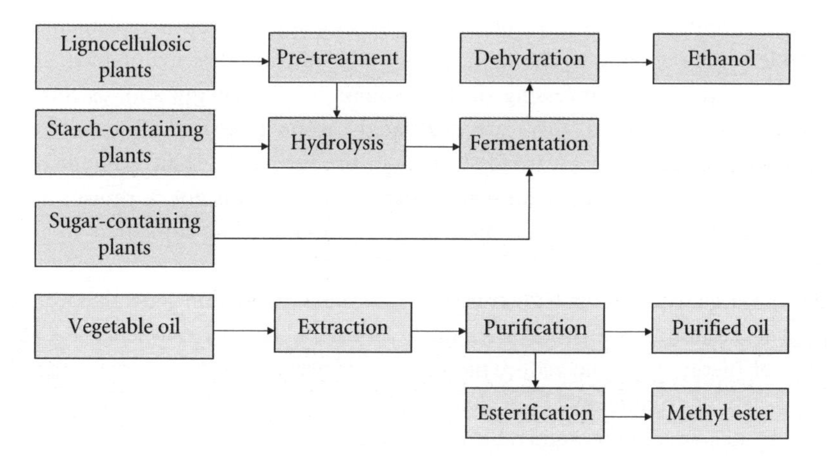

Fig. 7.9 Main processes involved in producing liquid biofuels from biomass.

expensive both in cost and energy. Currently production of oil appears uneconomic from algae though research is ongoing.

EXAMPLE 7.4

A palm tree plantation produces 5 t of oil per hectare per year. (a) What area of plantation would be required to displace 5×10^{11} litres of petrol (US usage, gasoline) per year, which is the US annual consumption, with biodiesel? (b) What would be the amount of carbon emissions displaced? (Density of petrol is 0.73 kg per litre.)

(a) Petrol has a LHV of $\sim$43.5 MJ kg^{-1}, so there are $\sim$32 MJ per litre. The energy content of 5×10^{11} litres of petrol is therefore 1.6×10^{13} MJ. An area A ha of palm trees would produce $5000A$ kg of oil per year. The conversion efficiency to biodiesel is close to 100% and biodiesel has a LHV of $\sim$38 MJ kg^{-1}. The efficiency of diesel cars is $\sim$4/3 times that of gasoline cars. Thus the area required is given by

$$5000 \times 38 \times A \times 4/3 = 1.6 \times 10^{13} \text{ so } A = 63 \times 10^6 \text{ ha} = 0.63 \times 10^6 \text{ square kilometres.}$$

(b) Assume that petrol combusts like octane. Equation (7.17) shows that 114 kg of octane produces 352 kg of CO_2. This is equivalent to $(12/44)352 = 96$ kg of C. So 1 kg of octane produces 0.84 kg of carbon on combustion. Petrol has an FER of 0.83, so 1 kg of petrol produces roughly 1 kg of carbon. A volume of 5×10^{11} litres of petrol has a mass of 3.7×10^{11} kg. This amount would produce $\sim$ 0.37 Gt of carbon. The mass of biodiesel is $0.37\ (43.5/38)\ 3/4 = 0.32$ Gt. Taking the FER of biodiesel as 3.2, 0.1 Gt of carbon from fossil fuels will be produced. So:

$$C_{\text{displaced}} = 0.37 - 0.1 = 0.27 \text{ Gt of carbon per year.}$$

7.4 Environmental impact of biomass

Biomass is a carbon-neutral source of energy provided the biocrop is replanted. It is also a sustainable source as long as the land quality (soil nutrients) is maintained. Irrigation, fertilizers, harvesting, and processing of the biomass require energy and, as this is typically derived from fossil fuels, there are associated carbon emissions. But these can be a small fraction of those given off by fossil fuels in producing the same amount of energy. The analysis of the overall emissions and energy requirements for a process (life cycle analysis) is generally favourable for biomass energy production compared to fossil fuel alternatives.

Besides the large reduction in CO_2 emissions, biomass combustion generally produces low emissions; the combustion of wood gives much less SO_2 than coal and hence less acid rain. Large areas of energy crops may reduce biodiversity. However, forestry energy crops can have a greater variety of wildlife and flora than arable or pasture land. It should be noted that biocrops, like all crops, can be vulnerable in bad weather.

Various crops have been identified as suitable as energy crops for combustion such as willow and miscanthus grass. The FER values for these crops are good. The area required to provide a significant fraction of the energy demand of a country, however, is large with roughly an output of 5 kW ha^{-1} being currently achieved. A 5 GW power plant would require an area of ~100 km × 100 km. In a European study ~10–15% of the projected demand in the EU in 2020 could be met by such energy crops. In the United States it has been estimated that, by using all the conservation reserve programme lands, which amount to about 7.5% of the cropland area, about 2 EJ y^{-1} could be produced from planting switch grass. The total cropland area of the US is ~179 × 10^6 ha, which is equivalent to ~20% of the US land area. An amount of 2EJ y^{-1} is about 2% of the primary energy consumption of the US and would suggest that, as in the EU, about 10% might be produced in the US without significant impact on agricultural production.

Energy crops grown for liquid biofuels would also require significant areas to provide sufficient fuel to displace gasoline or diesel. In the US the 2 EJ y^{-1} of switchgrass from ~7.5% of the US cropland area would produce ~1 EJ y^{-1} of ethanol, while ~38 EJ y^{-1} of petroleum imports are projected for the US by 2020. In the UK the most productive crop that can be grown is rapeseed and that yields ~1.5 t ha^{-1} of biodiesel. The amount of petroleum products used for transport in the UK in 2004 was ~37.6 × 10^6 t so some 26 × 10^6 ha would be required to displace all fossil fuels. But there are only 4.5 × 10^6 ha of cropland out of ~18.4 × 10^6 ha of agricultural land in the UK, so only a small fraction could be displaced.

7.5 Economics and potential of biomass

Combustion and gasification of biomass is typically the most economical competitive use of biomass, with gasification having the greater potential with its use of high temperature gas turbines. In Europe biomass combustion and gasification could provide up to ~15% of the projected demand for electricity by 2020. The development of fluidized-bed combustion plants with higher efficiencies has resulted in greater use of biomass and waste products in power and heat generation. As the cost of transport of biomass is expensive, because its energy density is typically quite low, small biopower plants are often more appropriate that can utilize locally produced feedstock. Currently forestry residues in the US are the main biofuel for over 6 GW of power plant. An associated economic benefit is that biocrops give rural employment as well as energy security. Liquid biofuels can be cost-effective, particularly with oil at ~$50 per barrel (2005 average price) (see Table 7.2). The oil price is currently (July 2006) ~US$75 per barrel, so liquid biofuels are even more cost-effective. However, the land area makes it unlikely that liquid biofuels will be able to replace fossil fuels at much more than a ~20% level.

Europe has promoted biofuels as a way of reducing its greenhouse gas emissions in response to international agreements. In 2003 the EU introduced a biofuels directive which set a target that 5.75% of all fuel for petrol and diesel engines must be from renewable sources. Biodiesel is Europe's main biofuel and tax exemptions and national targets are increasing demand. Diesel blended with biodiesel is being introduced and in the UK the reduction in duty by 20 pence/litre in 2002 helped investment.

Table 7.2 Relative cost of petrol (US, gasoline), bioethanol, diesel, and biodiesel

	Petrol	Bioethanol	Diesel	Biodiesel
Price (US$/litre)	0.32*	0.24–0.48	0.32*	0.33–0.59
Fossil energy ratio	0.83:1	1.34–4.6†:1	0.83:1	3.2:1
Relative energy density (%)	100	66	100	90

*Oil at US$38 per barrel (2004).
†Using enzymatic hydrolysis of biomass.

The target of 5.75% will create a demand for over 9 million tonnes of biodiesel, while current refining capacity is only just over 2 million tonnes. Rapeseed provides 80% of the biodiesel feedstock and more agricultural land will be required to meet the target of 5.75%. This land may be provided by set-aside land if it is suitable. Rapeseed is a relatively expensive crop as it requires frequent rotation and considerable amounts of fossil-fuel based fertilizer. So it may be difficult to meet demand economically using rapeseed. But the use of higher yielding crops such as jatropha that are grown in developing countries may be a cost-effective way of meeting demand, although, as the feedstock would be imported, it would not provide energy security.

The potential of the biomass energy supply in the US is shown in Fig. 7.10. For comparison the total primary energy consumption in the US in 2003 was 104 EJ. The energy supply in the US in 2002 was 6% from renewables. These were mainly biomass (46%) and hydropower (46%), with wind 2%, geothermal 5%, and solar 1%.

As the cost of non-biomass energy rises, different biomass supplies become increasingly competitive. At $60 a barrel energy from oil costs ~$10/GJ. But although competitive at these oil prices, the impact on agricultural production could be significant, for if biomass were to supply ~25% (~25 EJ) of the primary energy needs of the US then ~65% of US

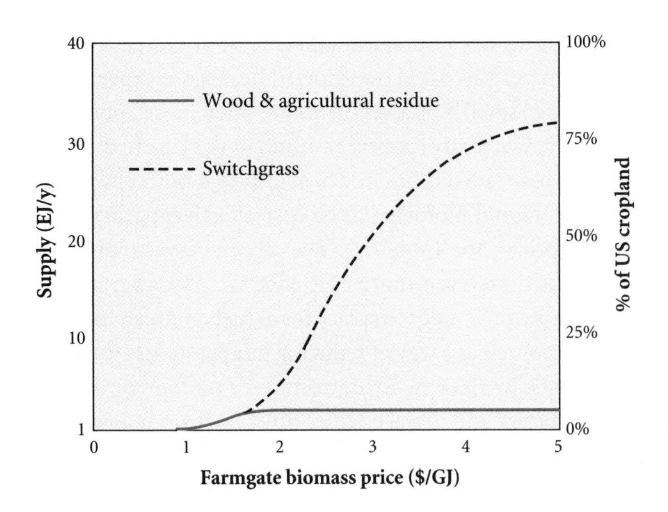

Fig. 7.10 Potential US biomass supply.

cropland would be required. The development of economically competitive liquid biofuels would reduce CO_2 emissions and the dependence of the US on foreign oil, 60% of which is imported, and in addition benefit the rural economy.

7.6 **Outlook**

Biomass has the potential to provide 10–20% of the primary energy needs of developed countries and a larger percentage in developing countries where improvements in the efficiency of biomass use could have a significant impact. Biomass can be stored, unlike other renewables, and can have low associated carbon dioxide emissions, which will help countries meet their international obligations on global warming reduction. Biomass can provide energy security and, if used for liquid biofuel production, can reduce dependence on foreign imports.

Its development aids rural economies and its use locally avoids high transport costs. The introduction of tax incentives and in particular a percentage share of energy production or use, e.g. the EU biofuel directive, has helped the development of biomass technologies. Liquid biofuels can help reduce our dependence on the fossil fuels but improved mileage per litre is important. Biofuels could be targeted at aviation where ~5% of CO_2 emissions occur. There are no substitutes for hydrocarbon fuels that can be used in aeroplanes but there are for cars. Hydrogen either directly or as methanol in fuel cells may well be able to displace fossil fuels in the future. Reserving biofuels for aeroplanes would then be sensible. The CO_2 from aeroplanes is well mixed in the atmosphere so biofuels would eliminate that source of greenhouse gas. Genetic engineering of bioenergy crops may increase yield and allow more harvests per year, but with present technology the potential contribution from biomass to projected global energy needs would appear to be limited to around 20–30%.

SUMMARY

- Plants store solar energy via photosynthesis as carbohydrates and as oils, with energy densities of ~16 MJ kg^{-1} and ~38 MJ kg^{-1}, respectively.
- The plant or biomass yield in the field is ~10 tonne ha^{-1} y^{-1}, which is equivalent to ~5 kW ha^{-1} or 0.5 MW per square kilometre.
- Biomass provides ~12% of the global power usage, ~1 TW in 2002, mainly for cooking and heating in the developing countries.
- On a small scale anaerobic digestion is widely used. On a larger scale combustion and gasification of biomass is utilized to provide both heat and electricity, with the potential to provide in Europe 10–20% of the demand.
- The conversion of biomass to liquid biofuels is increasingly important with the need to reduce fossil fuels. There is already large-scale production of ethanol though fermentation of sugar-containing plants in N. and S. America, and the production of biodiesel in Europe is increasing.

- The development of ethanol production from cellulose-based plants using enzyme-hydrolysis is being actively pursued as a means of providing large-scale economic production.
- Biomass is a significant source of carbon-neutral energy. But the large area of land required to produce a significant quantity of power or of biofuel limits its contribution. At ~0.5 MW per square kilometre, it unlikely that biomass will provide more than 20–30% of our global energy requirements.

FURTHER READING

Boyle, G. (ed.) (2004). *Renewable energy.* Oxford University Press, Oxford.

Useful overview of bioenergy.

Twidell, J. and Weir, T. (2006). *Renewable energy resources.* Taylor and Francis, London.

Provides good detail about all of the biomass processes.

Shepherd, W. and Shepherd, D.W. (2003). *Energy studies.* Imperial College Press, London.

Useful overview of biomass.

WEB LINKS

www.eere.energy.gov/biomass/ Source of information on biomass technologies.

www.nrel.gov Information on renewables in the USA.

www.dti.gov.uk/files/file22065.pdf Biomass potential in the UK.

LIST OF MAIN SYMBOLS

AD	anaerobic digestion	LCA	life cycle analysis
MSW	municipal solid waste	FER	fossil fuel energy replacement ratio
OSW	organic solid waste	FAME	fatty acid methyl esters (biodiesel)
CCGT	combined cycle gas turbine plant		

EXERCISES

7.1 Estimate the land area required to grow willow that would provide 1 GW of power in a region where the annual solar radiation is 1500 kWh m^{-2}.

7.2 A household disposes of 300 kg of domestic waste a year, 90% of which is carbohydrates and 10% of which is inert, in an anaerobic digester. The family uses the methane for cooking on an open fire. Estimate the number of litres of water that could be boiled annually. (Specific heat of water is 4.2 kJ kg^{-1}°C^{-1}.)

7.3 Estimate the annual reduction in CO_2 emissions in tonnes if the average number of miles travelled by cars in the USA per litre of petrol were increased by 30%. Take petrol to be 100% octane, density 0.73 kg/litre.

7.4 Estimate the heat of combustion of oleic acid, whose composition is $C_{18}H_{34}O_2$.

7.5 India has a land area of 2.97×10^6 square kilometres, 57% of which is cropland. If 5% of India's cropland were dedicated to producing jatropha plants, estimate the annual production of biodiesel. Compare your estimate with the global use of oil for transport and comment.

7.6 Discuss whether the UK should increase its biomass supply for heat and power or for liquid biofuel production.

7.7 Why is the USA promoting bioethanol production by enzymatic hydrolysis of cellulosic feedstock?

7.8 If all cars in the USA ran with petrol (gasoline) blended with 20% bioethanol produced by enzymatic hydrolysis of cellulosic feedstock, what would be the annual reduction in CO_2 production in tonnes?

7.9 Estimate what size of forest would need to be planted to absorb the carbon dioxide produced by a 3 GW_e coal-fired power station. Is this a practical way to combat greenhouse gas emissions?

7.10 Discuss the extent to which energy derived from the following sources is carbon-neutral: (a) short rotation coppice willow; (b) jatropha; (c) corn; (d) sugarcane.

7.11 (a) Estimate the annual reduction in CO_2 emissions if a 1 GW_e coal-fired power station were replaced by a MSW plant. (b) How much waste would be required per year by the MSW plant?

7.12 A 1 GW_e coal-fired power station is converted to use biomass as fuel. Estimate the area of land required for energy crops. (a) What is the reduction in CO_2 and in C emissions per year? (b) What area would be required for an annual reduction of 1 Gt of carbon?

7.13 Calculate the amount of carbon emitted when 1 litre of biodiesel is burnt. Take the composition of biodiesel to be 100% cetane (hexadecane) $C_{16}H_{34}$.

7.14 Estimate the number of miles per gallon (imperial) of petrol (gasoline) when a car travels at (a) 55 mph and (b) 70mph. Assume that the resistance to motion is predominantly air resistance, the drag coefficient is 0.4, and the cross-sectional area of the car is 3 m^2.

8 Energy from fission

List of Topics

- ☐ History of fission and fusion
- ☐ Binding energy and stability
- ☐ Neutron-induced fission
- ☐ Energy from fission
- ☐ Chain reactions
- ☐ Moderators
- ☐ Core design, control, and stability
- ☐ Reactor power output
- ☐ Fission waste products
- ☐ Radiation shielding
- ☐ Fast breeder reactors
- ☐ Present day nuclear reactors
- ☐ Economics of nuclear power
- ☐ Safety and environmental impact
- ☐ Public opinion and outlook

Introduction

Nuclear power is associated in some people's minds with nuclear weapons and nuclear waste, but we will see that it is an abundant source of carbon-free energy that could play a big role in combating global warming. There are two forms of nuclear energy—one from controlling fission: the reaction used in the first 'atomic' bombs; the other from controlling fusion: the energy source in stars. Fusion power is now only at the prototype stage, but it holds the promise of almost unlimited power.

Commercial nuclear power plants are fission reactors, most of which use uranium for fuel, an element that occurs in many parts of the world, with Canada and Australia currently the main producers. Compared with the amounts of coal, oil, or gas required to fuel a conventional power station, remarkably small amounts of uranium are needed for a nuclear reactor—roughly 1 tonne of uranium will deliver an equivalent amount of energy as 20 000 tonnes of coal.

Uranium (U) is approximately as common as tin or zinc. It is in many rocks and in the sea. The average concentration in the Earth's crust is 2.8 ppm. Granite contains about 4 ppm U while the sea has $\sim$0.003 ppm, which corresponds to 4000 Mt. At US $130 per kg U or less, the known reserves are 3.3 Mt, corresponding to 50 years operation at present consumption, with an estimated additional 10.7 Mt recoverable. High-grade ore containing $\sim$2% U is the cheapest to mine.

In this chapter we will describe nuclear power from fission and in Chapter 9 the progress on obtaining power from fusion. We will see that both come from converting part of the mass of nuclei into energy. In fission we will find that the process is initiated by neutrons and then yields more neutrons, giving rise to the possibility of a chain reaction that has to be controlled safely in a nuclear reactor.

We will explain how a chain reaction is brought about, how it is controlled, what is the power output from a reactor, and what are the fission waste products. The chapter ends with a discussion of the worldwide implementation of nuclear power, its economics, safety, environmental impact, and the effect of public opinion on its use.

8.1 Binding energy and stability of nuclei

In order to understand why energy is released in the fission of heavy nuclei or in the fusion of light nuclei, we need to consider the relationship between the mass and the stability of nuclei. A nucleus consists of protons and neutrons (nucleons) bound together by a short-range attractive force. Its mass is less than the sum of the masses of its constituent nucleons—the size of the difference ΔM gives the total binding energy B_E through Einstein's relation $B_E = \Delta M c^2$. This energy B_E is the energy that would be required to pull apart the nucleus into its constituent nucleons and determines whether a nucleus is stable or unstable. To illustrate this we will consider the element carbon whose nucleus has six protons.

There are two stable isotopes: ^{12}C, which is the most abundant, and ^{13}C. The isotope ^{14}C is unstable and decays to ^{14}N, with a neutron changing into a proton, with the emission of an electron and an anti-neutrino

$$^{14}\mathrm{C} \rightarrow {}^{14}\mathrm{N} + \mathrm{e}^- + \bar{\nu}.$$

The half-life of this beta decay is 5730 years and the relative amount of ^{14}C compared to ^{12}C is used in **radiocarbon dating** of archaeological objects. The reason ^{14}C is unstable is that ^{14}N is more tightly bound with an equal number of protons and neutrons than ^{14}C with eight neutrons and six protons. The lighter isotope ^{11}C is also unstable and decays to ^{11}B. ^{11}C is less bound than ^{11}B as it has more electrostatic repulsion with six rather than five protons. In this beta decay a proton changes to a neutron and a positron and neutrino are emitted.

Beta decay can leave a nucleus in an excited state that, being less bound, can decay to the ground state of the nucleus with the emission of a γ-ray (or γ-rays); just as an excited atom can decay with the emission of a photon (or photons). An excited state of a nucleus can also decay by transferring its energy to a bound electron causing the electron to be emitted. This latter process is called **internal conversion**.

The balance between minimizing the electrostatic repulsion and trying to equalize the number of neutrons and protons to make a nucleus more bound, determines the ratio of neutrons to protons and results in each element having at most only a few stable isotopes, as shown in Fig. 8.1. We will now look at how the nuclear binding energy varies across the periodic table.

The nuclear force between nucleons is short range and attractive, unless the separation between the nucleons is very small ($\lesssim$1 fm) when it becomes repulsive. This is like the force between molecules in a water droplet. Nucleons are therefore on average the same distance apart and interact primarily with their nearest neighbours. So we expect the total binding energy of a nucleus to be approximately proportional to the number of nucleons A, or the

Box 8.1 History of nuclear fission and fusion

In the mid-nineteenth century scientists were baffled by what was powering the Sun: any chemical process gave the age of the Sun as only a few thousand years. In the late 1850s Kelvin and Helmholtz suggested that the source was gravitational potential energy, which gave the age of the Sun as 20 million years. Though this was much more realistic it disagreed with Darwin's estimate in the *Origin of species* (1859) for the age of the Earth. This estimate was 300 million years based on erosion of the 'Weald' in the south of England, which Darwin took as the timescale for evolution. The problem was not resolved until 1920 when Eddington proposed the fusion of hydrogen to helium as the source of energy. However, the mechanism of fusion was not fully understood until quantum mechanical tunnelling was discovered in the late 1920s.

The exceedingly hot hydrogen plasma in the core of the Sun, where the fusion reactions occur, is contained by the large gravitational forces arising from the enormous mass of the Sun. Since the 1930s numerous attempts have been made to contain a hot plasma using magnetic fields with the eventual aim of producing fusion. These experiments have now reached that goal, but there are still significant technical advances required and it is some decades from being a commercial proposition.

Fission was observed, but not immediately recognized, shortly after the discovery of the neutron by Chadwick in 1932. Over the next few years many different nuclei were bombarded with neutrons. In 1934 Fermi and collaborators found that several radioactive nuclei were produced following neutron bombardment of uranium. Initially some of these were thought to be transuranic elements (i.e. elements with atomic numbers greater than 92), but Noddack speculated (correctly) that these nuclei might be isotopes of known lighter elements arising from the splitting apart ('fission') of uranium nuclei by the neutrons.

However, it was not until 1938 that the radiochemists Hahn and Strassman established that some of the nuclei produced by neutron bombardment of uranium were radioactive isotopes of barium, rather than isotopes of the chemically similar radium, as they had thought initially. This possibility had been dismissed earlier as physically impossible, as only α (helium nucleus) and β (electron or positron) decays were known at that time. But early in 1939 Meitner and Frisch explained this phenomenon by noting that a uranium nucleus would behave rather like a liquid drop and that the capture of a neutron caused the nucleus to oscillate and divide into two smaller nuclei.

In the fission process two to three neutrons are released together with a considerable amount of energy, giving rise to the possibility of an explosive chain reaction through further neutron-induced fission. In 1942 Fermi and Szilard demonstrated that this chain reaction could be controlled in a nuclear reactor but it was not until 1956 that the first prototype nuclear power station was built to utilize the energy released by fission reactions.

binding energy per nucleon, $B_E/A \equiv b(A)$, to be approximately constant. A plot of $b(A)$ versus A is shown in Fig. 8.2 and, while it is roughly constant above $A \sim 12$, it has a maximum near iron, Fe ($A \sim 60$).

The reason for this maximum is that nuclei would be most tightly bound if the number of neutrons N equalled the number of protons Z, were it not for the electrostatic repulsion

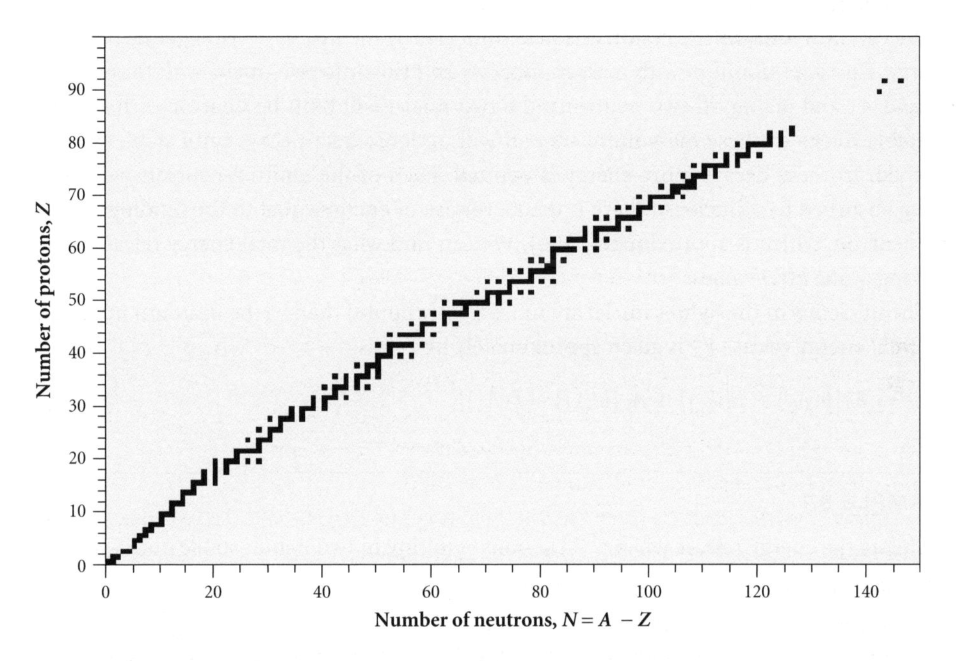

Fig. 8.1 Chart of stable nuclei.

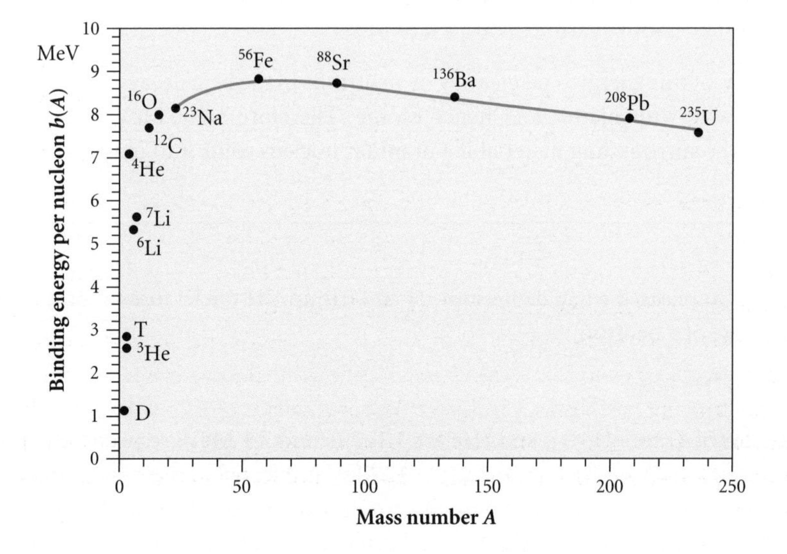

Fig. 8.2 Binding energy per nucleon $b(A)$ as a function of A.

between the protons, which increases with increasing Z. This causes heavy nuclei such as uranium to have more neutrons than protons (see Fig. 8.1), and causes the fall in $b(A)$ seen above $A \sim 60$. Below $A \sim 60$ the effect of the increase in the relative number of nucleons on the surface of the nucleus, which are less tightly bound as they have fewer neighbouring nucleons than those within, outweighs the reduced electrostatic repulsion from fewer protons and $b(A)$ reduces.

We can now see why the fission of a heavy nucleus or the fusion of two light nuclei releases energy. Consider uranium with mass number A_1 splitting into two nuclei with mass numbers A_2 and A_3 and giving off two neutrons: the two nuclei will both be neutron-rich compared to stable nuclei of these mass numbers and will undergo beta-decays until stable nuclei are reached. In these decays more energy is emitted. Each of the emitted neutrons will end up being absorbed by a nucleus. There is then a release of energy equal to the binding energy of the neutron, which is approximately $b(A)$. We can find what the total energy release E_R is by looking at the $b(A)$ values.

The nucleons in the lighter nuclei are more tightly bound than in the uranium nucleus and the total energy release E_R is given approximately by

$$E_R = A_2\{b(A_2) - b(A_1)\} + A_3\{b(A_3) - b(A_1)\}. \tag{8.1}$$

EXAMPLE 8.1

Calculate the energy release when ^{235}U fissions resulting in two lighter stable nuclei with mass numbers 140 and 93.

The value of $b(A)$ for uranium ($A = 235$) from Fig. 8.2 is $\sim$7.6 MeV, while for the two lighter stable nuclei with mass numbers of $A \sim 140$ and $A \sim 93$ it is $\sim$8.35 and $\sim$8.7 MeV, respectively. Substituting these values in eqn (8.1) gives

$$E_R = 140(8.35 - 7.6) + 93(8.7 - 7.6) \approx 210\ \text{MeV}.$$

About 10 MeV of this energy is taken away by neutrinos in the beta decays. Neutrinos interact extremely weakly with matter and hence escape. Therefore $\sim$200 MeV of energy will be deposited in the surrounding material if a uranium nucleus splits into mass 93 and mass 140 nuclei.

Energy is also released when deuterium ^2H and tritium ^3H nuclei fuse to form helium ^4He with the release of a neutron

$$^2H + {}^3H \rightarrow {}^4He + n.$$

The values for $b(A)$ for ^2H, ^3H, and ^4He are 1.1, 2.6, and 7.1 MeV, respectively, so approximately $(4 \times 7.1) - (2 \times 1.1) - (3 \times 2.6) \approx 18$ MeV is released in this fusion process.

We will now look at fission and how this process is harnessed to produce power before describing in Chapter 9 the progress that has been made in obtaining fusion power.

8.2 Fission

Although energy is released by the fission of uranium, the natural occurrence of this process (called **spontaneous fission**) is very rare. This is because uranium is stable with respect to small deformations from its equilibrium shape, as illustrated in Fig. 8.3.

Such deformations increase the surface area of the nucleus and there is a consequent loss in binding energy, which is not offset by the decrease in electrostatic repulsion arising from the increased separation of charge. The result is a barrier of a form indicated in Fig. 8.3. Classically such a nucleus would be stable, as a ball would be in a dip in the top of a mound, but quantum mechanically decay by fission can take place through a process called **tunnelling**.

This is the same process that occurs in α decay. For $Z = 92$ the barrier is sufficiently low at ~6 MeV that the process is just detectable for ^{238}U, with a half-life for spontaneous fission of $\sim 10^{16}$ y. Its half-life for α decay is comparatively much shorter at 4.5×10^9 y so fission is very rare.

8.2.1 Neutron-induced fission

The probability of fission of uranium is increased enormously when a uranium nucleus captures a neutron. For a ^{235}U nucleus this capture produces an excited ^{236}U* nucleus, $n + {}^{235}U \rightarrow {}^{236}U^*$, whose energy is above the height of the fission barrier and can therefore fission promptly. However, for $n + {}^{238}U \rightarrow {}^{239}U^*$, the excited ^{239}U* nucleus is ~1 MeV below the top of the barrier, whose height is ~6 MeV. This difference in excitation arises because a neutron in ^{236}U is slightly more strongly bound than in ^{239}U. In ^{236}U, which has 92 protons, all the neutrons are paired off, while in ^{239}U there is one unpaired neutron and this is less tightly bound than a paired-off neutron.

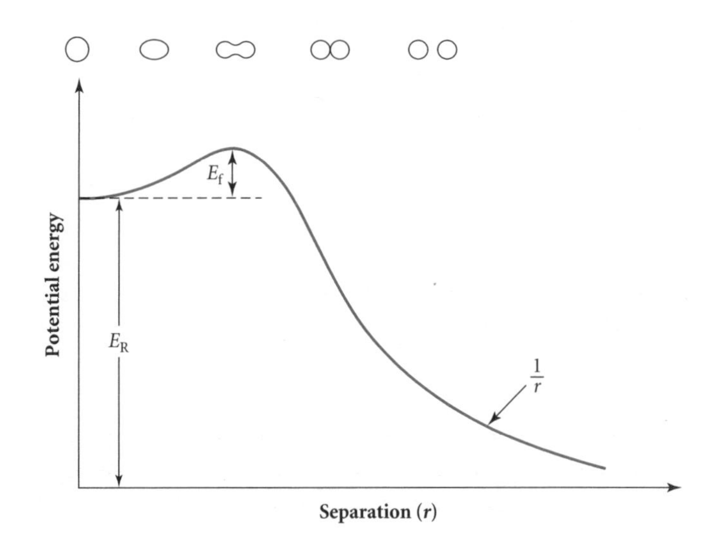

Fig. 8.3 Fission barrier.

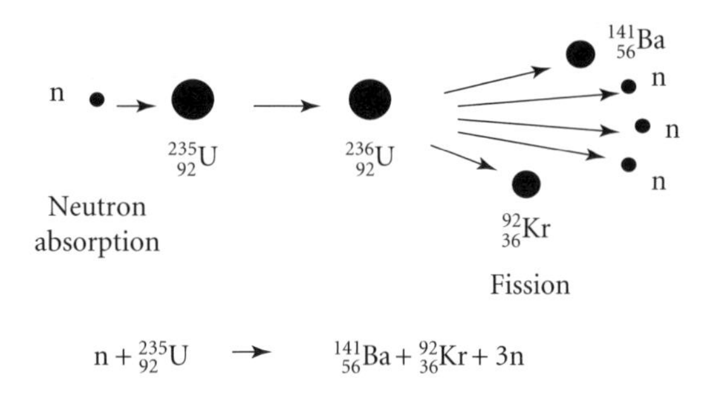

Fig. 8.4 Neutron-induced fission of ^{235}U producing ^{141}Ba and ^{92}Kr.

Figure 8.4 shows an example of neutron-induced fission of ^{235}U. Typically two or three neutrons are released promptly. This is because the barium and krypton isotopes that are produced initially are very neutron-rich compared with the stable isotopes of these elements, the most abundant of which are ^{138}Ba and ^{84}Kr, respectively (see Fig. 8.1). These very neutron-rich nuclei are produced in states that emit neutrons promptly, leaving neutron-rich isotopes of barium and krypton that beta decay to stable isotopes.
Some of these isotopes decay to excited states of nuclei which are unstable to neutron emission (see Fig. 8.5). This process is called beta-delayed neutron emission because the neutron is emitted only after a beta decay, and happens about 0.6% of the time in the neutron-induced fission of uranium. We will see in Section 8.4.3 that these **beta-delayed neutrons** are very important in the control of the chain reaction in a nuclear reactor.

8.2.2 Energy release in fission

We have estimated the energy release in the fission of uranium from the $b(A)$ values in Fig. 8.2 to be about 200 MeV $\equiv 3.2 \times 10^{-11}$ J. This is about 50 million times more than that released in a chemical combustion reaction such as

$$C + O_2 \rightarrow CO_2 + 4.2\ \text{eV} \equiv 7 \times 10^{-19}\ \text{J}.$$

A carbon atom is much lighter than a uranium atom so 1 tonne of ^{235}U is equivalent as an energy source to ~2.3 million tonnes of carbon. Coal produces about 75% less energy, and only 0.7% of natural uranium is ^{235}U, 99.3% being ^{238}U so, if only ^{235}U is used for fission,

1 tonne of uranium is actually equivalent in energy to about 20 000 tonnes of coal.

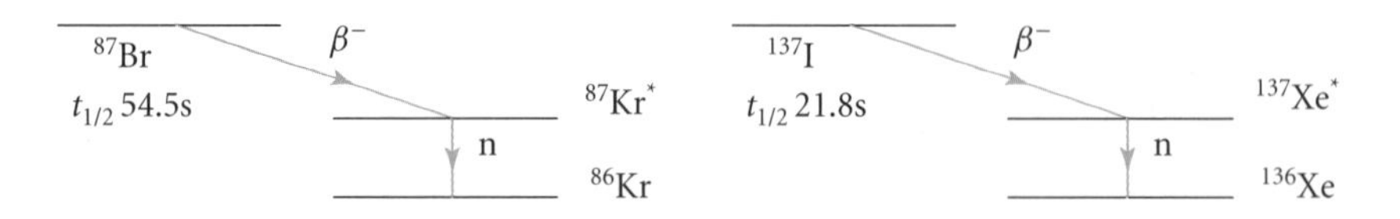

Fig. 8.5 Examples of beta-delayed neutron emitters.

The energy from fission can be divided into **prompt release**, and **delayed release** following the beta decay of the neutron-rich nuclei produced in the fission. The distribution of energy released is shown in Table 8.1.

In a reactor the antineutrinos escape because their interaction with matter is exceedingly weak. The neutrons that do not induce fission are captured by nuclei producing nuclei in excited states that decay by gamma emission. These γ-rays give another $\sim$5 MeV, making the total energy absorbed as heat per fission close to 200 MeV.

EXAMPLE 8.2

How much energy is released when 1 kg of uranium enriched to 3% in ^{235}U is consumed in a nuclear reactor?

One mole of uranium corresponds to 238 g so the number of uranium nuclei N_U in 1 kg is given by

$$N_U = 1000 \times 6 \times 10^{23}/238 = 25.2 \times 10^{23}.$$

The energy release per fission of ^{235}U is 200 MeV. As 3% of the uranium nuclei are ^{235}U, the total energy release E_T will be

$$E_T = 0.03 \times 25.2 \times 10^{23} \times 200 \times 10^6 \times 1.6 \times 10^{-19}\text{J} = 2420 \text{ GJ}.$$

On average $\sim$2.4 neutrons are emitted in the neutron-induced fission of ^{235}U with a broad range of energies about a mean energy of $\sim$2 MeV (see Fig. 8.6). The release of more than one neutron in the neutron-induced fission of ^{235}U, as in the reaction illustrated in Fig. 8.4, opens up the possibility of a **chain reaction**, which will occur if on average at least one of the neutrons released induces fission of other nuclei. Typically the fission is asymmetric as shown in Fig. 8.4. It is important to note that not all of the neutrons are emitted promptly: 0.65% are beta-delayed neutrons with a mean delay time of 13 seconds.

Table 8.1 Energy released by fission process

	Energy released per fission (MeV)
Prompt Release	
Fission products	168
Neutrons	5
γ-rays + internal conversion e^-	7
Delayed Release	
β-particles	8
γ-rays + internal conversion e^-	7
Antineutrinos	12

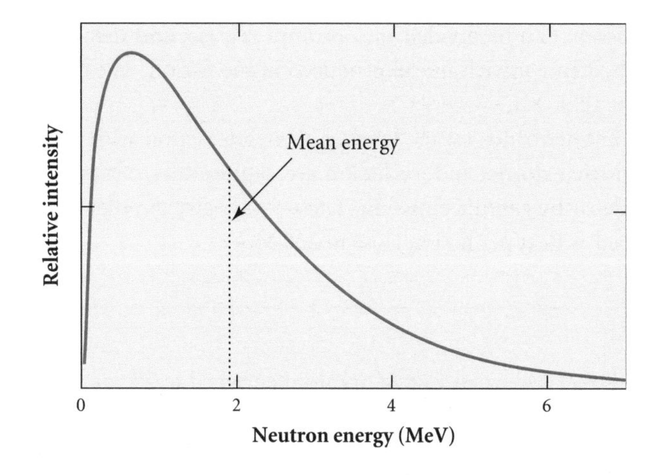

Fig. 8.6 Fission neutron energies.

We will now consider the conditions for a chain reaction to occur in uranium by looking at the relative probabilities for different neutron reactions on ^{235}U and ^{238}U.

8.2.3 **Chain reactions**

Whether a chain reaction actually occurs depends on the relative probability of neutron-induced fission compared to neutron loss. The dominant cause of loss is neutron capture, followed by gamma emission. The main reactions that neutrons with energies from a fraction of an eV (thermal neutrons) to several MeV undergo with uranium are scattering (both elastic and inelastic), capture, and induced fission. The probability that one of these reactions occurs can be described in terms of a cross-section σ. This can be visualized as the effective cross-sectional area within which a target nucleus and an incident neutron will interact and give rise to a particular reaction. Its units are barns (b) $\equiv 10^{-28}$ m^2. The cross-sectional area of a uranium nucleus is $\sim$2b. The value of σ can, however, be much larger than the cross-sectional area of a nucleus. This is a consequence of the wave-like properties of a neutron. The cross-section for any interaction is the total cross-section σ_t, and that for absorption equals the sum of the capture and neutron-induced fission cross-sections, i.e. $\sigma_a = \sigma_c + \sigma_f$.

Consider neutrons moving at a speed v through uranium where the number of ^{235}U nuclei per unit volume is n_f. If the cross-section for neutron-induced fission is σ_f, then in one second a neutron will sweep out a volume $\sigma_f\, v$ (see Fig. 8.7).

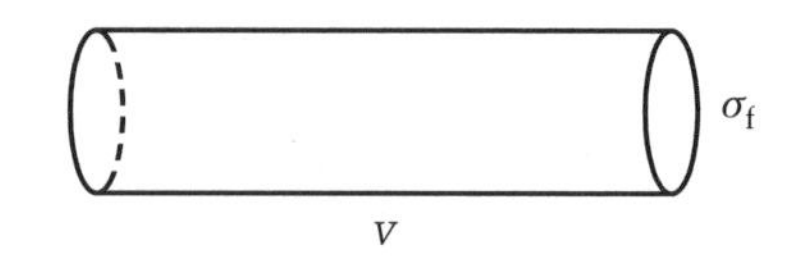

Fig. 8.7 Volume $\sigma_f V$ swept out by neutron in one second.

The number of ^{235}U nuclei enclosed within the volume swept through by this neutron per second is therefore $n_f \sigma_f v$ so, if there are n neutrons per unit volume within the uranium, then the fission reaction rate R_f per unit volume is given by

$$R_f = n_f \sigma_f v n \equiv \Sigma_f \varphi, \tag{8.2}$$

where

$$\Sigma_f = n_f \sigma_f \tag{8.3}$$

is called the macroscopic cross-section for neutron-induced fission of ^{235}U. Its units are m^{-1}. The product

$$nv \equiv \varphi \tag{8.4}$$

is called the **neutron flux**. The flux φ is very important as it determines the power output in a nuclear reactor. Its units are m^{-2} s^{-1} and a typical value for φ in a reactor is 10^{17} m^{-2} s^{-1}.

EXAMPLE 8.3

Calculate the macroscopic cross-section for fission in uranium that is enriched to 2% in ^{235}U. What is the energy produced per unit volume when the neutron flux is 10^{17} m^{-2} s^{-1}? The density of uranium is 18900 kg m^{-3} and the cross-section for neutron- induced fission of ^{235}U is 579 b.

One mole of uranium corresponds to 238 g, so the number density of uranium nuclei n is given by

$$n = (18900/0.238) \times 6 \times 10^{23} = 4.77 \times 10^{28}.$$

The amount of ^{235}U is 2% and the fission cross-section is 579 b, so the macroscopic cross-section Σ_f from eqn (8.3) is

$$\Sigma_f = 0.02 \times 4.77 \times 10^{28} \times 579 \times 10^{-28} = 55 \text{ m}^{-1}.$$

The reaction rate per unit volume R_f from eqn (8.2) is

$$R_f = \Sigma_f \varphi = 55 \times 10^{17} \text{m}^{-3} \text{s}^{-1}.$$

So the energy generated per second per unit volume P_f is R_f multiplied by the energy release per fission E_f, i.e.

$$P_f = R_f E_f = 55 \times 10^{17} \times 200 \times 10^{6} \times 1.6 \times 10^{-19} = 176 \text{ MW m}^{-3}.$$

For a medium containing a mixture of nuclei with number densities n_i and total cross-sections σ_t^i, the macroscopic total cross-section is given by

$$\Sigma_t = \sum_i n_i \sigma_t^i. \tag{8.5}$$

The average number of nuclei with number density n_i that a neutron interacts with after it has travelled a distance x is $n_i\sigma_t^i x$, so the total number of nuclei that a neutron interacts with is on average $\Sigma_t x$. Putting $\Sigma_t x = 1$ determines the **mean free path** λ between interactions in the medium as

$$\lambda = 1/\Sigma_t. \tag{8.6}$$

The macroscopic cross-sections determine the relative probabilities for a neutron giving rise to a particular reaction with different nuclei. For example, the ratio of $\Sigma_f^{235}$to Σ_f^{238} gives the relative probability that a neutron induces fission of ^{235}U rather than of ^{238}U.

We will now consider natural uranium, which has 99.28% ^{238}U and 0.72% ^{235}U, and look at what happens to the neutrons produced in fission, spontaneous or induced. The average number of neutrons emitted per fission, called ν, is ~2.4, and their energies range between ~0 and 10 MeV with a mean energy of ~2 MeV. In natural uranium these neutrons most likely scatter off ^{238}U and it is only when they have energies less than ~5 eV that neutron-induced fission of ^{235}U is more likely than capture by ^{238}U.

When the neutrons have energies between ~5 and 100 eV, sharp peaks (called **resonances**) occur in the cross-section for the capture of a neutron by ^{238}U, which correspond to excited states in ^{239}U being formed. At these peaks the probability of capture is close to one as the cross-section for capture is much larger than that for scattering, $\sigma_c \gg \sigma_s$. In this region the neutrons are only losing energy very slowly through elastic scattering off ^{238}U and the chance that their energy falls under a peak in the ^{238}U capture cross-section is high with the result that very few ($<1\%$) reach energies less than 5 eV and induce fission of ^{235}U.

Only a small percentage of the high energy neutrons induce fission of ^{238}U (~8%) and of ^{235}U (~2%). So the average number of neutrons left after each fission is the percentage that induces fission (~10%) times the average number emitted per fission (~2.4); i.e. only about 0.24 neutrons, and there is therefore no chain reaction in natural uranium.

We can conclude that in order to produce a self-sustaining chain reaction in uranium either the isotopic abundance of ^{235}U must be increased from its naturally occurring percentage of 0.72%, by a process called **enrichment**, so that the percentage of induced fission by fast neutrons of ^{235}U is larger, or the probability of capture by ^{238}U must be reduced.

Enrichment allows a chain reaction to be maintained with **fast neutrons**, where fast neutrons are those with energies greater than 100 keV. Alternatively, if nuclei with a low atomic number, called **moderators**, are added to the uranium, then the change in energy of a neutron following an elastic collision with a moderator nucleus can be sufficiently large that the chance of capture by ^{238}U in one of the resonance peaks can be significantly reduced. Once below the resonance region the probability of induced fission is large, and a chain reaction can be maintained with neutrons with energies ~0.05 eV. Neutrons with these energies are called thermal neutrons as they are at the same temperature as the uranium fuel.

Commercial nuclear reactors operate using thermal neutrons (obtained by the use of a moderator), and most use fuel enriched in ^{235}U to a few per cent. This enables a chain reaction to occur. As the time between a neutron from one fission inducing another fission is typically only ~1 ms then the chain reaction would be uncontrollable if it were not for the influence of the small fraction of **beta-delayed neutrons** that are also released. Their mean delay time of

$\sim$10 s allows sufficient time for the chain reaction to be controlled mechanically using **control rods**. Control rods contain nuclei with very high neutron absorption cross-sections (e.g. ^{10}B, ^{113}Cd) and the total number of neutrons inside the reactor can be controlled by continually adjusting how far the rods are inserted into the reactor core.

Box 8.2 Enrichment

The fraction of ^{235}U in natural uranium is only 0.72%. As the isotopes ^{235}U and ^{238}U are chemically identical, enrichment is achieved by taking advantage of their slight difference in mass. Several techniques have been demonstrated in the laboratory but only two have been developed commercially: **gaseous diffusion** and the **gas centrifuge process. Electromagnetic separation** was used in the Manhattan project (p. 6) but is not as economic: in this process uranium ions are accelerated through a potential drop and then deflected in a magnetic field around a circular path whose radius depends on the ion's mass. A process still under development is **laser isotope separation**. This method uses the isotopic shift of an atomic or molecular energy level to enable laser beams to selectively excite and then ionize ^{235}U, or molecules containing ^{235}U, which can then be collected.

Enrichment reduces the degree of mixing of the ^{235}U and ^{238}U isotopes. This produces a more ordered state with lower entropy, and this requires work. In a gaseous separation plant there is a feed and two outputs—the product and tail streams. The amount of separation required depends on the mass and enrichment of all three streams and is measured in **separative work units** or SWUs. For example, to produce 1 kilogram of 3% ^{235}U using natural uranium as feed requires 3.8 SWU if the tail enrichment is 0.25% or 5.0 SWU if it is 0.15%. SWUs have the dimensions of mass and the amount of work required depends on the number of SWUs and the process used.

About 120 000 SWU are required to enrich the annual fuel loading for a typical 1 GW pressurized water reactor (PWR). The gaseous diffusion process uses about 2500 kWh per SWU, while gas centrifuge plants only require about 50 kWh per SWU. This energy, if produced from hydrocarbon fuels, accounts for the main greenhouse gas impact of nuclear power. However, if gas centrifuge plants are used, it is equivalent to only 0.1% of the CO_2 produced from a coal-fired plant of the same power.

Gaseous diffusion

In this process uranium hexafluoride gas is passed under pressure down the centre of a porous tube, as shown in Fig. 8.8(a). On average the molecules containing ^{235}U atoms are moving slightly faster, since they are lighter, and thus make more collisions per second with the porous membrane than those containing ^{238}U. As a result the gas passing through the membrane gets enriched in ^{235}U. The amount of enrichment at each stage is very small and a cascade of over a thousand stages is required to produce a concentration of $\sim$3% ^{235}U. Each stage requires a compressor and a cooler to remove the compressive heating of the gas and these consume huge amounts of energy.

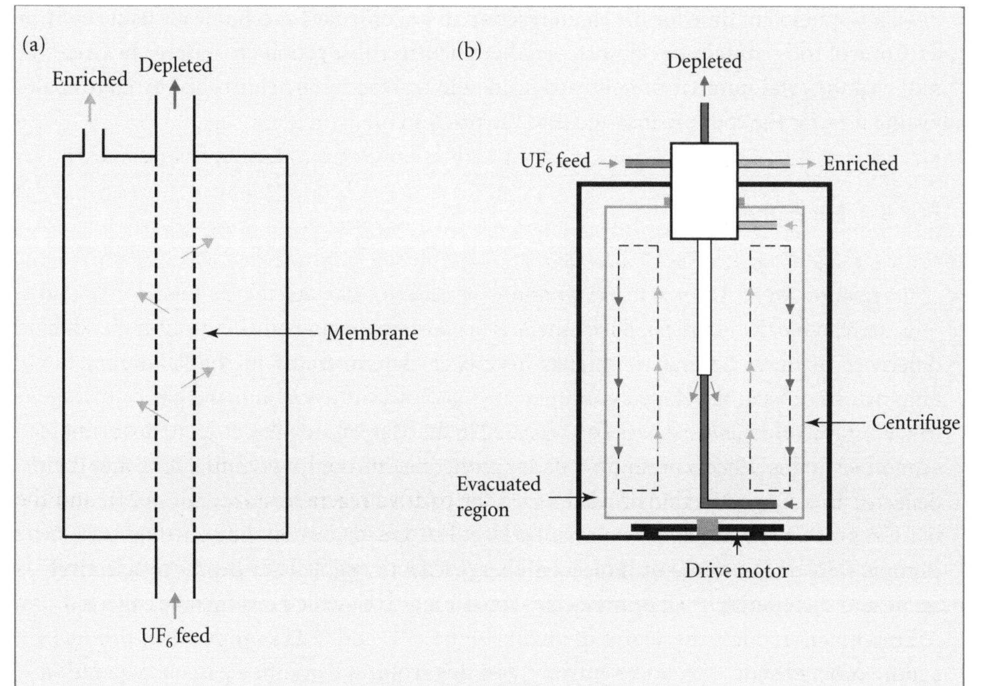

Fig. 8.8 (a) Gas diffusion stage. (b) Gas centrifuge.

Gas centrifuge

In the gas centrifuge process, uranium hexafluoride gas is injected near the axis of a very high speed centrifuge (see Fig 8.8b). The heavier molecules tend to be closer to the wall than the lighter molecules due to the larger centrifugal force and an axial flow causes more of the lighter molecules to leave through the top than the bottom tube. Although the capacity of a single centrifuge is much smaller than that of a single diffusion stage, the isotopic separation is much higher. As a result many centrifuges are arranged in parallel, but only some 20 stages are required to produce a concentration of $\sim$3% ^{235}U. The main advantage of the gas centrifuge process is that it requires only about 2% of the energy per SWU used in a gas diffusion plant.

We will now look at the layout of a typical thermal reactor before looking more closely at the fuel, moderator, design, control, and safety requirements of a reactor.

8.3 Thermal reactors

The fissile material (the fuel) is generally in the form of **fuel rods**, which allows for easy refuelling. The rods are immersed in a chemically inert fluid such as water, CO_2 or He, which is heated by the fuel. Shown in Fig. 8.9 is a schematic diagram of a pressurized water reactor (PWR). The PWR is the most common reactor and illustrates the main features of a nuclear reactor. The control rods are located above the core and the energy released by fission heats the fuel rods, which in turn heat the fluid. The hot fluid is pumped through the primary loop and releases its heat through the heat exchanger and the steam produced drives a turbine.

The fuel is surrounded by the moderator and the first aspect of the design of a reactor is determining the conditions required of the fuel and moderator for a chain reaction to occur. Whether the neutrons emitted following neutron-induced fission (called the **next generation of neutrons**) actually lead to a chain reaction in a piece of fissile material depends on the ratio of the number of neutrons producing fission in one generation to the number producing fission in the previous generation. This ratio is called the **multiplication constant** k, and if $k \geq 1$ a chain reaction is possible.

The multiplication constant k is determined by the relative probability of neutron-induced fission compared to neutron loss via other neutron reactions or by escaping from the reactor. The macroscopic cross-sections Σ determine the relative reaction rates (eqn (8.2)), and the size of the core determines the escape probability. For an infinite core no neutrons can escape and the multiplication constant is then called k_∞, which is greater than k. A chain reaction is possible if $k_\infty > 1$, and the size of core for which $k = 1$ is called the **critical size**. We will first consider the factors affecting k_∞.

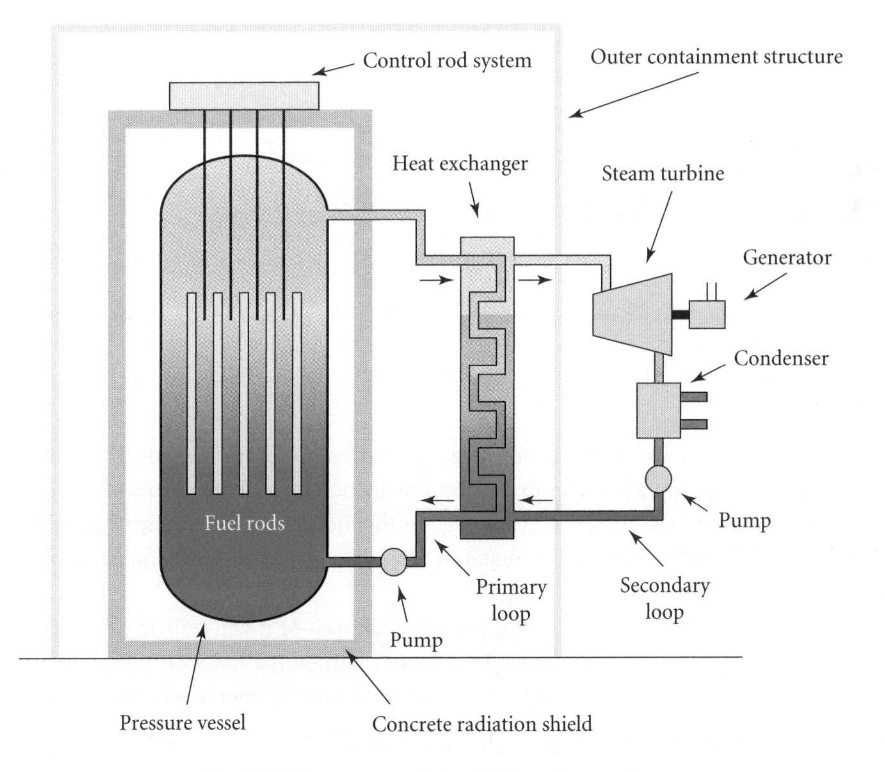

Fig. 8.9 Main components in a PWR nuclear reactor.

8.3.1 Four-factors formula for the multiplication constant k_∞

In a core of uranium that is sufficiently large that neutron loss is negligible, there are four factors that determine k_∞:

- fast fission factor ε;
- resonance escape probability p;
- thermal utilization factor f;
- number of neutrons produced per thermal neutron absorbed η.

Of the average initial number of fast neutrons emitted per fission, ν, the loss through capture is offset by the gain through fast neutron-induced fission of ^{238}U. For ^{235}U, $\nu = 2.42$. The ratio of the number below the ^{238}U threshold of $\sim$1 MeV to the initial number is called the **fast fission factor** ε. For low enrichment the fast fission factor ε is close to unity.

Below the ^{238}U threshold a fraction escape resonance capture on ^{238}U and manage to slow down to thermal energies. This **resonance escape probability** p depends strongly on the choice of material for the moderator.

The fraction f of thermal neutrons that are absorbed by U, rather than by the moderator or the fuel can, is called the **thermal utilization factor** f. It is given by the ratio of the macroscopic cross-section for absorption in the fuel (F) to that for absorption in the core (C)

$$f = \Sigma_a(\mathrm{F})/\Sigma_a(\mathrm{C}). \quad (8.7)$$

The ratio η of the number of fission neutrons produced to the number of thermal neutrons absorbed is less than the average number of neutrons ν emitted, since some of the neutrons are captured by ^{238}U and ^{235}U. This ratio is given by

$$\eta = \nu\Sigma_f(\mathrm{F})/\Sigma_a(\mathrm{F}). \quad (8.8)$$

Finally, combining all the above factors gives the **four factors** formula

$$k_\infty = \varepsilon p f \eta. \quad (8.9)$$

We will now consider how the choice of moderator affects the resonance escape probability p.

8.3.2 Moderators

The average energy of the neutrons produced by the fission reaction needs to be reduced by many orders of magnitude in order to reach thermal energies and the means of achieving this is called **moderation**. In a nuclear thermal reactor the fuel is generally in the form of fuel rods which are surrounded by moderating material. This arrangement of fuel and moderator gives a non-uniform core.

When a neutron elastically scatters off a nucleus of mass M it will transfer some energy ΔE to M, the amount depending on the angle of scatter and on the mass M: the larger the mass M the smaller the loss ΔE (Exercise 8.6). The average relative energy loss $\Delta E/E$ is called the **logarithmic decrement** ξ and is given for moderating nuclei with mass number A by

$$\Delta E/E \equiv \xi \approx 6/(3\mathrm{A}+1), \quad (8.10)$$

where the approximation is reasonable for $A \geq 12$. We can now see why neutrons slowing down in natural uranium by scattering off the uranium nuclei have only a small chance of not being captured by one of the ^{238}U resonances. Each peak in the capture probability occurring between ~5 and 100 eV extends over a small energy range, which is given by the width of the resonance. The peaks correspond to the formation of excited states in ^{239}U that are unstable, and the widths Γ are related to the mean lifetimes (τ) of the states through $\Gamma = h/(2\pi\tau)$, which is an example of the uncertainty principle. If the peak cross-section is at an energy E_0 then the cross-section is reduced by a factor of two at $E_0 \pm \Gamma/2$ and by a factor of five at $E_0 \pm \Gamma$.

One of the strong capture resonances is at a neutron energy of 6.7 eV, with a total width of 0.024 eV. The average neutron energy loss ΔE for a neutron with energy E when scattering from ^{238}U is given by ξE, so for a 6.7 eV neutron $\Delta E \approx 0.056$ eV by eqn (8.10). There is therefore a good chance that, while losing energy through elastic scattering from U nuclei, a neutron has an energy within Γ of the peak energy. As the peak cross-section is over 100 times the elastic scattering cross-section the neutron is likely to be captured. We therefore need to use low mass nuclei as a moderator, which will have a larger $\Delta E/E$, to reduce the chance of resonant capture.

To be an efficient moderator, both ξ and the probability of scattering (which is proportional to Σ_s) should be large. The moderator should also have a small total absorption cross-section, Σ_a, and the **moderating ratio** $\xi\Sigma_s/\Sigma_a$ (a useful figure of merit for a moderator) should be as large as possible.

An approximate expression for the resonance escape probability p for a reactor core, where elastic scattering is predominantly by the moderator and loss is predominantly by absorption by ^{238}U in the uranium fuel, is given by

$$p \cong \exp\{-2.4[n_f/(n_s\sigma_s)]^{1/2}/\xi\} \tag{8.11}$$

where n_f is the number density of fuel nuclei, n_s the number density of moderating nuclei, σ_s is the scattering cross-section in barns, and ξ is the logarithmic decrement of the moderator. In this expression p is closer to one for larger ξ and Σ_s and for smaller n_f, as expected.

We will now calculate k_∞ for a reactor core containing a mixture of enriched uranium and carbon in the form of graphite for which the nuclear cross-sections, densities, and values for ξ are given in Table 8.2.

Table 8.2 Nuclear and material properties*

Material	Density (kg m^{-3})	n (10^{28}m^{-3})	σ_f(b)	σ_s(b)	σ_a(b)	ξ	Σ_s(m^{-1})	Σ_a(m^{-1})
Graphite (C)	1600	8.23	–	4.7	0.0045	0.158	37.7	0.037
^{235}U	18700	4.79	579	10	680		47.9	3229
^{238}U	18900	4.79	–	8.3	2.72		39.8	13.0

*σ are at thermal energies: 0.025 eV

EXAMPLE 8.4

Calculate the multiplication constant k_∞ of a reactor core containing a mixture of uranium, enriched to 1.7% in ^{235}U, and graphite, with a ratio of $n_s/n_f = 500$, where n_s is the number density of the graphite moderator.

The logarithmic decrement of the mixture is essentially the same as that of graphite as $n_s \gg n_f$. Using the values for the cross-sections and logarithmic decrement given in Table 8.2 in eqn (8.11) gives the resonance escape probability as

$$p = \exp\{-2.4[n_f/(n_s\sigma_s)]^{1/2}/\xi\} = \exp\{-2.4[1/(500 \times 4.7)]^{1/2}/0.158\} = 0.731.$$

The thermal utilization factor f for this graphite moderated enriched uranium mixture is given by eqn (8.7) as

$$f = \Sigma_a(\text{fuel})/\{\Sigma_a(\text{fuel}) + \Sigma_a(\text{graphite})\}$$

where

$$\Sigma_a(\text{fuel}) = n_{235}\,\sigma_a^{235} + n_{238}\,\sigma_a^{238} = (0.017\,\sigma_a^{235} + 0.983\,\sigma_a^{238})n_f.$$

Applying eqn (8.3) to absorption in graphite gives

$$\Sigma_a(\text{graphite}) = n_s\sigma_a(\text{graphite}).$$

So the thermal utilization factor f is given by

$$f = (0.017\sigma_a^{235} + 0.983\sigma_a^{238})/\{(0.017\sigma_a^{235} + 0.983\sigma_a^{238}) + [n_s/n_f]\,\sigma_a(\text{graphite})\}.$$

Substituting in values from Table 8.2 and using $n_s/n_f = 500$ gives $f = 0.864$.
The ratio η of the number of fission neutrons produced to thermal neutrons absorbed is from eqn (8.8)

$$\eta = \nu\Sigma_f(\text{fuel})/\Sigma_a(\text{fuel})$$

where $\Sigma_f(\text{fuel}) = (0.017\sigma_f^{235})n_f$ and ν is 2.42. Substituting gives $\eta = 1.673$.
For this low enrichment and high n_s/n_f ratio $\varepsilon \approx 1$ so

$$k_\infty = \varepsilon p f \eta = 1.057.$$

This means that a chain reaction is possible and will be critical, i.e. $k = 1$, if the size of the reactor is such that the fractional loss $(k_\infty - 1)/k_\infty$, equals 0.054. This loss is the escape probability for neutrons escaping from the core.

8.3.3 Effect of finite core size

Neutrons are lost from the reactor core because a neutron will typically undergo many elastic scatterings before reacting. Thereby they diffuse away from their point of origin on average a distance that we will call δ. Those neutrons within δ of the outer surface of the core will therefore have a good chance of escaping. The fraction of neutrons that are within δ of the outer surface reduces with increasing core size. We show in Box 8.4 that the size of a core a_c required for a reactor to be just critical ($k = 1$) is given by

$$a_c = \frac{\pi\delta}{\sqrt{3(k_\infty - 1)}} \tag{8.12}$$

where

$$\delta = 1/\sqrt{\Sigma_a \Sigma_s}. \tag{8.13}$$

As the probability of absorption or scattering increases, i.e. Σ_a or Σ_s increases, then δ decreases as the neutron is likely to be absorbed in a shorter distance from where it was created. For larger k_∞ then the loss can be larger and hence a_c can be smaller. When the core is just critical, i.e. $k = 1$, then the probability that neutrons do not diffuse out of the core equals $1/k_\infty$.

We will now use eqs (8.12) and (8.13) to work out the critical size of a reactor core with a high ratio of moderator to fuel.

Box 8.3 Reactor core size

Consider a spherical core of radius a. Figure 8.10 shows the typical neutron flux distribution in the radial direction, and neutrons within a distance δ of the outer surface will have a good chance of escaping.

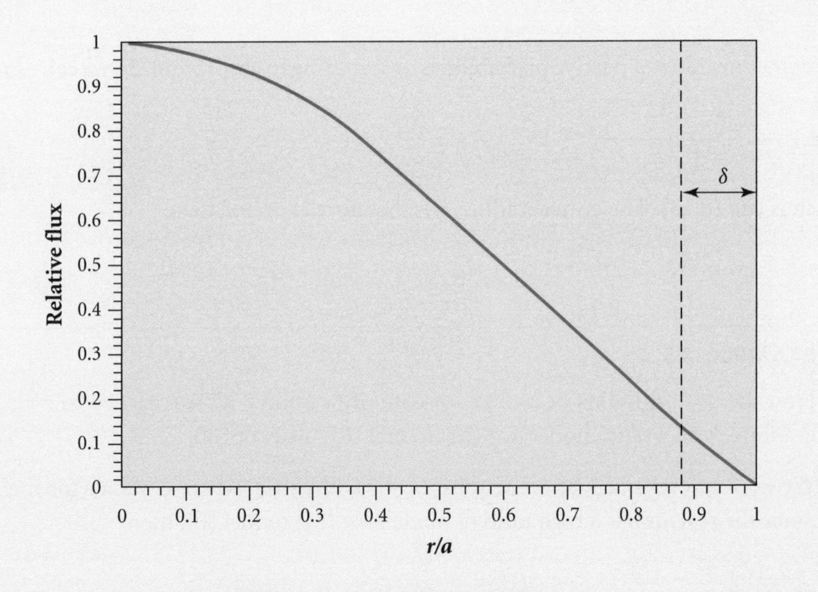

Fig. 8.10 Neutron radial flux distribution.

The flux at the surface is of order δ/a compared with that in the body of the reactor and the fractional volume of the outer layer from which neutrons can escape is also of order δ/a. The fractional loss of neutrons will therefore be $\sim(\delta/a)^2$. When critical, this loss equals the fractional gain of neutrons, so

$$(\delta/a_c)^2 \sim (k_\infty - 1)/k_\infty. \tag{8.14}$$

Rearranging, noting that $k_\infty \sim 1$, the critical radius a_c is then found to be

$$a_c \approx \frac{\delta}{\sqrt{k_\infty - 1}}. \tag{8.15}$$

A more precise relationship is that given by eqn 8.12.

$$a_c = \frac{\pi\delta}{\sqrt{3(k_\infty - 1)}}.$$

The size of the core is therefore smaller if the fuel is enriched (since this increases k_∞) and is proportional to the distance δ.

We can estimate δ by considering the path a neutron takes in a reactor. The neutron scatters many times before being absorbed and in the process makes a random walk. Each step is on average the mean free path λ, which for $\Sigma_s \gg \Sigma_a$ is

$$\lambda = 1/\Sigma_s,$$

using eqn (8.6) for λ. In a random walk the average distance travelled after n scatters is $\sqrt{n}\lambda$. The number n of elastic scatterings that the neutron makes before being absorbed is

$$n \sim \Sigma_s/\Sigma_a \tag{8.16}$$

as n equals the ratio of relative probabilities of scattering to absorption. So we can estimate δ by

$$\delta = \sqrt{n}\lambda = \left(\sqrt{\Sigma_s/\Sigma_a}\right) 1/\Sigma_s = 1/\sqrt{\Sigma_s\Sigma_a},$$

which is eqn (8.13). The critical radius a_c is therefore expected to be

$$a_c = \pi/\sqrt{[3(k_\infty - 1)\Sigma_a\Sigma_s]}.$$

EXAMPLE 8.5

Find the critical radius of a uniform core of uranium, 1.7% enriched in ^{235}U, and graphite, with a ratio moderator (M) to fuel (F) nuclei of 500.

We need to evaluate δ, so we require $\Sigma_a(C)$ and $\Sigma_s(C)$, where C stands for the core. Since the reactor has a high ratio of moderator (M) to fuel (F), then

$$\Sigma_s(C) \approx \Sigma_s(M).$$

The thermal utilization factor from eqn (8.7) is

$$f = \Sigma_a(F)/\Sigma_a(C).$$

Putting $\Sigma_a(C) = \Sigma_a(M) + \Sigma_a(F)$, we have

$$\Sigma_a(C) = \Sigma_a(M)/(1 - f).$$

From Example 8.4, the core has $k_\infty = 1.057$ and $f = 0.864$. Also, from Table 8.2, $\Sigma_s(M) = 37.7 \text{ m}^{-1}$ and $\Sigma_a(M) = 0.037 \text{ m}^{-1}$. Substituting into eqn (8.13) gives

$$\delta = 0.31 \text{ m}.$$

Putting this value into eqn (8.12) gives the critical radius as $a_c = 2.4$ m.

8.4 Thermal reactor designs

The first reactor was built in a squash court in Chicago in 1942 under the direction of Enrico Fermi as part of the **Manhattan Project**, which developed the first fission bomb. It was a graphite moderated pile containing natural uranium fuel in the form of rods embedded in graphite blocks. Although a uniform mixture of graphite and natural uranium cannot go critical, a non-uniform core containing fuel in the form of rods can go critical.

The development of graphite moderated reactors using natural uranium fuel was the approach adopted in the UK for the first generation of nuclear power stations, owing to the difficulty in producing uranium enriched in ^{235}U. These reactors were gas-cooled using high pressure carbon dioxide as the primary fluid and used uranium oxide fuel in a magnesium alloy container; hence their name **Magnox reactors**. They operated at a relatively low temperature of 400°C, which gave a low thermal efficiency of ~30%. In Canada a heavy-water moderated and cooled reactor (CANDU) was successfully developed. The use of non-enriched fuel, however, makes the core large and the pressurized light-water reactor developed in the USA, which uses enriched fuel and a more compact core, has been much more widely adopted. The techniques used to enrich uranium are described in Box 8.2.

8.4.1 Pressurized light water reactor (PWR)

The pressurized light water reactor (PWR) shown in Fig. 8.9 is the most widespread commercial reactor. Of a total of 439 nuclear reactors in 2003, 263 were PWRs. The PWR was initially developed for submarines since, unlike internal combustion engines, nuclear-powered submarines do not need oxygen and can therefore remain underwater for much longer. The heat from the reactor produces steam to drive a turbine and the relatively compact core proved a cost-effective design that could be scaled up to ~1 GW. The first prototype was operated in 1953.

In a PWR (see Fig. 8.9) water is circulated in the primary loop past the fuel rods at a high pressure of ~15 MPa to keep it in the liquid phase at a temperature of ~315°C. It is passed through a heat exchanger, where the water in the secondary loop, which is at a lower pressure of ~5 MPa, is heated to produce steam to drive the turbine, after which the steam is condensed and the water is returned to the heat exchanger. The high neutron flux in the core activates the cooling water and makes it radioactive. This radioactivity is kept within the primary loop and within the containment vessel. The thermal efficiency of a PWR is typically ~30%.

The water acts as a moderator as well as a coolant. It also absorbs neutrons. Should the pressure drop in the primary loop and the water start to boil, the creation of bubbles (voids) decreases the moderation and also the absorption. The effect on the moderation is the more significant and the chain reaction stops and the reactor is no longer critical. The moderation is also decreased if the core temperature rises, as this increases the Doppler broadening of the ^{238}U resonances (see Section 8.4.4), which decreases the resonance escape probability p. Hence there is a negative temperature coefficient of reactivity, which tends to stabilize the power output, since an increase in power causes the temperature to rise and the reactivity to fall, and vice versa.

Over a long period the high neutron flux causes embrittlement of the reactor vessel as the metal becomes less ductile and this affects the lifetime of the reactor. Corrosion in the steam-generating tubes must also be monitored. A loss-of-coolant accident (LOCA) in which the water in the primary loop is lost requires additional emergency cooling to be available. While the core would no longer be critical, the heat generated from the decay of the fission products in the fuel rods could cause the core to melt. This possibility has been termed the **China syndrome**, referring to the (mistaken) idea that the molten core would bore through the Earth from the US to China!

8.4.2 Core design

Putting the fuel in the form of rods increases the resonance escape probability p significantly. Part of this increase arises because some of the fast fission neutrons, which have a good chance of escaping the fuel rods, are thermalized by the moderator before interacting with another fuel rod and hence escape capture. However, the fuel rods are generally sufficiently close that this is not so significant. The greater effect arises because the flux of neutrons with energies in the resonance region is significantly reduced within a fuel rod due to the very strong absorption of these neutrons within the surface layer (<0.01 cm) of the rod. The average flux within the fuel is consequently considerably lower than in the fuel in a uniform mixture of fuel and moderator; hence the escape probability is higher.

A flux reduction also occurs within the fuel rods for thermal neutrons because of the large absorption cross-section. However, since the mean free path for thermal neutrons in natural uranium is about 2.5 cm, the effect on the thermal utilization factor f is much less than on the resonance escape probability p, with the result that the product fp is increased.

Fuel in the form of rods also increases the fast fission factor ε. The probability that the emitted fast neutrons induce fission in ^{238}U is larger because, typically, the fast neutrons travel through pure fuel for a distance of the order of the radius of the fuel rod before being moderated.

Besides these nuclear physics reasons for having fuel in the form of rods, there are excellent engineering reasons as well. The use of fuel rods allows good cooling, and hence good heat transfer, as well as easy refuelling. The control rods, which maintain the neutron flux at a constant value, also make the core non-uniform.

8.4.3 Reactor control

Whether using fast or thermal neutrons, it is essential to maintain the multiplication factor k close to unity, so that the neutron flux remains constant. In particular, as the fuel is 'burnt', k will decrease, so that the neutron absorption must be lowered. This is achieved by using electromagnetically operated control rods, which control the neutron flux in the core of the reactor. These contain nuclei with a high cross-section for neutron absorption, such as ^{10}B and ^{113}Cd, and can be lowered into the reactor core to reduce the flux or raised out of the core to increase the flux. The **reactivity** ρ, defined by $\rho \equiv (k-1)/k$, is generally used when describing the time dependence of the neutron population.

Control can also be maintained by altering the absorption of the coolant by adding a chemical containing a nucleus with a large neutron absorption cross-section (**chemical shim**). If water is the coolant (as in a PWR) then boric acid can be used. The effect on k is to alter the thermal utilization factor f. Its speed of response is slower than that of control rods, but it reduces the number of rods required.

Another method of control is to use **burnable poisons.** These are substances contain nuclei with very large thermal absorption cross-sections, such as gadolina (Gd_2O_3) or er (Er_2O_3), that are included in some of the fuel rods. Initially they absorb sufficient neutro that k is close to unity. As the fuel is consumed, the consequent decrease in k is partly offset the reduction in the amount of burnable poison nuclei ^{A}X through the absorption reacti $n + {}^{A}X \rightarrow {}^{A+1}X$. This method has been used to extend the time between refuelling for nuclear submarine reactor to over 10 years.

In a chain reaction there is a short period of time between the release of a neutron from a fission, and it initiating another fission and more neutrons. This period, called the generation time τ_g, for neutrons released promptly following fission is given by λ_a/u, where λ_a is the absorption mean free path ($= 1/\Sigma_a$) and u is the neutron speed. A typical value of τ_g for prompt thermal neutrons is $\sim 10^{-3}$s. If the number of neutrons per unit volume in the reactor core is n, then the build up of neutrons will be governed by the rate equation

$$dn/dt = (k-1)n/\tau_g. \tag{8.17}$$

The presence of the -1 term is because one neutron is absorbed in the chain reaction to produce the next generation, which has k neutrons, and this occurs on average every τ_g seconds. The population of neutrons therefore grows exponentially with a time constant

$$\tau = \tau_g/(k-1). \tag{8.18}$$

Hence, if k increases to 1.001, the neutron population increases by a factor of 20 in 3 seconds, which would make mechanical control exceedingly difficult. Fortunately, this is not what actually happens in practice due to the fact that a small percentage of neutrons, $\beta \sim 0.65\%$ for ^{235}U, are emitted following the beta decay of neutron-rich fission fragments with a mean delay time $\tau_d \sim 13$ s ($\tau = t_{1/2}/\ln 2$). The lifetimes and yields of beta-delayed neutron emitters are listed in Table 8.3.

Table 8.3 Half-lives and yields of beta-delayed neutron emitters following n $+^{235}$U

$t_{1/2}$	Yield (%)
55.7	0.0215
22.7	0.1424
6.22	0.1274
2.30	0.2568
0.61	0.0748
0.23	0.0273

For k near unity, it turns out that the generation time τ_g effectively depends on the delayed neutrons alone and equals the average neutron lifetime τ_a given by

$$\tau_a = \beta(\tau_d + \tau_p) + (1 - \beta)\tau_p = \beta\tau_d + \tau_p \tag{8.19}$$

where τ_p is the generation lifetime for prompt neutrons (~1 ms). The time constant for $k = 1.001$ therefore increases considerably to ~85 s and mechanical control of the reactor is possible. When starting up a reactor k is deliberately made sufficiently greater than unity that the flux increases to its operating value φ_o within a reasonable time.

EXAMPLE 8.6

Estimate what multiplication factor k is required for a thermal ^{235}U fuelled reactor to have a time constant of 300 s. How long would it take for the reactor output to change from 1 W to 1 GW?

Using Table 8.3 the mean half-life of the beta-delayed neutrons is 9.0 s, corresponding to a mean lifetime of $\tau_d = t_{1/2}/\ln 2 = 13.0$ s. Taking τ_p as 1 ms then the average neutron lifetime τ_a is given by eqn (8.19) as 85.5 ms. The generation time $\tau_g = \tau_a$, so the time constant τ is from eqn (8.18)

$$\tau = \tau_a/(k-1) \quad \text{so} \quad (k-1) = \tau_a/\tau = 0.0855/300 = 0.000285.$$

So $k = 1.000285$.

The power output depends on the flux of neutrons and hence on their number density. This density changes like $\exp(t/\tau)$ so the time t is given by

$$\exp(t/\tau) = 10^9 \quad \text{so} \quad t = 9\,\tau\ln(10) \approx 6220\text{ s} = 1.73\text{ h}.$$

8.4.4 Reactor stability

It is very important that the reactivity should decrease if the temperature of the core rises, so that the core temperature will be stable to small fluctuations in reactivity. In a thermal reactor this comes about through a broadening of the ^{238}U capture resonances with increasing temperature. The ^{238}U nuclei are vibrating and have a mean square speed that is proportional to the temperature of the uranium atoms. The neutron energy required to form an excited state of ^{239}U depends on the relative velocity of the neutron and ^{238}U nucleus and so has a spread Δ that increases with increasing temperature. This has the effect of increasing the neutron absorption rate. A simple example shows how this comes about.

Consider a capture resonance with a rectangular energy profile of width Δ and whose capture cross-section is four times the scattering cross-section. Let the neutron on average lose Δ at each scattering. Then typically every fast neutron will fall once under this resonance as it loses energy, and the chance it will be captured is $(1 - 1/5) = 4/5$. Now broaden the resonance to 2Δ. The total capture probability remains the same so the magnitude of the resonance is halved to twice the scattering cross-section. A neutron now on average lies twice under the resonance and its chance of being captured is $(1 - (1/3)^2) = 8/9$. Increasing the core temperature therefore decreases the resonance escape probability and hence decreases k.

8.4.5 Power output of a thermal reactor

The distribution of energy released in neutron-induced fission of ^{235}U is shown in Table 8.1. In a reactor the antineutrinos escape because their interaction with matter is exceedingly weak. The neutrons that do not induce fission are captured through (n,γ) interactions and the γ-rays give another ~5 MeV, making the total energy absorbed as heat per fission close to 200 MeV. This value translates to an energy output of 0.95 GWday per kilogram of ^{235}U, or about 1 GWday per kg ^{235}U. To produce 1 GW for a year requires 384 kg of ^{235}U (neglecting any contribution from conversion; see Section 8.5). However, this is thermal power, not electrical power. The thermal efficiency of a typical PWR is ~30% so about 1280 kg of ^{235}U per GW$_e$year is actually needed, where the subscript 'e' refers to electrical power.

The power generated in a reactor core depends on the neutron flux and the higher the flux the higher the power output. For a core of volume V containing N_{235} ^{235}U nuclei the power P is given by the fission reaction rate multiplied by the energy release per fission E_f. The reaction rate is the fission rate per unit volume R_f multiplied by the volume V. The rate R_f is given by eqn (8.2) as $\Sigma_f\varphi$, so the power P is

$$P = V\Sigma_f\varphi E_f \equiv N_{235}\sigma_f\varphi E_f \tag{8.20}$$

where φ is the neutron flux in the reactor and $E_f = 200$ MeV is the energy released per fission. (8.3) has been used to equate $V\Sigma_f$ with $N_{235}\sigma_f$. The power P can also be expressed in terms of the mass M_{235} of ^{235}U and flux φ as

$$P = 4.75\, M_{235}(\text{tonne})\, \varphi(10^{18}\ \text{m}^{-2}\text{s}^{-1})\ \text{GW}. \tag{8.21}$$

EXAMPLE 8.7

A thermal reactor has a core volume of 30 m^3, a macroscopic fission cross-section of 5 m^{-1}, and a neutron flux of 2×10^{17} m^{-2} s^{-1}. Calculate the power output, the mass of ^{235}U in the core, and the amount of ^{235}U consumed per year. Assume the power output remains constant.

The thermal reactor has: $\varphi \sim 2 \times 10^{17}$ neutrons s^{-1} m^{-2}, $\Sigma_f = 5\ m^{-1}$, and $V = 30\ m^3$, so from eqn (8.20) the power output P is

$$P = 30 \times 5 \times 2 \times 10^{17} \times 200 \times 10^6 \times 1.6 \times 10^{-19}\ W = 960\ MW.$$

From Table 8.2 σ_f^{235} is 579 b.

$$V\Sigma_f \equiv N_{235}\sigma_f, \quad \text{so} \quad N_{235} = 2.59 \times 10^{27}.$$

As 1 mole of ^{235}U is 235 g then N_{235} nuclei correspond to a mass m equal to

$$m = (N_{235}/6 \times 10^{23})(0.235) = 1014\ \text{kg of}\ ^{235}U.$$

The fission rate is given by $V\Sigma_f\varphi$, which equals 3×10^{19} fissions per second. Each fission means one ^{235}U nucleus lost, so in 1 month (2.63×10^6 s) the amount m_c of ^{235}U consumed is

$$m_c = (0.235)(3 \times 10^{19} \times 2.63 \times 10^6)/(6 \times 10^{23}) = 30.9\ \text{kg of}\ ^{235}U.$$

The **burn-up rate** is the number of kilograms per second used to produce the output power. In this case the burn-up rate is 1.17×10^{-5} kg s^{-1}.

8.4.6 Fission products

During the operation of a reactor there is a build-up of fission products within the fuel rods. Some are very long-lived actinides arising through successive neutron capture reactions on uranium. When the amount of fissile material in a fuel rod is insufficient to maintain criticality, the rod is removed and the remaining fissile material is extracted chemically (called **fuel reprocessing**). It is reutilized in new fuel and the waste products are separated for storage. The presence of these actinides means that the waste must be stored safely for many thousands of years (see Section 8.9).

Some of the fission products are volatile fission fragments and it is these that could be most easily released in the event of a major accident. The fission rate F is related to the thermal power output P in GW, assuming 200 MeV per fission, by

$$F = 3.13 \times 10^{19} P\ \text{fissions per second.} \tag{8.22}$$

If a fraction b_x, the fission yield, of fissions produce a nucleus x with a decay constant λ_x, then the decay rate R_x of the nucleus x after a time t such that $\lambda_x t \gg 1$ equals the production rate $b_x F$, i.e the activity R_x is given by

$$R_x = b_x F. \tag{8.23}$$

(This calculation neglects any contribution from a fission product decaying to the nucleus x.)

EXAMPLE 8.8

Estimate the activity of ^{133}Xe when in equilibrium in a 1 GW_e reactor. The fraction of fissions that produce ^{133}Xe is 0.0677 and the half-life of ^{133}Xe is 5.27 d. How long will it take after the reactor is shut down for the activity to decay to a rate of $10^6\ s^{-1}$ ($\equiv$ 1 MBq; see Box 8.5)?

An electrical power of 1 GW corresponds to about 3 GW thermal power. Using eqn (8.22), the fission rate $F = 9.39 \times 10^{19}\ \mathrm{s}^{-1}$. From eqn (8.23) the equilibrium activity R_x is

$$R_x = 0.0677 \times 9.39 \times 10^{19} = 6.4 \times 10^{18} = 6.4 \times 10^6\ \mathrm{TBq}.$$

Assuming production ceases on shutdown, then the activity R_t after a time t is

$$R_t = R_x \exp(-t/\tau)$$

where $\tau = t_{1/2}/\ln 2$. So the time t is given by

$$t = \tau \ln(R_x/R_t) = (5.27/\ln 2)\ln(6.4 \times 10^{18}/10^6) = 224\ \text{days}.$$

Table 8.4 lists three of the main noble gas and iodine fission products at the end of a $1\mathrm{GW_e}$ (electrical power) PWR plant fuel cycle. The activities in Table 8.4 are very large and in the case of a core meltdown some 20% of the ^{135}I in the fuel rods could be released into the reactor containment building and subsequently leak out into the atmosphere. The radioactive gas can then be dispersed by wind. Close to the reactor building the direct dose from the radioactive nuclides within the reactor building is dominant while farther away the internal thyroid dose from iodine isotopes is generally largest. These considerations are clearly important in determining where to site a reactor and what radiation shielding is required.

8.4.7 Radiation shielding

Shielding is required around the core to reduce the radiation dose to safe levels. (See discussion of radiation in Box 8.5.) Generally it is only necessary to shield against γ-rays and neutrons since α- and β-particles have very short ranges in matter. γ-rays of a few MeV passing through a concrete wall interact via the photoelectric effect, Compton scattering, or pair-production (the threshold is $2m_ec^2 \equiv 1.02$ MeV). While the intensity of an incident monoenergetic beam of γ-rays drops exponentially like $e^{-\mu x}$, where μ is the linear attenuation length of concrete and x is the thickness of the wall, the emergent beam will contain γ-rays of lower energy. These arise primarily through Compton scattering where a γ-ray scatters off and transfers energy to an electron. (μ equals the total macroscopic absorption cross-section, $\Sigma_t = n\sigma_t$.) The radiation dose is therefore not reduced by as much as $e^{-\mu x}$. This effect is parametrized by using a build-up factor $B(\mu x)$, so that the dose is given by

Table 8.4 Data on equilibrium amount of three fission products from a 1 $\mathrm{GW_e}$ reactor

Nuclide	Half-life	Fission yield (b_x)	Activity (TBq)
^{88}K	2.79 h	0.0364	2.52×10^6
^{133}Xe	5.27 d	0.0677	6.29×10^6
^{135}I	6.7 h	0.0639	5.55×10^6

Box 8.4 Radiation

Radiation affects tissues as it causes ionization, which breaks molecules apart and gives rise to free radicals, which can damage cells. The scale of the effect depends on the energy deposited per unit mass of tissue, the dose D, and on the type of radiation. Charged particles, such as α-particles cause relatively more damage than γ-rays or electrons depositing the same energy, since their energy loss per unit length (the **linear energy transfer**, LET) is higher. Likewise, neutrons transfer their energy in matter to nuclei, so their LET is also high. This difference in LET is taken into account by using a weighting factor, w, to give the equivalent radiation dose, $H = wD$. In SI units H is measured in sieverts (Sv) and the dose D in grays (Gy), with $1\ \text{Gy} \equiv 1\ \text{J kg}^{-1}$. An older unit for H that is still used is the rem $\equiv 10^{-2}$ Sv. The weighting factors for different types of radiation are given in Table 8.5.

Table 8.5 Radiation weighting factors

Radiation	Weighting factor w
X-rays, γ-rays, electrons	1
Neutrons	20*
α-particles, fission fragments, heavy nuclei	20

*Maximum value.

The dose received at a distance r in air from a radioactive source is proportional to $1/r^2$ and to the decay rate or **activity** of the source. The unit of activity is the becquerel (Bq), which is one disintegration per second. An older unit still in use is the curie (Ci), where $1\ \text{Ci} \equiv 3.7 \times 10^{10}$ Bq.

At a distance of r metres from a source of A MBq activity emitting γ-rays of energy E MeV, the flux of γ-rays F is

$$F = A/4\pi r^2 (10^6 \text{m}^{-2}\ \text{s}^{-1}).$$

An approximate expression for the equivalent dose rate D_{rate} in μSv h^{-1} is

$$\begin{aligned} D_{\text{rate}}(\mu\text{Sv h}^{-1}) &\approx A(\text{MBq}) \times E(\text{MeV})/6r^2(\text{m}^2) \\ &\approx 2F(10^6 \text{m}^{-2}\text{s}^{-1}) \times E(\text{MeV}) \end{aligned} \tag{8.24}$$

when E is in the range ~0.1 to ~5 MeV.

The equivalent dose gives a measure of the biological effect when the whole body is irradiated uniformly. If only certain organs receive a dose, the effect on the person is less and this is taken account in the effective dose. An example is the effective dose following radioactive iodine inhalation. Iodine concentrates in the thyroid and if the thyroid receives an equivalent dose of 20 μSv then the effective dose is 1 μSv.

Environmental radiation

There is naturally occurring radiation in the environment that comes from radioactive minerals and from cosmic rays. Uranium and thorium and their decay products, in particular radon, and the potassium isotope ^{40}K contribute most. Typical values for the annual effective dose in the UK from various sources are shown in Table 8.6.

Table 8.6 Typical radiation doses in the UK

Source	Average annual effective dose (μSv)
Cosmic rays	260
Food	300
Environmental*	1650
Medical	370
Miscellaneous	0.4
Fall-out	5
Occupational	8

*Higher in granite areas.

The cosmic ray contribution increases with altitude and is about three times higher at 2000 m. In an airliner at 10 000 m it is ~150 times larger than at sea level. The amount of radon in the air is higher in granite areas such as Cornwall, where the total annual dose is about 7800 μSv.

For people working with radioactive materials, such as in the nuclear power industry, there is the possibility of increased radiation exposure. The principal long-term risk from exposure to significant amounts of radiation is an increased risk of cancer. The limits set on the annual radiation dose and on the amounts of any radioactive isotopes that might be ingested or inhaled are such that receiving these amounts would not cause the worker a significant risk in comparison with other occupational risks. The current whole body annual dose limit for radiation workers in the UK is 20 mSv.

an effective flux F_{eff} of γ-rays of the incident energy given by

$$F_{\text{eff}} = B(\mu x)e^{-\mu x}. \tag{8.25}$$

For example, for a beam of 2 MeV γ-rays passing through 2 m of water, for which $\mu x \sim 10$, $B(\mu x)$ is ~10, so the build-up is significant.

EXAMPLE 8.9

A beam of 2 MeV γ-rays of intensity 10^8 m^{-2} s^{-1} is incident on 0.05 m thick lead shield. Calculate the attenuated flux and the effective flux behind the shield. Using eqn 8.24, estimate the dose rate behind the lead shield. The value of μ(2 MeV) for lead is 51.8 m^{-1}. The build-up factors $B(\mu x)$ of lead for 2 MeV γ-rays are $B(2) = 1.76$ and $B(4) = 2.41$.

The value of $\mu x = 51.8 \times 0.05 = 2.59$. The attenuated flux F_a is therefore

$$F_a = F\exp(-\mu x) = 10^8 \exp(-2.59) = 7.5 \times 10^6 \text{ m}^{-2} \text{ s}^{-1}.$$

Linearly interpolating between the values of $B(\mu x)$ for $\mu x = 2$ and 4 gives $B(2.59) = 1.95$. So the effective flux F_{eff} from eqn (8.25) is

$$F_{eff} = B(2.59)F_a = 1.46 \times 10^7 \text{ m}^{-2} \text{ s}^{-1}.$$

Substituting this effective flux into eqn (8.24) gives

$$D_{rate} \approx 2 \times F_{eff}(10^6 \text{m}^{-2}\text{s}^{-1}) \times E(\text{MeV}) = 2 \times 1.46 \times 10^1 \times 2 \approx 58 \ \mu\text{Sv h}^{-1}.$$

For neutrons, absorption cross-sections are much higher at low energies ($E_n \lesssim 1$ keV) than at high energies ($E_n \gtrsim 100$ keV), so good moderation is required of the fast fission neutrons. Water is a very effective neutron shield and boron can be added to the water to improve the absorption of low energy neutrons. Concrete is also effective due to its high hydrogen atom density (1/4 that of water). It can be cast into the required shape and is strong so it provides most of the shielding in a reactor.

8.5 Fast reactors

As mentioned above, a chain reaction can be maintained predominantly by fast neutrons ($E_n > 100$ keV) if the fuel is enriched. Making a fast reactor is, however, technically more challenging than making a thermal reactor, since it has a much higher energy density in the reactor core, and there are relatively few fast reactors in operation. However, when the supply of uranium eventually becomes more limited, fast reactors are likely to become particularly attractive for conserving uranium stocks since the conversion of fissile material is generally much higher than in a conventional fission reactor. For conversion, the core is surrounded by fertile material and the emitted neutrons convert it to fissile material. For example, when fuelled by ^{235}U some of the fission neutrons are absorbed by ^{238}U and produce ^{239}Pu, which is fissile, via the reactions:

$$n + {}^{238}\text{U} \rightarrow {}^{239}\text{U} \rightarrow {}^{239}\text{Np} \rightarrow {}^{239}\text{Pu} \tag{8.26}$$

where the last two reactions occur via β^- decay.

Conversion can be sufficient to yield more fissile material than used in the core. When this happens the reactor is called a **breeder reactor** and the utilization of uranium can be increased from ~2% to ~50%, an increase of a factor of ~25 in the size of the nuclear energy reserve. The details of breeding are described in Box 8.6. The predicted fossil and uranium (for thermal reactors) fuel reserves are $\sim 8 \times 10^{22}$ J and $\sim 4 \times 10^{21}$ J, respectively. Breeder reactors could therefore provide an enormous amount of energy, without any significant greenhouse gas emission.

In a fast reactor, to ensure that fission is maintained predominantly by fast neutrons, no moderating material (or very little) is used and the fuel is enriched. As a result the core is very compact. For a reasonable power output the energy density in the core must be high and this requires a coolant with excellent heat transfer properties. Also, the coolant must not moderate the neutrons, thereby favouring coolants not containing low atomic number nuclei. Sodium has been used in many designs. It has a high boiling point of 882°C at 1 atmosphere so the thermodynamic efficiency is high without requiring high pressures and a large pressure vessel. However, sodium is chemically very reactive and is also activated through neutron absorption producing ^{24}Na, which has a 15 h half-life, and so the coolant loop must be heavily shielded. As a result the design of a reliable economic sodium-cooled fast reactor has proved difficult.

Figure 8.11 shows a schematic design of a pool-type sodium-cooled fast breeder reactor. The core consists of a central region with fuel rods surrounded by rods containing fertile material. Since the sodium in the primary coolant loop becomes radioactive there is a secondary sodium loop that is used to generate the steam for the turbine. Although the reactor can be located partially underground to reduce the necessity for heavy shielding, access to all of the primary coolant components for repairs and maintenance is difficult because they are immersed in the sodium pool.

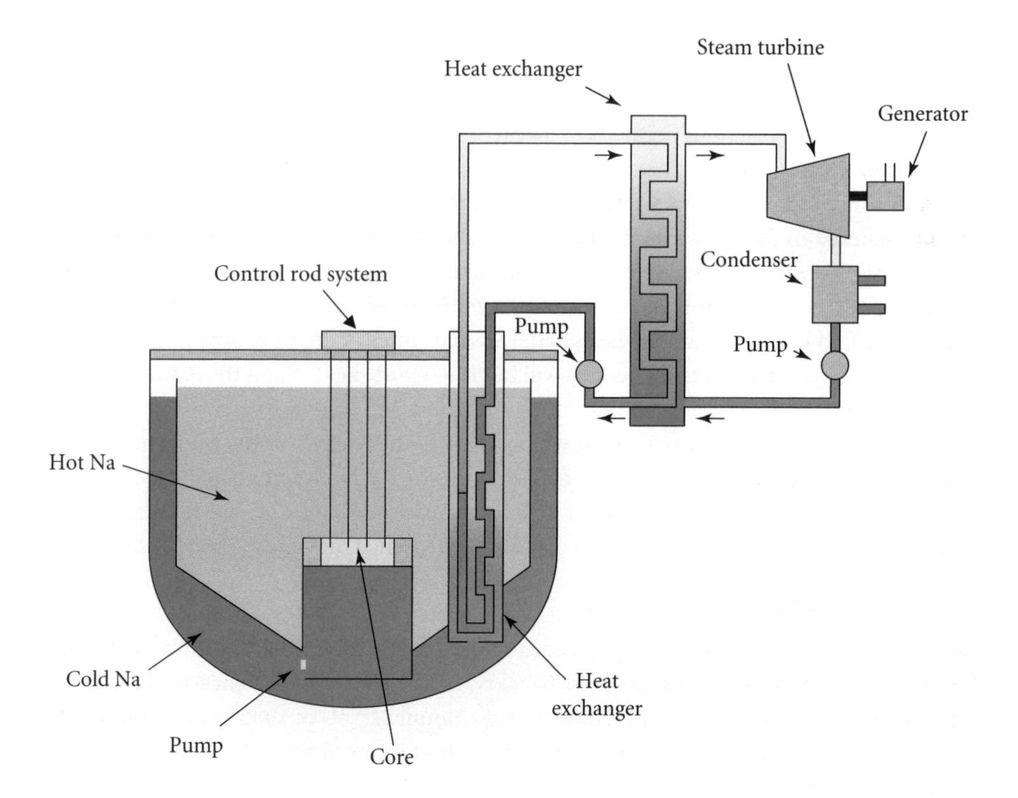

Fig. 8.11 A pool-type sodium cooled fast reactor.

Box 8.5 Breeders

In a reactor core η neutrons are emitted on average for every neutron absorbed by the fissile fuel. When fuelled by ^{235}U some of the fission neutrons are absorbed by ^{238}U and produce ^{239}Pu via the reactions (8.26). The ^{239}Pu can be used as fuel and, in a thermal reactor fuelled by ^{235}U, this can increase the percentage of uranium used to produce energy to ~2%. This production of fissile material from non-fissile nuclei is called **conversion**, and is quantified by the **conversion** or **breeding ratio** B, which is the average number of fissile nuclei produced per fissile nucleus consumed.

The breeding ratio B is related to η by $B < (\eta - 1)$, as one neutron must be absorbed by a fissile nucleus to maintain the chain reaction, i.e. keep the core critical.When $B > 1$ more fissile material is produced from fertile material than is consumed and the reactor is then called a **breeder reactor**. This increases the percentage of uranium used to produce power up to an economic limit of around 50%, representing a considerable increase in the potential amount of energy available compared with conventional fission nuclear power.

For breeding to be possible, $\eta > 2$. In practice, however, due to neutron leakage and neutron absorption by elements in the core other than the fuel, η must be greater than 2.2. For the fissile elements ^{233}U, ^{235}U, and ^{239}Pu, only ^{233}U could be used to breed with thermal neutrons, but, for fast neutrons (> 100 keV), all three fuels could be used. ^{233}U can be bred from ^{232}Th. The process is

$$n + {}^{232}\mathrm{Th} \rightarrow {}^{233}\mathrm{Th} \rightarrow {}^{233}\mathrm{Pa} \rightarrow {}^{233}\mathrm{U}$$

where the last two reactions occur via β^- decay. Thorium is about three times as abundant as uranium.

A measure of how good a reactor is at breeding is the time it takes for the amount of fissile material in the reactor to double; this is called the **doubling time** t_D. The breeding gain is $(B - 1)$, as one fissile nucleus is consumed for every B fissile nuclei produced. Consider a reactor operating at a constant power P_o (GW) with an initial mass of fissile material M_i. If m kilograms of fissile material are consumed per day per gigawatt of output then the gain in fissile material per day will be $(B - 1)mP_o$ kg d^{-1} and the doubling time t_D equals $M_i/[(B - 1)mP_o]$.

This estimate assumes that all the bred fuel is left in the reactor for the whole time. The reactor only needs an amount of fissile fuel M_i to produce P_o for a certain neutron flux and the excess could be removed and used in another reactor. As reactors are refuelled once or twice a year this mode of operation could be approached, in which case t_D becomes

$$t_D = M_i \ln 2 / [(B - 1)mP_o] \qquad (8.27)$$

(see Exercise 8.18). For example, consider a breeder reactor containing ^{235}U and ^{238}U which produces 1 GW, equivalent to 0.3 GW$_e$ if the thermal efficiency is 30%. This consumes about 1 kg/day of ^{235}U. For an initial amount of ^{235}U of 1000 kg and a breeding ratio of 1.1, the doubling time would be 27 y without removal of the bred fuel, and 19 y with removal.

8.6 Present-day nuclear reactors

Many reactor designs have been produced around the world, with different fuel, moderator, and coolant. The early prototype reactors, called generation I, were made during the 1950s and 1960s. Most of the reactors operating today are generation II reactors and were built during the 1970s and 1980s. The reactors operating in 2003 were PWRs (263), boiling water reactors BWRs (92), Russian graphite-moderated and water-cooled RBMK reactors (17), CANDU heavy water reactors (38), British Magnox and AGRs (26), and fast reactors (3). Many of the US light-water reactors were one-off designs, which led to duplication of effort and greater costs.

This prompted the development in the early 1990s of standardized advanced light-water reactor designs (generation III), which are simpler, requiring fewer components and less piping, are easier to build, and include additional passive emergency cooling systems. Passive cooling depends on gravity or temperature differences, rather than pumps, and as a result it is expected to be much more reliable and closer to a fail-safe system. A few of these generation III reactors have now been built. The advantage of standardization in design is seen in France where, over nearly two decades, 34 0.9-GW_e and 20 1.3-GW_e nuclear plants were built. They supply ~75% of France's electricity.

By the end of the 1990s it was recognized that there was a need for an international effort to decide upon improved designs that addressed public concerns over safety and were more economic. In 2003, the Generation IV International Forum, representing 10 countries, announced the selection of six reactor technologies. These were chosen for being safe, clean, cost-effective, fuel-efficient, secure, and resistant to proliferation. Some of these systems use a closed fuel cycle, in which reprocessing is carried out on-site and actinides (see Section 8.9) are recycled to minimize the amount of high-level waste that needs to be stored.

The very high temperature helium-cooled graphite moderated reactor (HTGR), shown in Fig. 8.12, uses a once-through uranium fuel cycle and operates at a temperature of 1000°C. This is considerably higher than in water-cooled reactors, which are limited by pressure to a maximum of ~300°C, and also higher than in liquid metal cooled reactors, which are limited by corrosion to ~600°C. A direct Brayton cycle gas turbine (see Chapter 2, Section 2.9) could be used to give high thermal efficiency as shown in Fig. 8.12, or the reactor could be used for thermochemical hydrogen production. The high operating temperature allows the graphite to constantly anneal the damage caused by the fast neutrons and thereby avoid any build-up of stored energy. There is also a strong negative temperature coefficient of reactivity and the chain reaction will stop if there is a loss of the helium coolant gas (LOCA).

For a moderate power HTGR of 150 MW_e the core temperature would rise following a LOCA to 1600°C—the rise eventually being limited by radiation cooling. The design of the fuel would prevent any significant loss of fission products at this temperature. The fuel is in the form of small spheres of mixed fissile and fertile material contained in thin carbon and silicon carbide shells. As a result the reactor is called a Pebble Bed reactor. The relatively small power output makes the inventory of fission products low and it would be further reduced by continuous refuelling. The uranium fuel would be quite highly enriched at ~8% and would not need to be reprocessed. The main safety concern is to avoid any air contact with the hot

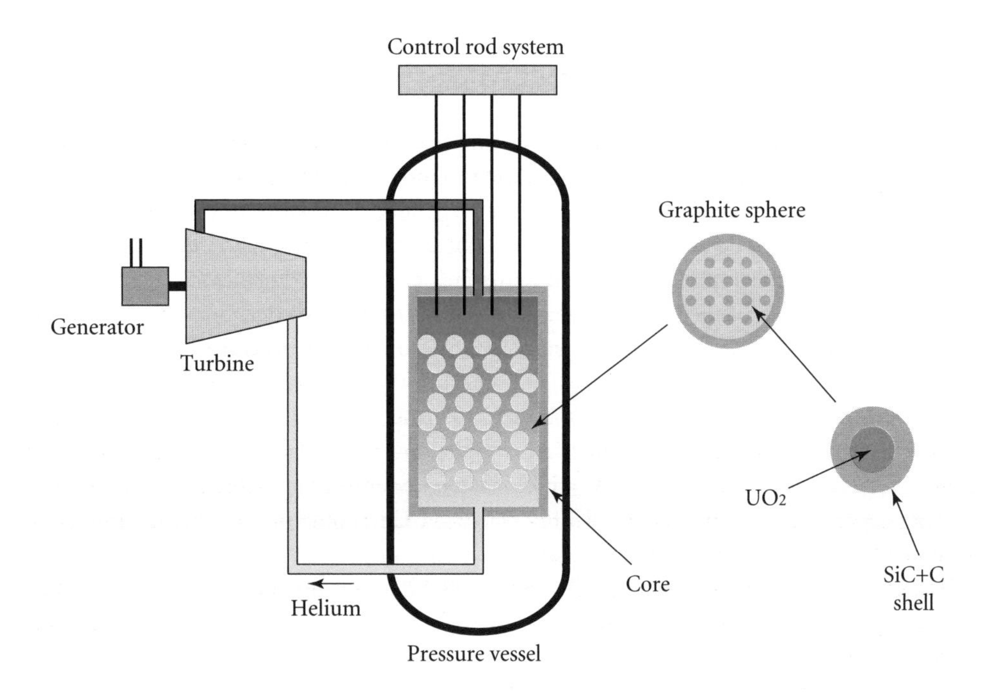

Fig. 8.12 Very high temperature gas cooled reactor.

graphite, which could cause a fire. This form of fuel gives good fission product containment. Surrounding the core with an outer containment building would provide additional safety, but would make the plant more expensive.

Another fast reactor design uses the natural convection of liquid Pb or liquid Pb–Bi to cool the core and is based on 40 year's experience of Pb–Bi cooling in Russian submarine reactors. There is also a sodium-cooled fast reactor design similar to the design discussed in Section 8.5.

An unusual technology that has been chosen as a generation IV design is the molten salt reactor. The uranium fuel is dissolved in a sodium fluoride salt coolant which circulates through a graphite moderator. Fission products are removed continuously and the actinides are fully recycled, which reduces the high-level waste. The operating temperature is high enough for hydrogen production and there is passive cooling of the core, which improves safety. The other two systems selected are a supercritical water-cooled reactor and a very high temperature gas reactor.

Although mostly large reactors have been built up to now, some of the new designs are smaller. The latter better match the needs of less-developed countries. Smaller units also reduce the magnitude of the radioactivity inventory in a reactor. The generation IV reactor designs are planned for operation in 2020–30. Nuclear power is one way to provide for carbon dioxide free energy and for greater national energy security by reducing the dependence on fossil fuel. The concern about the effect of fossil fuel emissions on global warming has renewed interest in nuclear power. Electricity generation using nuclear power has only increased slowly over the last two decades. The slow growth is due to both economic considerations and to public concern over safety.

8.7 Safety of nuclear power

There are over 400 nuclear plants operating worldwide, most of them being LWRs, operating at ~90% of maximum annual output. Operating experience has led to improved capacity as well as improved safety. However, there is considerable public concern over the use of nuclear power, due in part to three serious accidents in commercial plants since the appearance of nuclear power in the 1950s. In 1952 there was a fire at a gas-cooled graphite-moderated reactor at Windscale in the UK. This was caused by the release of stored energy, associated with the displacement of carbon atoms in the graphite by neutron irradiation (known as the Wigner effect) with a limited release of radiation into the atmosphere. In 1979 there was a loss of cooling accident (LOCA) in a PWR at Three Mile Island in Pennsylvania in the USA, caused by both mechanical and human failures, resulting in a 20% core meltdown but only a small release of radioactivity. In 1986 there was an uncontrolled reactor power increase in a water-cooled graphite-moderated RMBK reactor at Chernobyl in the Ukraine, causing a steam explosion and a huge release of radioactivity.

The outer containment building at the Three Mile Island PWR contained nearly the entire radioactivity released from the partial melt-down of the core. However, the costs of the cleanup were enormous, almost $1 billion dollars and the loss of public confidence in nuclear power, particularly in the USA, was considerable. The Chernobyl disaster was much worse since a significant fraction of the core inventory of fission products was released. The accident at Chernobyl was largely due to the unusual reactor conditions created for a safety test, coupled with procedures not being followed and an inherent weakness in the design of the RBMK reactor, insofar as certain parts of the reactor had a positive feedback effect on the reactivity of the reactor when the amount of steam in the coolant water increased. (More steam reduced the average density and hence the neutron absorption of the water, which caused the reactivity to increase.)

About 50 people died as a direct result of the accident, most from high radiation doses (Chernobyl Forum (2006)). The radioactivity was carried great distances from the reactor. It is possible amongst the 600 000 people who received the most radiation exposure that there might be ⅔ of a per cent increase in cancer mortality. There has been an increase in thyroid cancer (most of whom were curable) amongst children in the exposed areas of Ukraine, Belarus, and Russia. However, apart from this increase in thyroid cancer, there has been no clear evidence from epidemiological studies for a radiation-induced increase in cancer deaths or mortality. Health, though, was significantly affected by the anxiety and the massive relocation caused by the accident. While the long-term effects of the Chernobyl disaster may not be as bad as had been feared, the effect on public confidence in nuclear power has been very significant.

Following the Chernobyl accident there has been increased international cooperation on safety and reactor design issues to reduce the chance of a similar accident. In the USA there have been almost 3000 reactor years of operation during which time there has been only one serious accident (at Three Mile Island). Based on probabilistic risk assessments a similar frequency estimate of one core damage incident every 10 000 reactor years has been made.

New reactor designs with both passive and active safety features will reduce this probability. There have also been improvements in operator training and in instilling the importance of safety at work. Obtaining a risk of 1 in 100 000 reactor years is thought to be possible and should help to improve public confidence in nuclear power.

There is another aspect to reactor safety and that is the ability of a reactor to withstand a terrorist attack. This has received particular attention following the 9/11 attacks. Fortunately, the strength of the construction in nuclear power plants would provide considerable protection from the effects of the crash of a hijacked aircraft or of a truck bomb. The spent fuel, which is initially stored on-site, is in reinforced concrete pools and so is also protected. However, the security of nuclear power plants has been increased as a result of 9/11.

8.8 Economics of nuclear power

Nuclear reactor plants require large capital costs and those operating today were originally state-owned or run by regulated utility monopolies. Under these arrangements the financial risks associated with capital expenditure could be more easily absorbed within a large state-protected framework (or borne by the consumers rather than by the suppliers, with a consequent reduction in the cost of capital), and projects with a long-term payback could be funded as part of the national infrastructure for long-term economic security. A competitive electricity supply market, however, tends to favour less capitally intensive energy sources and ones with shorter construction times. In the 1980s and 1990s in the USA, construction costs were higher than expected for nuclear plants and there were regulatory and political difficulties in obtaining site approval. During this period fossil fuel costs also decreased. All of these factors made nuclear power uncompetitive in the deregulated economic power market in the USA at that time.

Nuclear power, with its relatively low operating costs, can compete with alternative sources for 'base-load' (high load factor) power. A group at MIT has recently looked at the economics and role of nuclear power in the USA. Table 8.7 shows their estimated relative costs for electricity produced by nuclear, coal, and gas power plants. Since gas and coal prices are typically higher in Europe, and even more so in Japan and Korea, a similar comparison for these countries would favour nuclear more. However, nuclear power would only be competitive with fossil fuel power plants in the USA if the construction costs, construction time, and cost of borrowing were all reduced.

The MIT study considered the effect of placing a tax on carbon emissions to reflect the cost of achieving a reduction in global CO_2 emission. The lowest tax would pay for a reduction of US emissions by 10^9 tonnes per year. The higher taxes cover the range of estimated costs of carbon sequestration (capture, transport, and storage), but these are currently very uncertain. When such taxes are imposed nuclear energy becomes much more competitive. For the higher rates, however, the MIT study points out that it may prove more economic to generate electricity using a combined cycle gas turbine (CCGT) plant burning gas obtained from the gasification of coal. Relatively cheap coal can be used, the CCGT plant has good thermal efficiency, and it may be cheaper to capture CO_2.

Table 8.7 Relative costs of electricity in the USA (2003)

	US cents/kW$_e$hour
(25-year capital recovery, 85% lifetime capacity factor)	
Nuclear	7.0
Coal	4.4
Gas	4.1
Nuclear costs with	
construction costs reduced by 25%	5.8
and construction time reduced by 12 months	5.6
and cost of capital ≡ coal and gas	4.7

	Cost (US cents/kW$_e$hour) with carbon tax (per tonne of carbon) of		
	$50	$100	$200
Coal	5.6	6.8	9.2
Gas	4.6	5.1	6.2

8.9 Environmental impact of nuclear power

The principal environmental advantage of nuclear power is the very small amount of associated CO_2 emission, which makes it a very good energy source in the light of global warming. The main environmental considerations are over the siting of nuclear reactors, in particular, the seismology, meteorology, geology, risk of flooding, and population distribution in the area of the reactor. Meteorology is important since, in the event of an accident, the effect of the prevailing weather conditions on the dispersal of radioactivity has to be assessed. In addition, the land requirements, the effect of thermal discharges to the environment, and in particular the storage and the disposal of waste have to be considered.

The waste generated in a nuclear reactor is classified into **high-**, **intermediate-** and **low-level waste**. High-level waste is the main concern and comprises ~25 tonnes of spent fuel per year for a 1 GW$_e$ PWR, which represents ~95% of the activity produced. It contains both transuranic and fission products: the transuranic waste contains long-lived isotopes of plutonium, americium, neptunium, and curium, which are all actinide elements, while after a few decades the fission products are mainly ^{90}Sr (and its daughter ^{90}Y) and ^{137}Cs, though ^{99}Te and ^{129}I are also quite abundant and potentially mobile.

The spent fuel is first stored for several years on-site, to allow the intense short-lived activity to decay. On-site it is kept in storage pools to remove the heat and provide shielding of the radiation. The spent fuel can then be stored or reprocessed to recover the uranium and plutonium and the remainder immobilized by vitrification, in which the waste is incorporated into borosilicate glass. Another method is to incorporate the waste in natural stable mineral lattices. The spent fuel or vitrified waste can then be placed in a corrosion-resistant can and stored in an underground repository.

Caverns in dry stable rock formations or in salt deposits have been proposed for these repositories. Whether this high-level waste should be retrievable is a subject of debate. It would enable the waste to be eventually disposed of as new technologies emerge in the future, such as accelerator-induced transmutation of the long-lived nuclides to shorter-lived ones. Sealing waste securely would protect it from terrorists, but the main concern is whether the waste would remain well-contained for thousands of years. In some sites there would be a risk in the long-term that the waste could leak into the ground water and contaminate drinking water. There is therefore the need to identify geologically stable sites. Deep bore holes have also been suggested as possible repositories but the same issue arises. The amount of waste generated during the early years of nuclear reactors, called legacy waste, is far greater than that generated by the current generation of reactors and also needs to be stored safely.

In the Oklo uranium mine in West Africa, a natural uranium reactor has been discovered. Two thousand million years ago the enrichment of the uranium was about 3%, due to the difference in the half-lives of ^{235}U (7×10^8y) and ^{238}U (5×10^9y). The presence of water in the ore provided the necessary moderation for the ore to become critical. The fission and transuranic products have remained within the ore, providing an example of long-term secure storage. However, the unavoidable question remains as to how certain we can be about the security of any proposed site.

8.10 Public opinion on nuclear power

In America public opinion is divided over expanding nuclear power, with recent surveys varying from 60% opposed to 55% in favour of expansion. A survey in 2003 concluded that public opposition in the USA was primarily focused on safety, waste disposal, and poor economics, with a large majority strongly opposed to a plant within 25 miles of their home and a majority feeling that nuclear waste could not be stored safely for many years. People expressed concern over global warming, but this did not translate into favouring nuclear over fossil fuels. It was the relative cost of alternative energy sources that was the major factor in influencing people as to which source (nuclear, coal, oil, solar, or wind) should be developed, with information about global warming having little influence.

In Europe a poll in 1997 showed that only 16% were in favour of nuclear power, while 40% were against because they felt it posed unacceptable risks. However, in 2001 over 50% agreed with keeping the nuclear option open if all the waste were safely managed, with only 25% disagreeing. There was, though, a lack of trust amongst EU citizens that the nuclear industry is open or provides accurate or sufficient information about radioactive waste. A survey in 2002 showed that almost 90% thought that global warming and climate change were serious problems, with about 75% thinking that fossil fuels contributed significantly to climate change. Worryingly, however, almost 50% thought that nuclear power was a significant contributor! As in the USA, the environment and cost were seen as important issues, with protection of the environment and the need to keep prices low the top priorities for energy policy.

In Japan, where currently 25% of the electricity is from nuclear energy, opinion polls conducted in the 1990s showed that the public was divided on the issue of nuclear safety, with about the same percentage saying it was safe as unsafe. Recently, a serious criticality accident in 1999 at a fuel fabrication plant in Tokaimura undermined public confidence in the government's handling of nuclear energy, which may impact on the Japanese government's plans for the expansion of nuclear power.

8.11 Outlook for nuclear power

The production of electricity from nuclear power has been increasing steadily but its share of the total world production has been decreasing slightly since ~1990 as is seen in Fig. 8.13. Table 8.8 gives some statistics on the actual and proposed number of reactors in countries in 2004 that are already generating a significant fraction of their electricity from nuclear power or propose to increase their fraction. As can be seen, there were no plans in 2004 to build any new reactors in the USA or in Europe—the proposed expansion is in China, India, Japan, Korea, and Russia. Germany in 2000 decided to phase out their nuclear reactors by ~2025 and to promote the development of renewable energy sources. In 2005 28% of Germany's electricity was provided by nuclear power. This difference in national policies reflects economic and environmental considerations and public attitudes, as discussed above, and also strategic and political considerations.

The facilities used to enrich uranium for use in nuclear reactors could be used to provide highly enriched weapons grade ^{235}U. The development of nuclear power in a country therefore has considerable political and security implications. The enrichment facilities found in Iraq in the 1990s had major international repercussions. The possibility of nuclear proliferation as a result of the expansion of nuclear power is a very important factor in determining its future.

The principal strategic considerations in favour of nuclear power are that it is a very compact fuel source with a high energy density that can be easily stockpiled. As there are diversified sources of supply, it is unlikely that the supply of fuel would be interrupted, making it a very secure energy resource. Within Europe it is currently the only major source of electricity that does not contribute to global warming.

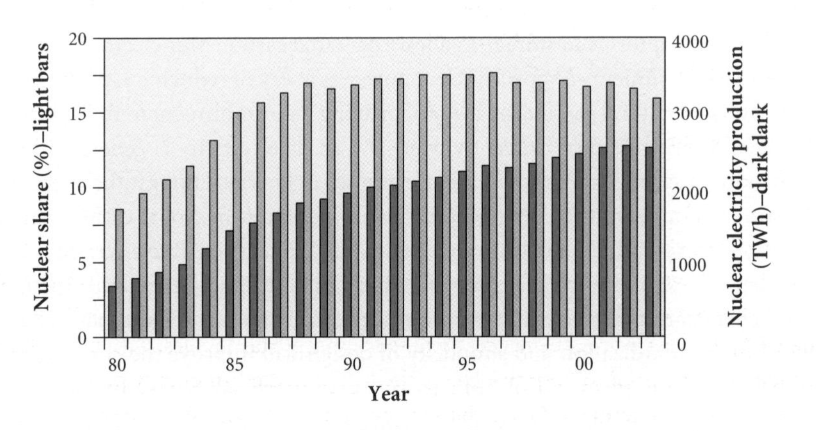

Fig. 8.13 Nuclear electricity production and share of total electricity production.

Table 8.8 Number of reactors and percentage of electricity generated in countries that use or propose to use nuclear power significantly

Country	Nuclear electricity generation (2005)		Number of reactors		
	Billion kWh	% Total electricity	Operating	Being built	Planned/ proposed
Belgium	45.3	56	7	0	–
Canada	86.8	15	18	0	2
China	50.3	2.0	10	5	24
France	430.9	79	59	0	2
Germany	154.6	31	17	0	0
India	15.7	2.8	15	8	24
Japan	280.7	29	55	1	12
Korea RO (South)	139.3	45	20	0	8
Russia	137.3	16	31	4	9
Sweden	69.5	45	10	0	0
United Kingdom	75.2	20	23	0	0
USA	780.5	19	103	1	13
WORLD	2626	16	441	27	153

In the UK, a government policy document (White Paper) published in 2003 identified nuclear power as an important source of carbon-free electricity. The government felt that current economics made it an unattractive option for new carbon-free generating plant, and that there were important aspects of nuclear waste still to be resolved.

However, it did not rule out the possibility of building new nuclear reactors in order to meet the national carbon emission targets, but felt that this would need full public consultation. The immediate plan to reduce carbon emissions was to increase renewable energy supplies, in particular wind power, and to improve energy efficiency. It also noted the importance of developing methods of carbon capture and storage to allow coal-fired carbon-free electricity generation.

The Japanese government supports nuclear power as a way of reducing Japan's dependence on imported oil and natural gas, and as a way of meeting its commitment to the Kyoto protocol on reducing carbon emissions. Currently over 25% of its electricity is generated by nuclear reactors, making it second to France in the use of nuclear energy amongst the G-8 nations.

The desire for both a secure energy supply and a reduction in global carbon emissions is causing nations to re-evaluate nuclear power as a source of carbon-free electricity. Currently there are plans to expand nuclear power in India, China, Korea, Japan, and Russia but the economic benefits are unattractive in the short term, except when oil and natural gas prices are high. More standardization and simplicity of design will improve the competitiveness of nuclear power and the inclusion of both passive and active safety systems will reduce the likelihood of accidents. Public involvement in considering the option of nuclear power will be important in regaining public confidence.

It will be more critical for any future reactor design to be very safe and economic than for it to be efficient. Public trust is essential and safe operation and waste disposal are paramount in any new reactor building programme. Reprocessing may not be required as fuel costs are not a major factor in reactor economics. The development of fast breeder reactors is less advanced than that of thermal reactors and the reserves of uranium are such that breeders are not essential in the short term—the UK stopped research on liquid-metal cooled fast reactors in 1992.

Internationally approved design and inspection would help build public confidence. France is considering an advanced PWR for its next generation of reactors, which is a design based on existing PWRs that have had many reactor years of experience. There are advantages in the new proposed HTGR design, which is efficient, relatively small, and has good passive safety features but there is much less operating experience with this type of reactor than with PWRs.

In parts of the world the political and security issues arising from possible nuclear proliferation and terrorist attacks will be a significant additional concern over the development of nuclear power. In conclusion, the question of how large a part nuclear power will play in providing carbon-free electricity and national energy security over the next 100 years is far from clear.

SUMMARY

- Nuclear power is a very compact source of essentially carbon-free electricity with 1 tonne of uranium equivalent in energy to 20 000 tonnes of coal.
- Uranium reserves are ~14 Mt, equivalent to ~200 years at the current global production of 350 GW_e.
- In PWRs, water is both the coolant and the moderator. The neutrons are moderated to thermal energies to allow the chain reaction to proceed with fuel only a few per cent enriched in ^{235}U.
- Beta-delayed neutrons allow mechanical control of the chain reaction by the insertion and withdrawal of control rods.
- The multiplication factor in the chain reaction is determined by $k_\infty = \varepsilon p f \eta$, and by the size of the reactor core.
- The power output in terms of the mass of ^{235}U and the flux φ in the core is

  $$P = 4.75 M_{235}(\text{tonne})\, \varphi\, (10^{18}\ \text{m}^{-2}\ \text{s}^{-1})\ \text{GW}.$$

- Fast reactors can utilize about 50% of the uranium fuel, rather than essentially only the ^{235}U component in a thermal reactor, but are technically difficult and currently uneconomic.
- Safety of new nuclear reactors with passive safety features is predicted to be good, but public confidence in nuclear power has been affected by the Chernobyl accident.
- Nuclear power is more expensive than from fossil fuels, but could be competitive if there were a carbon-tax on CO_2 emissions.
- Disposal of nuclear waste is a concern and a long term solution has not been decided upon.

- Nuclear power provides good energy security and very low carbon emissions, particularly important in combating global warming. But concern over waste disposal, safety, and nuclear proliferation make its future role uncertain.

FURTHER READING

Lamarsh, J. R. and Baratta, A. J. (2001). *Introduction to nuclear engineering*. Prentice-Hall, Englewood Cliffs, NJ. Good detailed description of nuclear reactors.

Lilley, J. (2001). *Nuclear physics—principles and applications*. Wiley, New York. Clear discussion of the physics of nuclear reactors.

WEB LINKS

web.mit.edu/nuclearpower/ MIT report on the Future of Nuclear Power, 2003.

www.uic.com.au Information on nuclear energy for electricity.

www.world-nuclear.org World Nuclear Association.

LIST OF MAIN SYMBOLS

$b(A)$	binding energy per nucleon
σ_x	cross-section for reaction x
φ	neutron flux
Σ_x	macroscopic cross-section for reaction x
λ	mean free path
ν	average number of neutrons emitted per fission
k	multiplication constant
k_∞	multiplication constant for no neutron loss
ε	fast fission factor
p	resonance escape probability
f	thermal utilization factor
η	number of neutrons produced per thermal neutron absorbed
ξ	logarithmic decrement
τ	mean lifetime
Γ	width of excited state
τ_g	generation time
μ	linear attenuation coefficient
B	breeding ratio
GW_e	gigawatt electrical power

EXERCISES

8.1 Calculate the energy released in the neutron-induced fission of ^{235}U into two nuclei with mass numbers 114 and 118 together with the emission of three neutrons.

8.2 How many tonnes of uranium fuel enriched to 3% in ^{235}U is equivalent to 100 000 barrels of oil?

8.3 Estimate the amount of natural uranium required annually to provide 10% of the primary global power consumption of 14 TW.

8.4 The global production of electricity by nuclear power in 2001 was about 2500 TWh. Estimate the annual consumption of uranium.

8.5* The isotope ^{157}Gd has a thermal neutron capture cross-section of 255 000 barns. Compare this value with (a) the cross-sectional area of the Gd nucleus and (b) the magnitude of $\pi\lambda^2$, where λ is the de Broglie wavelength of the thermal neutron.
(A thermal neutron has a kinetic energy of 0.025 eV. $\lambda = h/p$, where p is the momentum of the neutron. The radius R of a nucleus with A nucleons $\approx 1.2 \times A^{1/3}$ fm.)

8.6* Derive the expression for the ratio r of the initial to the final neutron energy for a single scatter through an angle θ in the centre of mass frame:

$$r = E_1/E_0 = 1 - \alpha \sin^2(\theta/2)$$

where $\alpha = 4mM/(m+M)^2$.

8.7* Show that the relation for the logarithmic decrement $\xi \equiv \ln(E_1/E_0)$ is approximately equivalent to $\Delta E/E = -\xi$, where $\Delta E = E_0 - E_1$. The time Δt between scatterings is given by $\Delta t = 1/(u\Sigma_s)$, where Σ_s is the elastic scattering macroscopic cross-section and u is the speed of the neutron. Show that the rate of loss of energy is approximately given by

$$\mathrm{d}E/\mathrm{d}t = -\xi u \Sigma_s E.$$

Assuming that Σ_s is independent of energy, show that the time t_t to thermalize a neutron of a few MeV is given by

$$t_t \approx 2t/\xi,$$

where t is the time between collisions at thermal energies given by $t = 1/(u_t\Sigma_s)$, with u_t the mean speed of thermal neutrons.

8.8* Neutrons scatter down from an energy interval A: $E \rightarrow E + \Delta E$ to B: $E - \Delta E \rightarrow E$, where $\Delta E = -\xi E$ is the average energy lost in scattering. The flux in A is φ and in B is $\varphi' = \varphi - (\partial\varphi/\partial E)\,\Delta E$. By considering the rate of scattering and of absorption, show that

$$\Sigma_s\varphi = \Sigma_a\varphi' + \Sigma_s\varphi'.$$

Hence show that

$$\varphi_l = \varphi_h \exp\left(-\int_{E_l}^{E_h} \frac{\Sigma_a\,\mathrm{d}E}{\xi E(\Sigma_s + \Sigma_a)}\right),$$

where φ_l is the flux at E_l, and φ_h is the flux at E_h. The exponential term is the probability p that neutrons escape capture.

8.9* The diffusion equation for a single group of neutrons is

$$\nabla\cdot\mathbf{j} = -\partial n/\partial t$$

where $\mathbf{j} = -(\lambda/3)\nabla\varphi$. The mean free path $\lambda = 1/\Sigma_s$. The production rate of neutrons per unit volume is the rate from induced fission minus the rate from absorption. So

$$\partial n/\partial t = (\nu\Sigma_f - \Sigma_a)\varphi.$$

For a spherical core the flux φ satisfies

$$(1/r^2)(\mathrm{d}/\mathrm{d}r)(r^2\mathrm{d}\varphi/\mathrm{d}r) = -3\Sigma_s(\nu\Sigma_f - \Sigma_a)\varphi \equiv -B^2\varphi.$$

Show that a solution to this equation is

$$\varphi = (A/r)\sin Br$$

where A is a constant. For a spherical core of radius a, the flux φ is zero to a good approximation at $r = a$. In this single group approximation the neutron multiplication factor $k_\infty = \nu\Sigma_f/\Sigma_a$. Show that the smallest value of a that gives $\varphi = 0$ at $r = a$ is

$$a_c = \frac{\pi\delta}{\sqrt{3(k_\infty - 1)}}$$

where $\delta = 1/\sqrt{\Sigma_a\Sigma_s}$.

8.10 Consider a spherical reactor core containing a homogeneous mixture of graphite and uranium enriched to 2.0% in ^{235}U and with a ratio of graphite to uranium n_s/n_f of 600. (a) Find p, f, η, and k_∞ assuming $\varepsilon = 1$. Use eqn (8.12) to calculate the critical radius a_c.

8.11* After the shutdown of a nuclear reactor, show that the number of ^{135}Xe nuclei n_p and ^{135}I nuclei n_I satisfy

$$dn_p/dt = \lambda_I n_I - \lambda_p n_p; \qquad dn_I/dt = -\lambda_I n_I.$$

Show that the number of ^{135}Xe nuclei grows to a maximum $\sim$11 hours after shutdown. Neglect the equilibrium number of ^{135}Xe compared with the equilibrium number N_o of ^{135}I present in the core prior to shutdown.
If the maximum amount of ^{135}Xe that can be compensated for by removing control rods is $0.2N_o$, how long is the reactor out of action after the shutdown? (For ^{135}I, $t_{1/2} = 6.7$ h; for ^{135}Xe, $t_{1/2} = 9.2$ h.)

8.12* A PWR plant releases radioactivity at rate R from a vent h metres above the ground while a wind of speed v is blowing. Downwind at a distance x from the plant the concentration χ of radioactivity is given by

$$\chi = (R/\pi v\sigma_x\sigma_z)\exp\{-h^2/(2\sigma_z^2)\}$$

where σ_y and σ_z, the horizontal and vertical **dispersion coefficients**, are functions of x. Both σ_y and σ_z are very approximately proportional to x for $10^2 < x < 2 \times 10^3$ m in reasonably stable atmospheric conditions, and $\sigma_y = 50$ m and $\sigma_z = 20$ m at $x = 10^3$ m. The external γ-ray dose rate dH/dt received at x is given by

$$dH/dt = 0.07\chi\langle E_\gamma\rangle \text{ mSv s}^{-1}$$

where χ is in MBq m^{-3}, and $\langle E_\gamma\rangle$ is the average energy in MeV of the γ-rays emitted per disintegration.
The radioactive gas ^{135}Xe ($\langle E_\gamma\rangle = 0.246$ MeV) is released for 10 minutes at a rate of 100 MBq per second from a vent 40 m above ground into a wind with $v = 3$ m s^{-1}. (a) Calculate the total external γ-ray dose H received downwind at a distance of 10^3 m. (b)* At what distance would the dose be a maximum?

8.13 Neglecting the time for fast neutrons to thermalize, the generation time τ_g is given by $\{u\Sigma_a(\text{core})\}^{-1}$, where u is the mean speed of a thermal neutron. Estimate τ_g for a graphite-moderated core that has a thermal utilization factor f of 0.88.

8.14 The value of $\beta\tau_d$ for the delayed neutrons from the fissile nucleus ^{239}Pu is 0.0324. Calculate the time constant for control of a ^{239}Pu fuelled reactor when $k = 1.0005$. Assume $\tau_g = 1$ ms.

8.15 A PWR nuclear reactor is generating 1 GW_e of electrical power. The reactor core contains 100 tonnes of uranium and the neutron flux density is 4×10^{17} neutrons s^{-1} m^{-2}. Estimate the enrichment in ^{235}U of the fuel.

8.16 The fission yield of the nuclide ^{88}Kr ($t_{1/2} = 2.79$ h) is 0.0364. Estimate the activity in the core from ^{88}Kr one day after a 2 GW_e (electrical power) reactor is shutdown. (Assume the production of ^{88}Kr ceases on shutdown.)

8.17 A beam of 3 MeV γ-rays of flux 10^9 m^{-2} s^{-1} is incident on a 1 m thick water shield. The linear attenuation coefficient μ of water for 3 MeV γ-rays is 3.96 m^{-1}, and the build-up factor for the 1 m thick shield is 3.55. Calculate (a) the effective γ-ray flux; (b) the effective dose rate D_{rate} (μSv h^{-1}).

8.18* In a breeder reactor show that, if the excess bred fuel is removed and used in another reactor, then the doubling time t_D is given by

$$t_D = M_i \ln 2/[(B-1)mP_o].$$

A 2 GW breeder reactor contains ^{235}U surrounded by ^{232}Th. For an initial amount of ^{235}U of 1000 kg and a breeding ratio of 0.05 to breed ^{233}U, calculate the doubling time with and without removal of the bred fuel.

8.19 (a) Estimate the volume of spent fuel generated annually by a 1 GW_e PWR (without reprocessing). The fuel is uranium dioxide enriched to 3% in ^{235}U and the density of the fuel is 4×10^3 kg m^{-3}. (b) Reprocessing reduces the volume of waste by about a factor of four. Estimate and comment on the volume of waste generated annually with reprocessing by PWRs with a total output of 0.9×10^6 GWh of electricity (~25% of the US annual consumption.)

8.20 Devise a demonstration (practical or computer) that illustrates how moderators increase the resonance escape probability in a reactor.

8.21 Discuss the advantages of one of the proposed generation IV reactors compared with an existing PWR.

8.22 Summarize the main economic considerations that affect the choice of nuclear power compared with an alternative source of power.

8.23 Discuss *critically* the statement that the environmental risks associated with nuclear power are far less than those associated with global warming.

8.24 Does public opinion preclude the expansion of nuclear power?

8.25 Describe the options for the storage of nuclear waste. Which one should be pursued?

8.26 Is nuclear power the only sufficiently developed large scale power source that could make a significant impact on global warming?

9 Energy from fusion

List of Topics

- ☐ Magnetic confinement
- ☐ D–T fusion reactor
- ☐ Fuel resources
- ☐ Lawson criterion
- ☐ Plasmas
- ☐ Charged particle motion
- ☐ Magnetic mirror
- ☐ Tokamak
- ☐ Plasma equilibrium
- ☐ Energy confinement
- ☐ Divertor tokamak
- ☐ Inertial confinement

Introduction

We have seen at the beginning of Chapter 8 that the increase in binding energy with mass number for light nuclei (Fig. 8.2) means that energy is released in fusion. The fusion of two light nuclei to form a heavier nucleus results in the release of energy, which is the source of energy in stars. In many stars, including our Sun, the energy results from a series of reactions that convert hydrogen into helium (the p–p chain), in which the initial step is the fusion of two hydrogen nuclei (protons) to form deuterium. This process is a weak interaction, as one of the protons changes to a neutron with the emission of a positron and a neutrino, and is the rate-determining step in the 'burning' of hydrogen.

In order for two protons to react they must have sufficient energy to overcome their mutual electrostatic (Coulomb) repulsion. The electrostatic potential energy of two protons is shown in Fig. 9.1(a) as a function of their separation. According to classical mechanics, protons require more kinetic energy than the Coulomb barrier (~700 keV) in order to fuse. However, quantum mechanical tunnelling enables the reaction to take place for the much lower proton energies that occur at the centre of the Sun (~1 keV). The protons have a Maxwellian energy distribution, $N(E) \propto E \exp(-E/kT)$, so the strong energy dependence means that the small high energy tail of the distribution contributes most to the fusion reaction rate in the Sun (Fig. 9.1(b)).

The energy release in the fusion of hydrogen to form helium in the p–p chain is about 25 MeV. This is some 10^6 times larger than in a typical chemical reaction involving electrons. This is a similar yield to that in fission and again there is no release of greenhouse gases. However, the fusion of hydrogen is totally impractical for a fusion reactor since the reaction rate is too low as it only involves the weak interaction. There are, though, other fusion reactions that occur via the strong interaction with a similar energy yield and with a reaction rate that can be made fast enough for a practical reactor. In particular, the fusion of deuterium and tritium to form helium

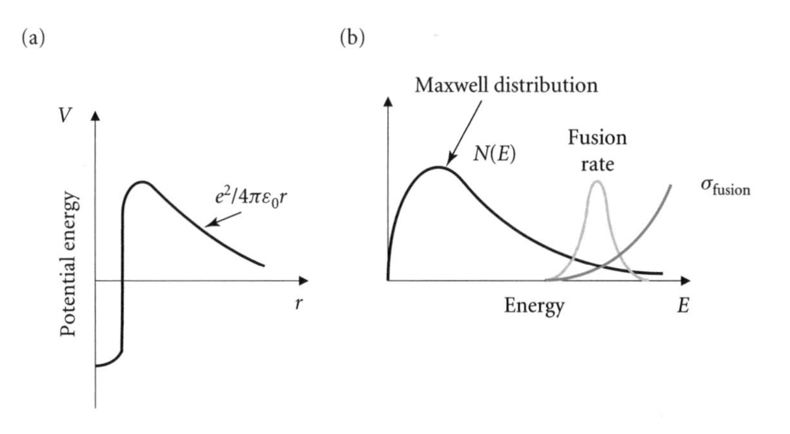

Fig. 9.1 (a) Coulomb barrier. (b) Fusion rate.

and a neutron has the highest rate and releases 17.6 MeV. This has the potential to provide the world's energy needs for thousands of years. The energy density is enormous—only 100 kg of deuterium (D) and 150 kg of tritium (T) are required to produce about 1 GW of electricity for 1 year.

For the fusion of deuterium and tritium to occur at a sufficient rate the nuclei must have energies $\sim$10 keV, corresponding to a temperature $\sim 10^{8\circ}$C. At these temperatures deuterium and tritium are highly ionized and form a gas of charged particles, called a **plasma**. No material can survive at such temperatures, so the plasma must be kept away from any containing walls. In the Sun the hydrogen plasma is so massive that it is contained by the huge gravitational force of attraction that balances the outward force resulting from the plasma pressure. In a terrestrial reactor the plasma must be contained by some other means. One method is to contain the plasma using magnetic fields, called **magnetic confinement**. Another is to compress the plasma by means of huge implosive forces, called **inertial confinement**. We will concentrate principally on magnetic confinement, as it is more developed.

9.1 Magnetic confinement

The earliest experiments were carried out in the USA in the late 1930s. The basic idea of using a toroidal plasma fusion reactor heated by radiofrequency (rf) waves was patented by Thompson and Blackman as early as 1946 in the UK. During the first decade of the Cold War, fusion research was conducted in secret. By the time it was de-classified in 1958 the mirror, theta and z-pinch machines and the stellarator had all been proposed. Progress in obtaining high temperature confinement, however, was slow until a breakthrough was made in 1968 in the Soviet Union. This advance was made using a tokamak machine (Tamm, Sakharov), and since then most research has concentrated on this type of magnetic confinement. Although the plasma in a tokamak can be contained by the magnetic field, obtaining stability for significant periods has proved exceedingly difficult. It is only recently that the conditions necessary for a fusion reactor have been achieved, after over 50 years of research.

9.2 D–T fusion reactor

The fusion reaction of deuterium (D $\equiv$ ^{2}H) and tritium (T $\equiv$ ^{3}H) forms a ^{4}He (α-particle) plus a neutron

$$D + T \rightarrow {}^4He + n + 17.6\,MeV. \tag{9.1}$$

(The nucleus of a deuterium atom is called a deuteron and that of a tritium atom a triton).

The 17.6 MeV of energy released is shared by the α-particle (3.5 MeV) and the neutron (14.1 MeV). Deuterium is stable but tritium is unstable with a half-life of 12.3 years; however, tritium can be produced using the emitted neutrons. The energetic neutrons, which take ~80% of the fusion reaction energy, can be stopped in lithium external to the plasma containment vessel. The energy deposited would be used as a source of heat for a thermal power station and the neutrons would produce tritium in the lithium via the following two reactions

$$n + {}^6Li \rightarrow T + \alpha + 4.78\,MeV, \tag{9.2}$$

$$n + {}^7Li \rightarrow T + \alpha + n - 2.87\,MeV. \tag{9.3}$$

The charged alpha particles from the D + T reaction provide a source of heat to maintain the plasma at a temperature of 10^{8}°C, which must be kept away from the vessel walls. There is a loss of energy by the plasma through radiation, particle diffusion, and heat conduction to the walls. For steady state operation this loss would be balanced by the heat generation in the bulk of the plasma from the alpha particles. There would also be some additional heating and fuel injection to maintain optimum running conditions. The ejected particles would be diverted away from the main body of the plasma, and thereby exhaust the helium ('ash') and heat from the plasma.

A schematic design for a D–T fusion reactor is shown in Fig. 9.2. The fast neutrons (14.1 MeV) pass through the plasma containment wall and are stopped in a lithium blanket,

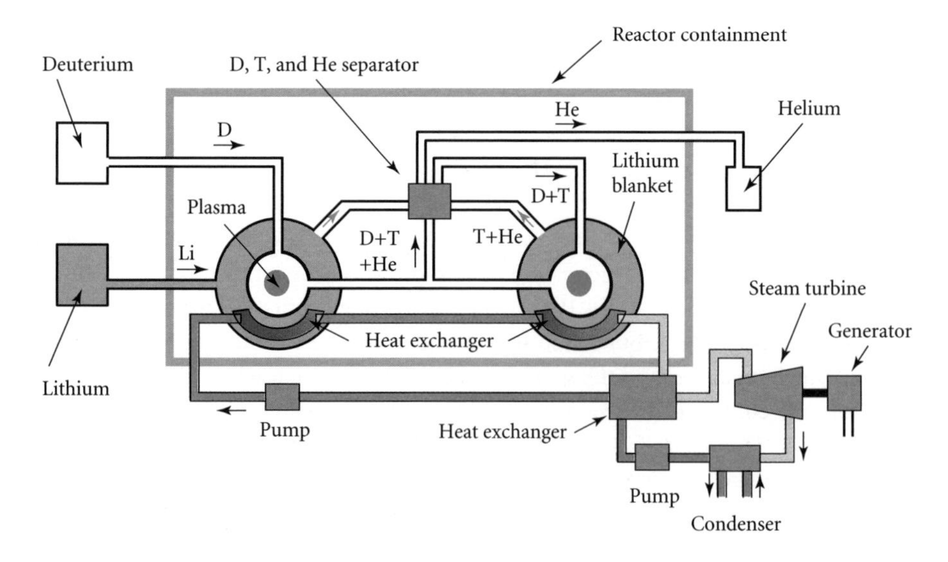

Fig. 9.2 Schematic layout of a fusion reactor.

where they produce tritium. The natural isotopic abundance of the lithium isotopes is 7.4% ^{6}Li and 92.6% ^{7}Li. Enriching the percentage of ^{6}Li enhances the tritium breeding ratio (TBR), which needs to be greater than unity for self-sufficiency. The ^{7}Li reaction produces a neutron which can make another T nucleus. Nuclei with significant (n, 2n) reactions, such as Be and Pb, could also be added to the lithium blanket to increase T production. Making the blanket material in the form of pellets aids the diffusion of the tritium, which is purged by helium gas.

The 14.1 MeV neutrons from the D + T reaction interact with the walls of the plasma chamber and other components. Their interaction can produce radioactive nuclei, i.e activate the materials. A low activation stainless steel has been developed in which Cr, W, and Ti replace Mo, Ni, and Nb, which allows the steel to be recycled after 50–100 years. Ceramic and fibre-composite materials are also being examined. The level of waste is significantly less than that from a fission reactor and there would not be a large nuclear waste disposal problem. Also, the tritium does not pose a serious radiation risk.

9.2.1 D–T fuel resources

The fuels for D–T fusion are deuterium and lithium. Deuterium is very plentiful as there are about 35 g per cubic metre of sea water, or 1 in 6500 hydrogen atoms are deuterium. The reserves of lithium are estimated as greater than 10^{6} tonnes in the earth. Less easily available are the amounts in the sea where there are more than 10^{11} tonnes, at a concentration of Li^{+} of about 175 mg per cubic metre. The energy density of the fuel for a fusion reactor is enormous—only 100 kg of D and 150 kg of tritium, equivalent to ~300 kg of ^{6}Li, are required to produce ~1 GW of electricity for 1 year (Example 9.1). One tonne (~1200 litres) of oil has the same energy content as 50 mg of D plus 150 mg of ^{6}Li. The cost of deuterium is 1 euro/g and that of lithium is 20 euro/kg, so the contribution of the cost of the fuel to the cost of electricity is negligible at 0.003 euro cents/kWh. The reserves of lithium in the earth alone would provide the world with 10 TWy of energy for ~1000 years.

EXAMPLE 9.1

What is the energy released when 100 kg of deuterium and 150 kg of tritium are consumed in 1 year in a fusion reactor? If the reactor's thermal plant is 35% efficient, find the average continuous electrical power output over the year.

100 kg of deuterium and 150 kg of tritium each correspond to 5×10^{4} moles. The energy release E is therefore

$$
\begin{aligned}
E &= 5 \times 10^{4} \times 6 \times 10^{23} \times 17.6 \times 10^{6} = 5.28 \times 10^{35}\ \text{eV} \\
&= 5.28 \times 1.6 \times 10^{16}\ \text{J} = 8.45 \times 10^{16}\ \text{J}.
\end{aligned}
$$

An amount of energy E in a thermal plant of efficiency ε can provide a continuous electrical power output P for a time t given by

$$\varepsilon E = Pt \quad \text{i.e.} \quad P = \varepsilon E/t = 0.35 \times 8.45 \times 10^{16}/(3.15 \times 10^{7}) = 0.94\ \text{GW}.$$

9.2.2 Characteristics of the D + T → ^{4}He + n reaction

The reaction rate for a flux ϕ of deuterons interacting with N tritons equals $\phi N\sigma$ (Chapter 8, eqn (8.2)). Hence the reactivity R, which is the reaction rate per unit volume, is given by

$$R = \phi n_T \sigma = n_D n_T \langle v\sigma \rangle, \tag{9.4}$$

as the flux ϕ of tritons equals $n_D v$, where n_D and n_T are the number of deuterons and tritons per unit volume, respectively. The average value $\langle v\sigma \rangle$ is the product of the relative velocity of deuterons and tritons and the fusion cross-section σ, averaged over the velocity distributions of the particles. Figure 9.3 shows that $\langle v\sigma \rangle$ is of similar magnitude for the D + T ((eqn 9.1)), D + ^{3}He, and D + D reactions:

$$\mathrm{D + T \to {}^4He + n + 17.6\,MeV},$$

$$\mathrm{D + {}^3He \to {}^4He + p + 18.3\,MeV}, \tag{9.5}$$

$$\mathrm{D + D \to T + p + 4.0\,MeV \text{ or } \to {}^3He + n + 3.3\,MeV}, \tag{9.6}$$

but that at a given temperature (T) the D + T fusion reaction has the highest rate of reaction. Over the temperature range corresponding to 10–20 keV the value of $\langle v\sigma \rangle$ for D + T is approximately proportional to T^2. This strong temperature dependence comes from the sharp increase with velocity of the probability of penetrating the Coulomb barrier.

For a D–T fusion reactor the fusion power P_{fusion} produced by the plasma is divided between the α-particles P_α (20%) and the neutrons P_n (80%). The α-particles are stopped within the plasma, so P_α heats the plasma. The neutrons escape the plasma and pass through the containment walls. Outside they are stopped in the lithium blanket, where tritium is produced, and thereby generate heat for the fluid for the power turbine.

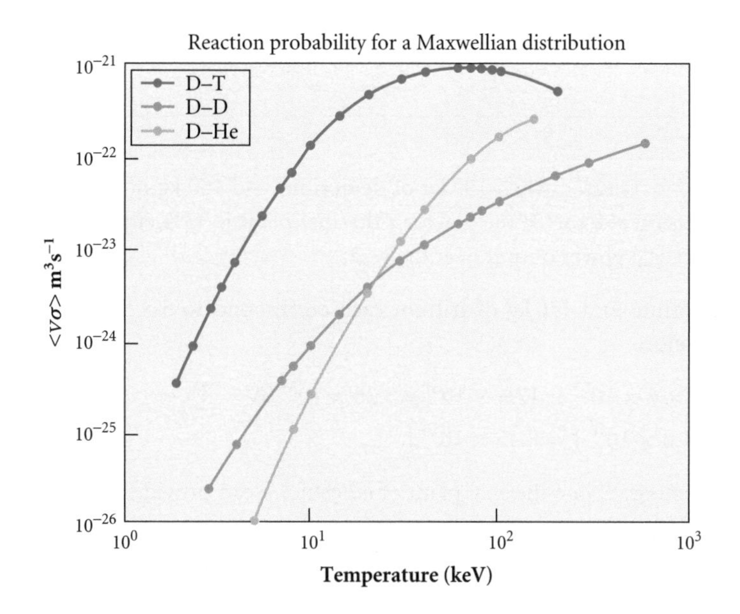

Fig. 9.3 Rate of fusion reactions as a function of temperature.

External power P_{ext} may be required in addition to P_α to compensate for losses from the plasma and to optimize the plasma conditions. The power loss P_{loss} arises from particles and heat that diffuse from the plasma centre to the walls of the container. In addition there is radiation (bremsstrahlung and synchrotron) together with line radiation from impurities emitted by the containing walls. The power loss P_{loss} is related to the **energy containment time** τ_E and the total plasma energy W by

$$P_{loss} = W/\tau_E \tag{9.7}$$

where τ_E is the time for plasma energy to be lost to the walls when the plasma is in its operating state but with no energy input. (N.B. τ_E is not the same as the plasma duration (burn) time: P_α and P_{ext} heat the plasma during a burn so the duration time is longer.)

A quality factor Q is defined by

$$Q = P_{fusion}/P_{ext}. \tag{9.8}$$

Break-even is defined as $Q = 1$ and **ignition** as $Q = \infty$. Ignition corresponds to $P_\alpha = P_{loss}$ and $P_{ext} = 0$. We show in Derivation 9.1, when deriving the Lawson criterion, that for a 50–50 mixture of D and T the requirement for ignition ($Q = \infty$) is

$$n\tau_E T \geq 3 \times 10^{21}\ \text{m}^{-3}\text{s keV}\ (T \text{ in keV}). \tag{9.9}$$

Temperatures are often expressed as an energy in keV, e.g. T is equivalent to $kT/|e|$, so 10^8 $K \sim 10^4$ eV or $\sim$10 keV.

For $T \sim 10$ keV and $n \sim 10^{20}$ m^{-3} the Lawson condition requires a confinement time $\tau_E \geq \sim 3$ s.

Derivation 9.1 Lawson criterion

For a D–T reactor the rate of change in the total plasma energy W is given by

$$dW/dt = P_\alpha + P_{ext} - P_{loss},$$

so under equilibrium conditions ($dW/dt = 0$)

$$P_{input} \equiv P_\alpha + P_{ext} = P_{loss},$$

and $\tau_E \equiv W/P_{input}$. Therefore

$$P_\alpha + P_{ext} = W/\tau_E. \tag{9.10}$$

Defining

$$f_\alpha \equiv P_\alpha/P_{fusion} = E_\alpha/E_{fusion} \tag{9.11}$$

and noting from eqn (9.8) that $P_{ext} \equiv P_{fusion}/Q$, then (9.10) is equivalent to

$$(f_\alpha + 1/Q)P_{fusion} = W/\tau_E \tag{9.12}$$

the reaction rate per unit volume R, reactivity, is given by eqn (9.4)

$$R = n_D n_T \langle v\sigma \rangle.$$

Each reaction produces energy E_{fusion}, where

$$D + T \rightarrow \alpha + n + E_{\text{fusion}},$$

and

$$E_{\text{fusion}} = E_\alpha + E_n = 17.6 \text{ MeV}.$$

The power released is reaction rate multiplied by energy released per reaction, i.e.

$$P_{\text{fusion}} = RV \times E_{\text{fusion}} \tag{9.13}$$

where V is the volume of the plasma.

For a 50–50 mixture of D and T then $n_D = n_T = n/2$. Plasma neutrality means that the electron and ion densities are equal, so that $n_e = n$. Hence the total kinetic energy per unit volume W/V is given by

$$W/V = 2n(\tfrac{3}{2}kT) = 3nkT \tag{9.14}$$

assuming the electron and ions are in thermal equilibrium (i.e. at the same temperature). Using eqn (9.4) with $n_D = n_T = \frac{1}{2}n$ then P_{fusion}, eqn (9.13), becomes

$$P_{\text{fusion}} = \tfrac{1}{4}n^2\langle v\sigma\rangle E_{\text{fusion}} V. \tag{9.15}$$

Substituting eqns (9.14) and (9.15) into equation (9.12) gives

$$\tfrac{1}{4}(f_\alpha + 1/Q)n^2\langle v\sigma\rangle E_{\text{fusion}} = 3nkT/\tau_E.$$

The **break-even** condition ($Q = 1$) can be written as

$$n\tau_E \geq 12kT/\{(1 + f_\alpha)\langle v\sigma\rangle E_{\text{fusion}}\} \equiv f(T) \text{ only.} \tag{9.16}$$

This is known as the **Lawson criterion**.

In the operating temperature region of 10–20 keV the value of $\langle v\sigma\rangle$ is approximately proportional to T^2 (see Fig. 9.3), i.e.

$$\langle v\sigma\rangle \sim 1.2 \times 10^{-24} T^2 \text{ m}^{-3} \text{ s}^{-1} \tag{9.17}$$

where T is in keV so this condition can then be expressed as one on the triple product $n\tau_E T$. The requirement for **ignition** ($Q = \infty$) is

$$n\tau_E T \geq 3 \times 10^{21} \text{ m}^{-3} \text{ s keV } (T \text{ in keV}). \tag{9.9}$$

For $T \sim 10$ keV and $n \sim 10^{20}$ m^{-3} then $\tau_E \geq \sim 3$ s.

With no external heating, equilibrium requires $P_\alpha = P_{\text{loss}}$. However, due to diffusion, the loss will be greater than just that from bremsstrahlung due to electron–ion collisions. The temperature dependence of the reactivity R_b for bremsstrahlung interactions is $T^{1/2}$ which is weaker than that for fusion, which depends on T^2. This means that at sufficiently high temperatures $P_\alpha > P_b$. The operating temperature for D–T fusion plasmas is ~10 keV, which corresponds to a temperature of ~10^{8}°C. At these temperatures deuterium and tritium are almost completely ionized and form a very hot plasma. This plasma has to be contained in the reactor away from the walls.

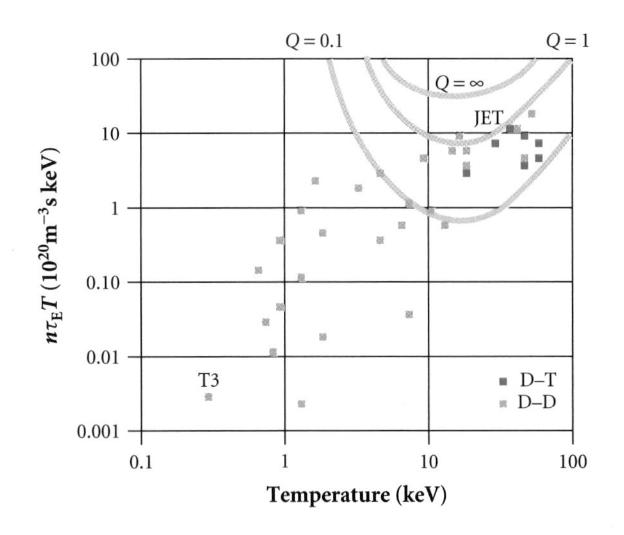

Fig. 9.4 Progress on increasing $n\tau_E T$ since the Russian tokamak T3 in 1968.

9.3 Performance of tokamaks

Figure 9.4 shows the progress in obtaining better confinement since 1968, when the results from tokamak T3 were announced. The increase of $n\tau_E T$ in the following 35 years has been four orders of magnitude, with the Joint European Torus (JET) obtaining a $Q \sim 0.6$ and $\tau_E \sim 1$ s in 1997 with a D–T plasma.

Energy confinement in the TFR tokamak in France in the mid-1970s was 20 ms at $T = 1$ keV. We will assume that the plasma conditions were such that heat loss was mainly via particle diffusion and heat conduction. Then, as these are diffusive processes, the energy containment time τ_E is proportional to the square of the diameter (or to the cross-sectional area of the toroid). In TFR the area was 0.13 m^2 while in JET it was 8 m^2. This increase in area would then predict a confinement time of 1.2 s, which is close to that achieved by JET. However, to achieve plasma conditions where the energy transport is primarily diffusive and the plasma is stable has proved very difficult. To understand why, we need to consider the characteristics of plasmas.

9.4 Plasmas

In the centre of the Sun hydrogen is highly ionized and consists of a gas of charged particles, hydrogen ions, and electrons, called a **plasma**. The long-range electrostatic interactions between the charged particles cause the plasma to exhibit long-range collective effects. (This can also occur when a gas is only partly ionized and contains neutral as well as charged particles.) The average kinetic energy of the particles is related to the temperature T of the gas by

$$\tfrac{1}{2}m\langle v^2\rangle = \tfrac{3}{2}kT \qquad (9.18)$$

where m is the mass of a particle and v is its speed. A typical particle velocity is therefore $\sim(kT/m)^{1/2}$. The temperature is not necessarily the same for ions as for electrons. For a plasma obtained from ionizing neutral atoms, there will be a roughly equal number of positive and negative charge carriers per unit volume.

Any time-independent external electric field is shielded from the interior of a plasma by a polarization layer of charge (known as a plasma sheath), the thickness of which is of the order of the Debye length

$$\lambda_D = (\varepsilon_0 kT/ne^2)^{1/2}, \tag{9.19}$$

and is typically a fraction of a millimetre for plasma machines. The typical distance that particles are separated in a plasma is of order

$$r_s \sim n^{-1/3}. \tag{9.20}$$

The relative kinetic energy of a pair of charged particle is about kT, so the distance apart r_c when their electrostatic potential energy $U_c = e^2/(4\pi\varepsilon_0 r_c)$ equals their relative kinetic energy is given by

$$r_c \sim e^2/(4\pi\varepsilon_0 kT). \tag{9.21}$$

If their typical distance apart is much greater than r_c i.e. $r_s/r_c \gg 1$, then binary Coulomb interactions are rare. It is then a good approximation to treat such hot diffuse plasmas as **collisionless**; these plasmas are also called **weakly coupled**.

The pressure p exerted by the plasma is given by

$$p = nkT. \tag{9.22}$$

EXAMPLE 9.2

A plasma has a temperature of 10 keV and a number density $n = 10^{20}$ m^{-3}, typical for fusion. Calculate the plasma pressure, the Debye length, and the ratio r_s/r_c.

Using eqn (9.22) for p and (9.19) for λ_D,

$$p = 10^{20} \times 10^4 \times 1.6 \times 10^{-19} = 1.6 \times 10^5 \text{ Pa} = 1.6 \text{ bar},$$
$$\lambda_D = (8.85 \times 10^{-12} \times 10^4 \times 1.6 \times 10^{-19}/\{10^{20}(1.6 \times 10^{-19})^2\})^{1/2} = 0.7 \times 10^{-4} = 0.07 \text{ mm}.$$

The distances r_s and r_c are given by eqns (9.20) and (9.21) so

$$r_s/r_c = (4\pi\varepsilon_0 kT)n^{-1/3}/e^2 \equiv 4\pi\lambda_D^2 n^{2/3} = 4\pi(0.7 \times 10^{-4})^2 \times 10^{40/3} = 1.3 \times 10^6.$$

Hence $r_s/r_c \gg 1$, so the plasma is weakly coupled (or collisionless).

In a magnetized weakly coupled plasma, the charged particles spiral quite freely along the magnetic field lines with a radius of gyration ρ. In a fusion plasma the dimensionless **magnetization parameter**

$$\delta \equiv \rho/L, \tag{9.23}$$

where L is the typical distance over which the field is essentially constant, is very small. The magnetized plasma is therefore highly anisotropic, with motion parallel and perpendicular to the field being very different. Small perturbations can also set up an enormous variety of motions, both oscillatory and turbulent, some of which are **unstable** and lead to exponential growth. This has been a subject of intensive research for many years and not all forms of instability are yet understood.

For magnetic confinement, though, a necessary condition is that the tightly spiralling particles must remain within the magnetic fields generated externally and by their own motion. To understand how the magnetic field in a Tokamak can confine a hot diffuse plasma, we need to consider first the motion of charged particles in magnetic and electric fields. We will consider later the stability of the confined plasma.

9.5 Charged particle motion in *E* and *B* fields

The force experienced by a charged particle of mass m in an electric field $\boldsymbol{E}$ and a magnetic field $\boldsymbol{B}$ is the Lorentz force $\boldsymbol{F}$ given by

$$\boldsymbol{F} = q\boldsymbol{E} + q\boldsymbol{v} \times \boldsymbol{B}, \tag{9.24}$$

where q is the charge and $\boldsymbol{v}$ is the velocity of the particle. In a region where $\boldsymbol{E}$ is zero and $\boldsymbol{B}$ is uniform, then the particle experiences a constant force $\boldsymbol{v}_\perp \times \boldsymbol{B}$, where $\boldsymbol{v}_\perp$ is the component of velocity perpendicular to $\boldsymbol{B}$. This force causes the particle to gyrate about a magnetic field line at an angular frequency ω_c (the cyclotron frequency), forming a helical path with a radius ρ (Larmor radius) where

$$\rho = mv_\perp/qB \tag{9.25}$$

and

$$\omega_c = qB/m. \tag{9.26}$$

Since in a plasma at a temperature T the mean kinetic energy $\frac{1}{2}mv^2$ equals $\frac{3}{2}kT$, only one component of velocity is parallel to $\boldsymbol{B}$, so $mv_\perp^2 = 2kT$. For a temperature of 10 keV and a B field of 5 tesla, typical values in a fusion plasma are

Deuterons: $\rho = 4.1$ mm, $\omega_c = 239 \times 10^6$ rad/s $= 38$ MHz,

Electrons: $\rho = 67$ µm, $\omega_c = 879 \times 10^9$ rad/s $= 40$ GHz.

When a charged particle is moving in a non-uniform magnetic field it will spiral along the magnetic field lines provided that the field varies sufficiently slowly that $\boldsymbol{B}$ is essentially uniform over a distance of order ρ. A particle moving into a region of higher field (Fig. 9.5) experiences a decelerating force since there is a radial component of $\boldsymbol{B}$.

The decelerating force is in the direction of varying field and is given by (see Exercise 9.5)

$$F_z = -\mu \partial B/\partial z \tag{9.27}$$

where

$$\mu \equiv \tfrac{1}{2}v_\perp \rho q \tag{9.28}$$

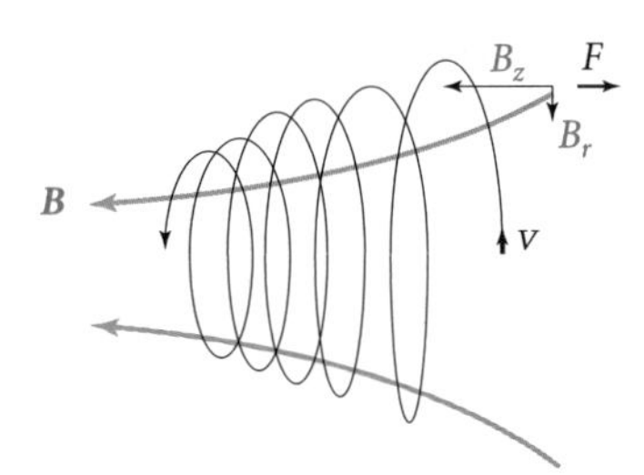

Fig. 9.5 Charged particle moving in a non-uniform axially symmetric magnetic field.

is the magnetic moment associated with the spiralling motion of the charged particle. As the particle does not experience a force in the azimuthal direction (since $\boldsymbol{B}$ is slowly varying), the angular momentum of the particle will be conserved, i.e. $mv_{\perp}\rho$ is constant, so as ρ decreases $v_{\perp}$ increases. Hence the magnetic moment μ is conserved; it is called an **adiabatic invariant**. The kinetic energy of the particle is also invariant as the magnetic force does no work on the particle since it acts perpendicular to the motion of the particle.

The decelerating force causes the component of velocity, v_z parallel to the field to decrease as the particle spirals along the field line, and it can reverse and so confine the particle's motion. A field configuration that confines particles in this way is called a **magnetic mirror**. Figure 9.6 shows the magnetic field configuration in a magnetic mirror, where the magnetic field is a minimum midway between the two coils ($z = 0$) and a maximum close to the position of the coils ($z = \pm z_{\mathrm{m}}$). Whether a particle is reflected or passes through a coil as it spirals along the field lines depends on the pitch angle α of the spiralling particle, given by

$$\tan\alpha = v_{\perp}(0)/v_z(0). \tag{9.29}$$

All particles with $\alpha > \alpha_{\mathrm{m}}$, where

$$\sin^2\alpha_{\mathrm{m}} = B(0)/B(z_{\mathrm{m}}), \tag{9.30}$$

will be reflected and contained in the field between the coils (see Exercise 9.6).

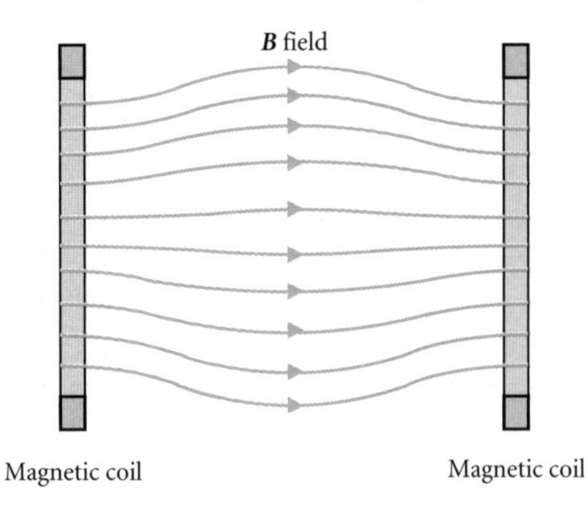

Fig. 9.6 A magnetic mirror configuration showing the magnetic field between the coils; the field extends beyond the coils.

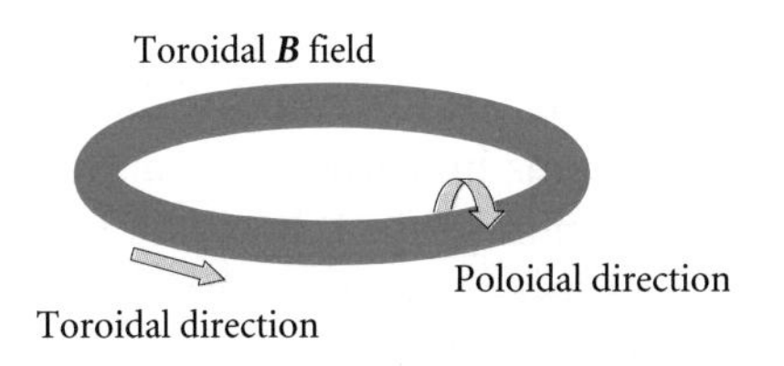

Fig. 9.7 Toroidal magnetic field.

Unfortunately, this magnetic mirror field configuration is not a good one for containing a plasma. Collisions occur within the plasma, which cause particles to change their pitch angle such that they are lost through the coils.

To avoid this loss we need to consider a field that has no ends, i.e a toroidal field as illustrated in Fig. 9.7, which shows the toroidal and poloidal directions. However, the curvature and non-uniformity of such a field cause the spiralling motion of a charged particle to drift away from the toroidal magnetic field lines. This drift is explained in Derivation 9.2; it can be counteracted by adding a poloidal field.

Derivation 9.2 Drift in toroidal and poloidal fields

We can see why drift occurs in a toroidal field by considering a particle initially moving with speed v in a circle of radius ρ in a uniform magnetic field $\boldsymbol{B}$. It then experiences a force $\boldsymbol{F}$ at right angles to $\boldsymbol{B}$ (see Fig. 9.8).

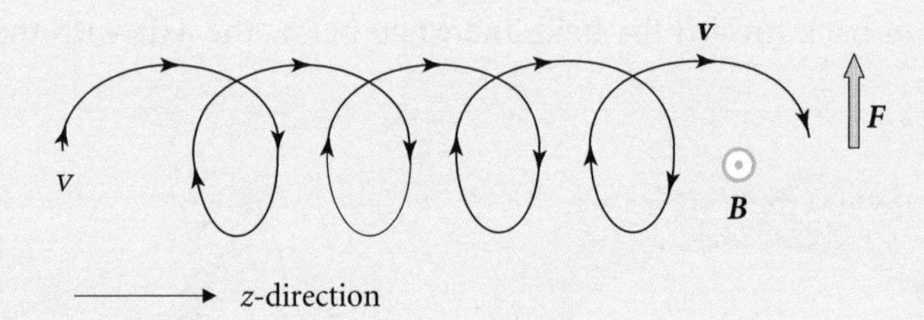

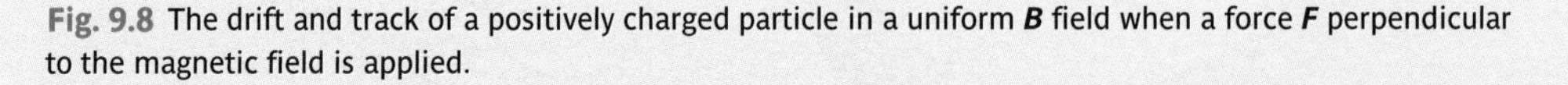
Fig. 9.8 The drift and track of a positively charged particle in a uniform ***B*** field when a force ***F*** perpendicular to the magnetic field is applied.

The particle, which starts on the left-hand side of Fig. 9.8, initially curves to the right. When it is moving in the z-direction the force $\boldsymbol{F}$ is in the opposite direction to the magnetic force $q\boldsymbol{v}\times\boldsymbol{B}$ so the curvature of its trajectory is less. When the particle has curved round and is travelling in the negative z-direction, the force $\boldsymbol{F}$ and the magnetic force are in the same direction and the curvature is greater. This causes a drift in the positive z-direction as shown.

The motion is equivalent to a circular motion plus a drift in the direction $\boldsymbol{F}\times\boldsymbol{B}$ with a drift velocity $v_{d\perp}$ such that the Lorentz force $qv_{d\perp}B$ balances the force F. The drift velocity produced by a constant force F is therefore

$$v_{d\perp} = F/qB.$$

The magnetic field in a toroidal field (Fig. 9.9), varies in the radial direction, so a spiralling charged particle experiences a force, as in the magnetic mirror configuration, that is also in this direction, i.e. radially. The centre of the helical path is circular so the particle is on average accelerating radially. It therefore experiences a further radial force. The result is that a spiralling particle in a toroidal field will drift in the direction $\boldsymbol{F} \times \boldsymbol{B}$, i.e. vertically as illustrated in Fig. 9.9(b), up or down depending on the sign of the particle's charge.

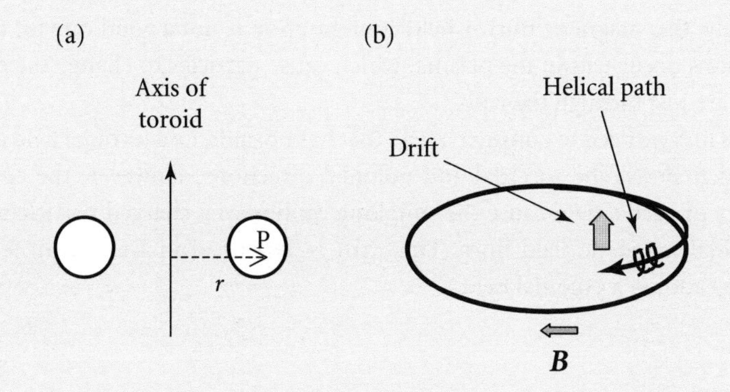

Fig. 9.9 (a) The drift of a charged particle in a toroidal magnetic field. (b) The magnetic field at a point P depends on $1/r$, where r is the distance from the axis of the toroid.

This drift can be counteracted by adding a poloidal component to the magnetic field producing a helical field, as illustrated in Fig. 9.10. The direction of the drift is shown by the arrows and its effect is to move the path up and away from the field line when the particle is above the axis but back toward the field line when below the axis with the result that the overall drift is zero.

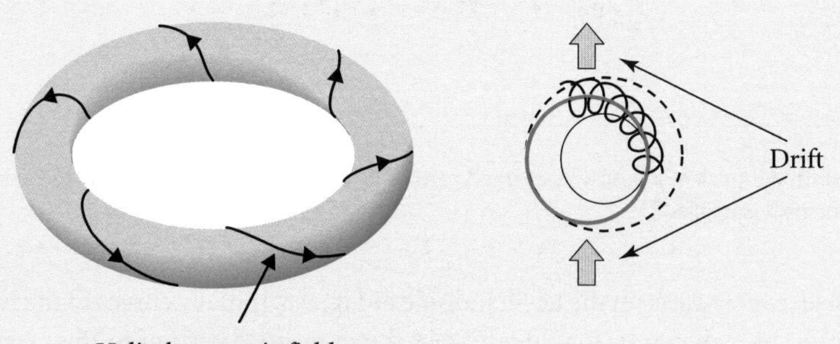

Fig. 9.10 A helical field produced by adding a poloidal field to a toroidal field and its effect on the drift of a particle's motion.

The helical toroidal magnetic field can be produced by a series of twisting magnetic coils in a configuration called a **stellarator** or by a **tokamak**. Fusion research with tokamaks is the most advanced and we will now describe the main features of a tokamak.

9.6 **Tokamaks**

In a tokamak, the toroidal magnetic field is produced by a series of toroidal field-coils (Fig. 9.11). For continuous operation in a fusion reactor superconducting coils would be used to reduce the power requirements for generating the large fields of several tesla required. The vertical field-coils are used to position and shape the plasma. Negative feedback is used to maintain the plasma in position.

To create the poloidal field a current is induced in the plasma; the resulting field is helical as shown in Fig. 9.11. The current is induced in the plasma by a time-varying large magnetic field that passes through the centre of the torus. The plasma acts like the secondary circuit in a transformer (see p. 280). If the flux linkage is Φ, then

$$V = -\mathrm{d}\Phi/\mathrm{d}t \quad \text{and} \quad I = V/R$$

where R is the resistance of the plasma. In the Joint European Torus (JET) at Culham in the UK a voltage of 10 V was required to establish the current since initially R was relatively high. Once the plasma was fully ionized by ohmic heating then a voltage of 0.3 V was sufficient to drive a 3.5 MA current. JET had a flux swing of 34 V s and was able to sustain a current for about a minute ($[\Phi_{\max} - \Phi_{\min}] = -\int V\mathrm{d}t$).

The need to apply an external field reduces the efficiency, or availability, of a fusion reactor. Fortunately a radial pressure gradient within the plasma causes a toroidal current to flow. As it is produced by the plasma itself, it is called a **bootstrap** current and reduces the amount of external current drive required to maintain the toroidal current.

In the early tokamak experiments the temperature of the plasma was raised by ohmic heating. However, the resistivity of the plasma decreases with increasing temperature, due to the increased ionization, like $T^{-3/2}$. The ohmic heating from the current required to give the necessary poloidal field raised the temperature to only a few 10^7 °C. At this temperature the power losses due to radiation and transport balanced the power input by ohmic heating. So we need to raise the temperature of the plasma by other means.

One method of raising the temperature is to inject a beam of neutral hydrogen ions into the plasma. Charged hydrogen ions are accelerated and then neutralized by passing them through

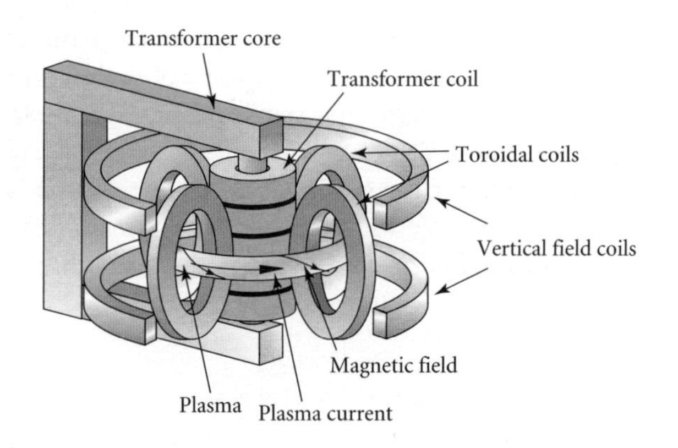

Fig. 9.11 Principal components of a tokamak and the confining helical magnetic field.

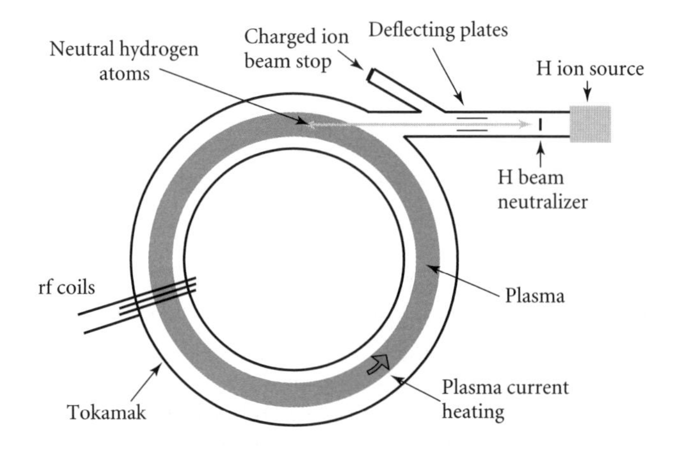

Fig. 9.12 Ohmic, rf, and neutral beam plasma heating techniques.

a neutral hydrogen gas. In the gas, charge exchange yields high energy (~80 keV) neutral hydrogen ions that can cross magnetic field lines. Once inside the plasma they transfer energy to the plasma particles and also become ionized. (D or T would be used in a fusion reactor.)

Another technique to heat the plasma is to use high (radio) frequency electromagnetic waves, a technique called **rf heating**. The oscillating electric field in the rf wave resonates with the cyclotron motion of the plasma particles, either ions or electrons, depending on the frequency. The frequencies are 20–50 MHz for the ion cyclotron resonance heating (ICRH) or 70–140 GHz for the electron cyclotron resonance heating (ECRH). Ohmic, rf, and neutral beam heating are illustrated in Fig. 9.12. Both the neutral beam and rf can also be used to drive a current within the plasma. Using high speed pellets of frozen D, a high central plasma density can also be obtained.

9.7 Plasma confinement

Confining the energy within a hot plasma requires that the plasma be held in stable equilibrium. There is also the need to reduce the rate of loss of energy from the plasma to the containing walls. Instabilities can be caused by gradients in the pressure or in the current. The tokamak magnetic configuration can provide macroscopic stability under certain running conditions, i.e. stability over distances of the order of the size of the tokamak. The rate of energy transport, however, depends in particular on turbulence occurring over smaller distances, which is only partially understood. We will first consider the conditions for equilibrium, then the stability of the equilibrium, before discussing the factors that limit the energy containment time.

9.7.1 Plasma equilibrium

In a hot plasma collective effects among the plasma particles make the plasma act like a fluid characterized by a temperature T, pressure p, and particle number density n. This is a useful approximation for low-frequency large-scale phenomena. The current $\boldsymbol{j}$ is given by the sum of the ion and electron currents:

$$\boldsymbol{j} = n_i e\boldsymbol{v}_i - n_e e\boldsymbol{v}_e = ne(\boldsymbol{v}_i - \boldsymbol{v}_e),$$

since $n_i = n_e = n$. The velocity of the fluid $\boldsymbol{v}$ is approximately equal to that of the ions $\boldsymbol{v}_i$, because the fluid is restrained by the inertia of the ions ($m_i \gg m_e$). The relatively small difference in ion and electron velocities gives rise to the current. This **single-fluid model** for the motion of a plasma in a magnetic field is called **magnetohydrodynamics** or MHD.

At higher frequencies the relative motions of the electrons and ions have to be taken into account and a **two-fluid model** is used. For even higher frequencies the effect of the distribution of particle velocities must be considered. For example, certain departures from a Maxwellian velocity distribution can be unstable and give rise to an exponential growth of small amplitude waves. There is an enormous variety of non-linear complex phenomena in magnetized plasmas. We will just consider the conditions for equilibrium in the MHD approximation.

9.7.2 Plasma equilibrium in a tokamak

When a current is induced in the plasma in a tokamak the current flows in a ring. A short section of the current can be approximated by a cylindrical section. The current gives rise to a very large poloidal magnetic field, which exerts an inward force on the plasma current—a **pinch effect**, as shown in Fig. 9.13. For equilibrium the plasma current must be large enough to generate sufficient magnetic force to balance the outward pressure of the plasma.

We can estimate the magnitude of this current I by finding the magnetic field B at the edge of the plasma which has a radius R. The magnetic field B due to the plasma current is $\mu_0 I/2\pi R$. Approximating the current to be all at a radius R, then the force on a length l of plasma is IBl. The force acts inwards over an area $2\pi Rl$ so the magnetic 'pressure' (i.e. force per unit area) p_B is

$$p_B \sim \mu_0 I^2 l/\{(2\pi R)^2 l\} = \mu_0 I^2/(2\pi R)^2 = B^2/\mu_0.$$

The pressure of the plasma p in the cylindrical section is nkT or $NkTl/(\pi R^2 l)$, where N is the number of charged particles per unit length and l is the length of the plasma section. The magnitude of current required to balance p is therefore given by

$$I^2 \sim 4\pi NkT/\mu_0.$$

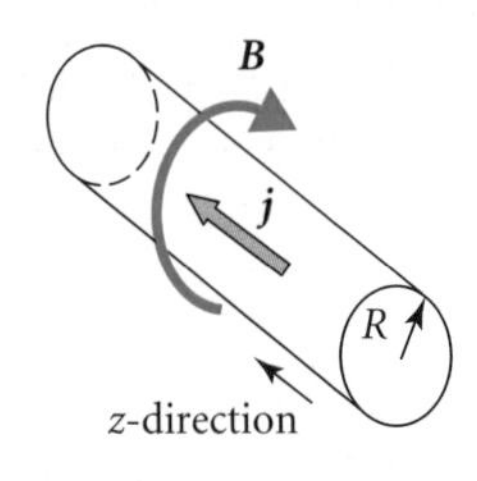

Fig. 9.13 *z*-pinch configuration.

The exact relationship is derived in Derivation 9.3 for the case of a cylindrical plasma, called a z-pinch, and is

$$I^2 = 16\pi NkT/\mu_0. \quad (9.31)$$

EXAMPLE 9.3

Estimate what current is required to contain a tokamak plasma at a temperature of 10^8 K, a number density $n = 10^{20}$, and a cross-sectional area of 1 m^2.

The linear number density $N = 10^{20}$. For equilibrium we require a current

$$I = (16\pi 10^{20} \times 1.38 \times 10^{-23} \times 10^8/4\pi 10^{-7})^{1/2} = 2.35 \text{ MA}.$$

This is the order of magnitude of the current required in a tokamak.

The exact relationship for the magnetic pressure in terms of the field B is $B^2/2\mu_0$. The ratio $\beta \equiv p/(B^2/2\mu_0)$ of the plasma pressure (which is equivalent to its kinetic energy density as $p = nkT$) to the magnetic pressure (which is the same as the magnetic energy density) is an important measure of the effect of the thermal motion on the confining ability of the magnetic field. This ratio, called the plasma beta, needs to be < 1 for confinement and typically $\beta \approx 0.2$.

Derivation 9.3 The z-pinch

Consider a short section of plasma carrying a current I as shown in Fig. 9.13. Equilibrium requires balancing the force arising from the pressure gradient ∇p within the plasma, with the magnetic force $\boldsymbol{j} \times \boldsymbol{B}$ arising from the plasma current density $\boldsymbol{j}$ in the magnetic field $\boldsymbol{B}$. For equilibrium the resultant force on an element of the plasma is zero, i.e.

$$-\nabla p + \boldsymbol{j} \times \boldsymbol{B} = 0. \quad (9.32)$$

We can derive another relationship from Maxwell's equations assuming the electrical conductivity of the plasma is sufficiently high that any electrical forces can be neglected. We will also assume, for low frequency effects, that the displacement current $\partial \boldsymbol{D}/\partial t$ can also be ignored. Then

$$\mu_o \boldsymbol{j} \approx \nabla \times \boldsymbol{B}. \quad (9.33)$$

This approximation of taking the conductivity as effectively infinite is called ideal MHD.

Rather than considering the current flowing in a tokamak, we analyse the simpler z-pinch configuration shown in Fig. 9.13. The axial current $\boldsymbol{j}$ gives rise to an azimuthal $\boldsymbol{B}$ field. The force per unit volume is $\boldsymbol{j} \times \boldsymbol{B}$. If the plasma pressure is p then

$$\mu_0 \nabla p = (\nabla \times \boldsymbol{B}) \times \boldsymbol{B} \ (\text{since } \nabla \times \boldsymbol{B} = \mu_0 \boldsymbol{j}), \tag{9.34}$$

$$\mu_0 \mathrm{d}p/\mathrm{d}r = -(B/r)\mathrm{d}(rB)/\mathrm{d}r,$$

$$\mu_0 \int r^2 (\mathrm{d}p/\mathrm{d}r)\mathrm{d}r = -\int (rB)\mathrm{d}(rB),$$

$$\mu_0 [r^2 p] - 2\mu_0 \int rp \, \mathrm{d}r = -\tfrac{1}{2}[(rB)^2]. \tag{9.35}$$

As $p(R) = 0$ at the boundary of the axial plasma current, the first term of eqn (9.35) is zero. Since $p = (n_+ + n_-)kT$ and $n_+ = n_- = n$, we have

$$2\mu_0 \int rp \, \mathrm{d}r = 2\mu_0 \, kT \int 2nr \, \mathrm{d}r = \tfrac{1}{2}\,[(rB)^2]_{r=R}.$$

Defining the total number of particles per unit length $N = \int 2\pi nr \, \mathrm{d}r$ we can write

$$2\mu_0 NkT = \tfrac{1}{2}\pi[(rB)^2]_{r=R}$$

From $\nabla \times \boldsymbol{B} = \mu_0 \mathbf{j}$ we have $(1/\mathrm{r})\mathrm{d}(rB)/\mathrm{d}r = \mu_0 j$ and

$$[rB]_{r=R} = (\mu_0/2\pi) \int 2\pi jr \, \mathrm{d}r = \mu_0 I/(2\pi).$$

Finally, we obtain the Bennett pinch relation (9.31)

$$I^2 = 16\pi NkT/\mu_0.$$

9.7.3 **Plasma stability**

In a z-pinch configuration the plasma is not contained at the ends, but using a toroidal field, as in a tokamak, avoids this problem. However, while the z-pinch is in equilibrium, it is not a stable equilibrium—it is unstable to sausage and kink instabilities, which are illustrated in Fig. 9.14.

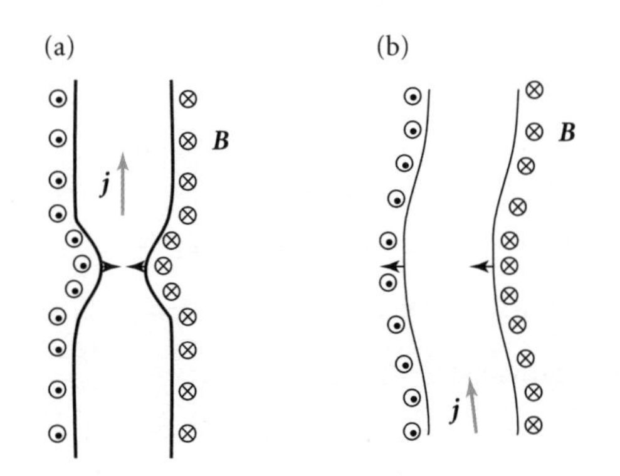

Fig. 9.14 (a) Sausage instability. (b) Kink instability.

In the sausage mode, Fig. 9.14(a), a small radial contraction of the plasma current causes the azimuthal magnetic field and the radially inward force to both increase, leading to further radial contraction. The volume of the plasma can remain constant by bulging out so there is no compensating force and the plasma column is highly unstable. Likewise in the kink mode (Fig. 9.14(b)) the magnetic field increases on the inward side, thereby increasing the force and the magnitude of the kink. The sausage instability is also seen in the break-up of a thin stream of water falling from a tap. Surface tension (Rayleigh) waves grow in amplitude as the jet falls and lead to the formation of droplets.

In a tokamak the toroidal component of magnetic field tends to stabilize the plasma to both sausage and kink instabilities. How the sausage instability is stabilized is explained in Derivation 9.4. The tokamak safety factor q is defined as the ratio of toroidal to poloidal turns of the magnetic field and is close to unity near the centre of the plasma. The variation of q with distance from the plasma centre ($S = (r/q)\mathrm{d}q/\mathrm{d}r$), called the **magnetic shear**, is an important parameter in determining the transport of energy and particles.

Derivation 9.4 Stabilization of the sausage instability in a tokamak

Consider a cylindrical section of plasma carrying a current I as in Fig. 9.13, but with a magnetic field B_z along the axis. The magnetic pressure p_B equals $B^2/2\mu_0$. The azimuthal component B_θ produces an inward pressure on the cylindrical plasma of magnitude $B_\theta^2/2\mu_0$. In the sausage instability p_B increases as the radius a of the plasma decreases, because $B_\theta = \mu_0 I/2\pi a$ increases as a gets smaller.

The axial field B_z produces an outward pressure on the plasma. A contraction of the plasma's radius causes B_z to increase as currents are induced in the plasma opposing the change. In the MHD approximation the plasma acts like a perfect conductor. In this limit the induced currents exactly maintain the magnetic flux through a section perpendicular to the axis of the plasma. The flux Φ equals the field multiplied by the cross-sectional area, i.e. $\Phi = B_z \pi a^2$. The current I is constant so both $B_z a^2$ and $B_\theta a$ are constant when the plasma radius changes. So $\partial(B_z a^2)/\partial a$ and $\partial(B_\theta a)\partial a$ are zero, i.e.

$$\partial B_\theta/\partial a = -B_\theta/a \quad \text{and} \quad \partial B_z/\partial a = -2B_z/a. \tag{9.36}$$

For stability we require the inward magnetic pressure p_B to increase if a increases. The pressure p_B can be expressed as

$$p_B = (B_\theta^2 - B_z^2)/2\mu_0. \tag{9.37}$$

Differentiating p_B with respect to a and using eqn (9.36) gives

$$\partial p_B/\partial a = (B_\theta \partial B_\theta/\partial a - B_z \partial B_z/\partial a)/\mu_0 = (2B_z^2 - B_\theta^2)/a\mu_0. \tag{9.38}$$

This expression is positive provided

$$B_z^2 > B_\theta{}^2/2, \tag{9.39}$$

i.e. the plasma is stable against a sausage instability provided the inequality (9.39) is satisfied.

9.7.4 **Energy confinement**

The energy confinement time τ_E is determined by the time taken for particles and energy (both kinetic and radiation) to be transported from the hot plasma to the outside wall. Although a fusion plasma is collisionless to a good approximation, some collisions do occur. A typical deuteron ion in a plasma at a temperature of 10 keV has a velocity of ~1000 km/s, so in a pulse of a few seconds duration the deuteron travels a great distance. Plasma particles spiralling along a magnetic field will scatter and transfer energy to other particles over a distance of about ρ—the Larmor radius. This diffusion process, called **classical transport**, will have a diffusion coefficient of order $\rho^2/\tau_\perp$, where $\tau_\perp$ is the characteristic time between collisions. There is a larger contribution (called **neoclassical transport**) from trapped particles. Particles with only a small component of velocity in the direction of the helical $\boldsymbol{B}$ field in the tokamak can be reflected through the mirror effect as they move to a smaller radius where the field is greater as it depends on r^{-1} (see Fig. 9.15).

The characteristic size of these banana-shaped trapped orbits is greater than ρ and so they contribute more to diffusion losses. However, transport is observed in many operating configurations to be much faster, where it is caused predominantly by micro- and macroturbulence. Microturbulence involves fluctuations over distances of the order of 0.1 mm to 10 mm, due to density and temperature gradients. These fluctuations form unstable waves that transport energy across field lines from the hot centre to the cooler edge of the plasma and then to the containing wall. The process is not well understood and much research effort is being directed towards modelling turbulence.

The control of macroturbulence, involving disturbances over distances of the order of 0.1 m to 1 m, has been largely achieved in tokamaks by choosing particular operating conditions. A significant advance in improving confinement came with the discovery of an H-mode (high confinement mode) in the ASDEX tokamak in Germany. It was associated with the use of a divertor in which the magnetic field is altered to keep the main interactions of the plasma with any material away from the central plasma region.

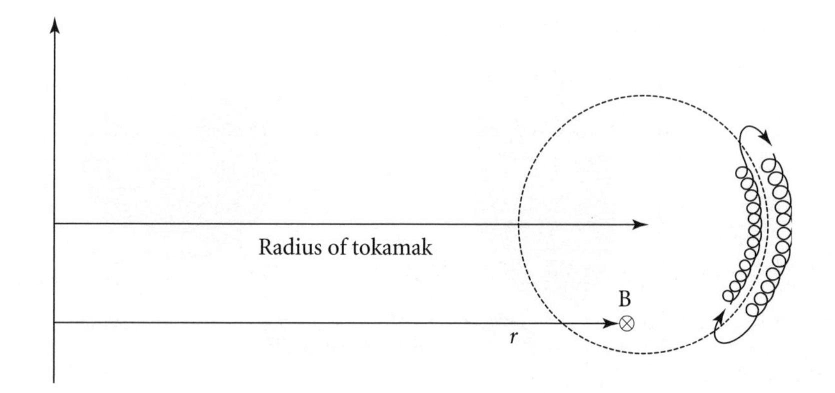

Fig. 9.15 Trapped particle in a banana orbit.

9.8 Divertor tokamaks

Improved performance was obtained in Russia in the early tokamak experiments by introducing an artificial boundary, known as a **limiter**. It was used to define the radial extent of the plasma, as illustrated in Fig. 9.16. Diffusing particles once they pass the last closed flux surface rapidly spiral along the field lines and deposit their energy on the limiter. The limiter can be cleaner, have a higher melting point than the containing wall, and also have a lower Z. The latter is important as high Z impurity atoms in the plasma cause a high radiative loss through line radiation. Low Z atoms are fully ionized in the hot centre of the plasma and can only radiate by bremsstrahlung, which is much weaker than line radiation. However, it is difficult to keep the deposited power density low enough to avoid significant melting or sublimation of the limiter material.

By elongating the field vertically in a tokamak it is possible to form a cross-over, called a **separatrix**, in the magnetic field configuration near the wall. This configuration diverts the plasma away from the main central region on to **divertor plates**, as shown in Fig. 9.17.

The plasma is incident over a larger area which helps reduce the power density. In addition, gas is added near the plates to increase the radiative loss and, through charge exchange,

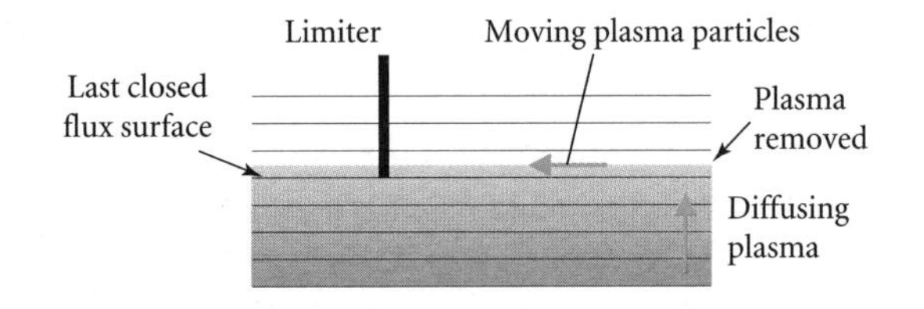

Fig. 9.16 Limiter in a tokamak.

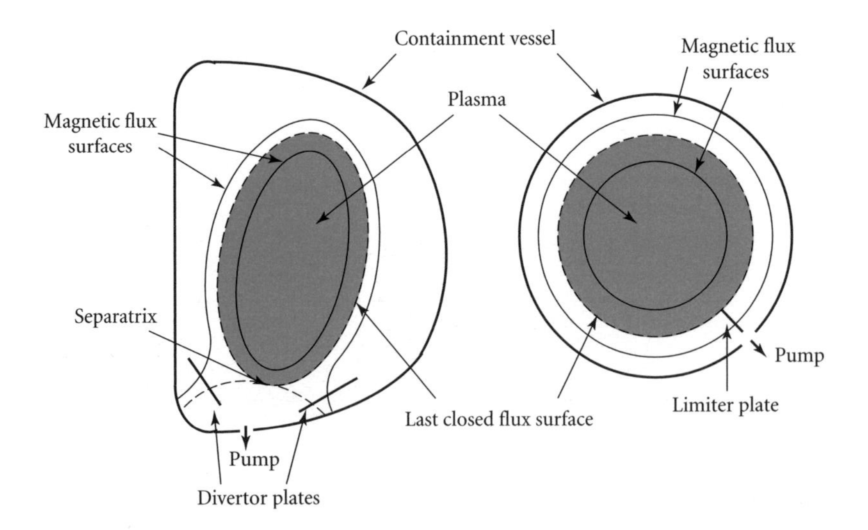

Fig. 9.17 Limiter and divertor tokamak configurations.

neutralize the energetic charged particles. Both the radiation and the neutral particles spread the heat load. The temperature of the plasma must be sufficiently low, of the order of eVs, for the gas, such as nitrogen, to retain some of its atomic electrons, and thereby emit line radiation. If the temperature is too high, the gas atoms are nearly fully ionized and can only radiate through bremsstrahlung, a much weaker process than line radiation.

The divertor needs to be made from a low Z material so that any evaporated atoms are fully ionized in the hot plasma. Carbon with its very high sublimation temperature of 3825°C and beryllium have been used.

9.8.1 H-mode confinement

When sufficient plasma heating was supplied in the ASDEX divertor tokamak, a high confinement mode was found with τ_E approximately twice as large as in the normal configuration (low confinement L-mode). The H-mode develops as turbulence at the plasma edge is suppressed, which gives rise to a transport barrier. The pressure distribution in different modes is shown in Fig. 9.18. The L-mode can be affected by periodic disruptions, called **sawtooth oscillations**, which reduce the stored energy but expel impurities from the plasma core. These are associated with the formation of magnetic islands in the field distribution in which magnetic energy is converted to kinetic energy. (This process, called **magnetic reconnection**, is important in solar flares.) They can be stabilized by additional plasma heating.

The H-mode shows a sharp pressure drop near the edge of the plasma, and over an extended region transport is close to neoclassical (Section 9.7.4). Above a critical pressure gradient

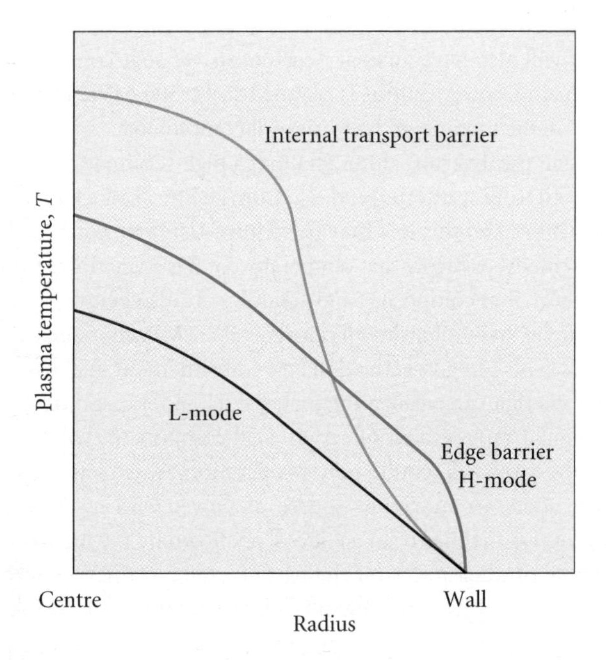

Fig. 9.18 Radial pressure distribution for different modes in a tokamak.

MHD instabilities develop, giving edge localized modes (ELM). These modes reduce the edge pressure periodically and expel particles and energy out of the plasma.

Improved confinement can also be achieved by generating a radial current distribution in the plasma to give negative magnetic shear. This occurs when the centre of the plasma has less current flowing than in the outer part of the plasma. Negative magnetic shear can reduce ion turbulent transport and set up an internal transport barrier (ITB), resulting in higher central temperatures. Feedback can also be used to control instabilities, for example, by using ECRH to change the temperature and therefore the resistivity of the plasma, and hence the current flow.

9.9 Outlook for controlled fusion

The successful culmination of the experiments at JET in 1997, when fusion with a 50–50 D–T plasma produced 16 MW of power corresponding to a $Q \sim 0.6$, made the prospects of a commercial fusion reactor by 2050 much more feasible. The H-mode provides a fusion plasma that can be confined and from which the output heat and particles can be extracted. However, considerable research is required on the materials to be used, particularly for the divertor, and on avoiding excessive heat loads from the ELM instabilities.

These problems will be addressed in the International Thermonuclear Reactor (ITER), which will be a tokamak designed to give a $Q \sim 10$ and provide data upon which a demonstration reactor (DEMO) can be built. ITER will produce significant power under quasi-stationary conditions, investigate α-particle heating and particle wall interactions and test the tritium breeder lithium blanket design. An international collaboration of the EC, Japan, United States, and the Russian Federation is involved. ITER is planned to have an inductive drive that will last for 8 minutes. It will also have auxiliary current drives that could extend operation up to 30 minutes. The fusion power output is planned to be 500 MW. In particular ITER will provide information on the lifetime of the torus wall components.

The use of graphite in the divertor, although it has a high sublimation temperature shows a high erosion rate—both from sputtering and also from chemical attack to form hydrocarbons. Graphite also traps tritium and this loss may be serious. Using tungsten with a melting point of 3695°C may be a remedy, as below that temperature it has a very low erosion rate and does not trap tritium. Carbon fibre composites and ceramics are also being considered.

In a fusion reactor neutron bombardment can deposit $\sim 2\ \mathrm{MW\ m^{-2}}$ and cause transmutation of structural materials. This can give activation and embrittlement due to α production. It can also cause lattice defects that can weaken the material. High temperatures, though, can cause partial annealing. Regular replacement of certain wall components will be necessary and the radioactivity within the torus will require remote operation, which has been pioneered at JET.

Controlled fusion offers an enormous source of power with essentially no greenhouse emissions and the progress that has been achieved, as shown by the increase of four orders of magnitude in the triple product $n\tau T$ over the past 40 years, justifies a continued investment in fusion research. Research is also ongoing into **inertial confinement fusion** and this method is described in Box 9.1. Though the timescale for a commercial reactor may still be several decades from fruition, the long-term benefit to future society cannot be overstated.

Box 9.1 Inertial confinement fusion (ICF)

Principle of ICF

Another approach to obtaining controlled fusion is to use lasers to heat up small spheres of deuterium and tritium to very high temperatures and pressures. By focusing the laser light so that the spheres are uniformly illuminated, it is predicted that the density can be raised sufficiently for fusion to take place. The process is illustrated in Fig. 9.19.

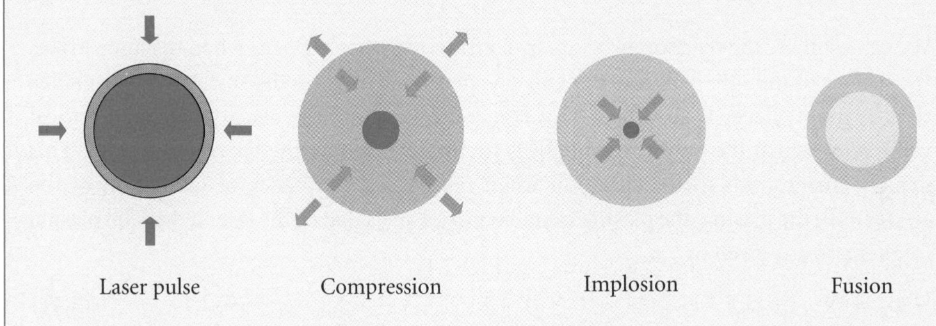

Fig. 9.19 Laser light focused on a D–T spherical shell causing implosion and fusion.

When pulses of laser light strike the surface of the D–T spheres a plasma is created. The hot plasma expands rapidly into the vacuum surrounding the spheres. This expansion causes a reaction pressure on the surface (Newton's third law) of the spheres that compresses them to very small radii. This compression raises the density by a large factor and initiates fusion. If the final radius is r_f then the plasma confinement time can be estimated as the time for the sphere's volume to double. The expansion is limited to that of the speed of sound in the plasma u_s so the confinement time τ_E is given by

$$\tau_E \sim r_f/4u_s. \tag{9.40}$$

The plasma is confined by its own inertia, hence the name **inertial confinement fusion.**

ICF operating conditions

In inertial confinement fusion (ICF) the Lawson criterion eqn (9.9) can be satisfied by having a very large number density n compared with that in magnetic confinement fusion (MCF). The confinement time τ_E required is correspondingly much smaller. For break-even at a temperature of 10 keV the requirement on the product $n\tau_E$ is given by eqn (9.9) as

$$n\tau_E > 5 \times 10^{19}\ \text{m}^{-3}\ \text{s}.$$

Substituting for τ_E from eqn (9.40) gives a minimum condition on the product nr_f. At break-even the energy supplied by the laser equals that produced in fusion, which is $E_{\text{fusion}} \times n4\pi r_f^3/3$. The minimum energy required in each laser pulse is therefore proportional to $1/n^2$.

Lasers can currently produce pulses of about 1 MJ energy. This energy translates to a minimum number density required that corresponds to a minimum pressure of order 10 Gbar. We will now look at what pressure is created when a laser is focused on a sphere.

Reaction pressure

We can estimate the reaction pressure by considering what happens when the laser strikes the surface of the sphere in more detail. The plasma density on the surface of the spheres rises as more energy is absorbed. This rise continues until the density reaches a critical value ρ_c at which the unabsorbed light is totally reflected by the charged particles. This density corresponds to the situation when the natural frequency of oscillation of the electrons in the plasma, the plasma frequency ω_p, equals that of the laser light. The plasma frequency ω_p is given by

$$\omega_p^2 = ne^2/(m_e\varepsilon_0) \tag{9.41}$$

where n is the number density of electrons and m_e is the mass of an electron. As there are two electrons per D–T molecule, the connection between ρ_c and the corresponding n_c is

$$\rho_c = 2.5 n_c m_p \tag{9.42}$$

where m_p is the mass of a proton.

At this critical density ρ_c, the pressure in the plasma is p_c and the speed of sound u_s is given by

$$u_s \sim (p_c/\rho_c)^{1/2}. \tag{9.43}$$

The reaction pressure p_a, arising as the material of the sphere is ionized (a process called **ablation**), is made up of two components: the plasma pressure p_c and the reaction from the flow of material away from the surface. As the mass flow per second per square metre away is $\rho_c u_s$, the rate of change of momentum per square metre, i.e. the pressure, from the flow of material is $\rho_c u_s^2$. The two components are of the same magnitude, so $p_c \sim \rho_c u_s^2$ and the total pressure p_a is

$$p_a \sim 2\rho_c u_s^2 \sim 2p_c. \tag{9.44}$$

The energy per second per square metre, W, absorbed by the plasma from the laser light raises the temperature and pressure of the plasma and gives energy to the ejected material. These are of comparable magnitude so $W \sim \rho_c u_s^3$ and

$$p_a \sim 2\rho_c^{1/3} W^{2/3}. \tag{9.45}$$

EXAMPLE 9.4

Estimate the total pressure p_a generated on D–T sphere when using a laser with $W = 10^{19}$ W m^{-2} and $\lambda = 1$ μm. Assume half of the laser light energy is absorbed.

For $\lambda = 1$ μm, $\omega_c = 2\pi c/\lambda = 1.885 \times 10^{15}$ rad s^{-1}. Substituting in eqn (9.41) gives $n_c = 1.12 \times 10^{27}$ m^{-3}. From eqn (9.42) $\rho_c = 4.7$ kg m^{-3} so, using eqn (9.45) $p_a \sim 9.8 \times 10^{12}$Pa = 98 Mbar.

We can see from Example 9.4 that pressures of only ~100 Mbar can be created by ablation. Fortunately this pressure is amplified by the implosion of the sphere.

Pressure amplification from implosion

The spheres can be made in the form of a thin spherical shell of plastic of thickness Δr surrounding the D–T fuel. The ablation pressure from the laser pulse causes the shell to implode raising its pressure. There is also a shock wave created that can initiate fusion if the pressure is high enough. For laser pulses with an energy of about 1 MJ, this pressure is about 20 Gbar. The shock wave creates a very high temperature pulse right at the centre of the compressed plasma that starts the burn.

We can estimate the pressure gain roughly using a simple model. The ablation pressure acts for a distance of order r_0, where r_0 is the radius of the shell. The energy gained will be about $(p_a 4\pi r_0{}^2)r_0$, (force multiplied by distance). This kinetic energy is transformed into the internal energy of the plasma formed from the shell material. The plasma occupies a volume about that of the original shell, $4\pi r_0{}^2\Delta r$, and its pressure acts on the D–T fuel.

As the pressure of a gas equals the internal energy per unit volume (N.B. $p = nkT$), then the final plasma pressure p_f is given by

$$p_f \sim (p_a 4\pi r_0^2)r_0/(4\pi r_0^2 \Delta r) \sim p_a(r_0/\Delta r).$$

The ablation pressure has been increased by the aspect ratio $r_0/\Delta r$. For an aspect ratio $r_0/\Delta r$ of 200 then, a final plasma pressure of ~20 Gbar would be achieved for $p_a \sim 100$ Mbar.

There is a limitation to the size of aspect ratio $r_0/\Delta r$ that can be used because of a hydrodynamic instability known as the **Rayleigh–Taylor instability**. This instability occurs when two fluids in contact have density and pressure gradients of opposite sign. An example would be a film of water floating on a layer of oil. The pressure in the oil layer would be higher than in the film of water, as the weight of water acts on the oil. However, the oil has a lower density than water. The result is that the layers are unstable and the oil would rise on top of the water if the layers were disturbed.

A similar situation occurs in the implosion of the spherical shell. The density of the shell is higher than the plasma on its surface caused by the laser pulse. But the pressure in the plasma is much higher than in the shell. As a result any imperfections in the spherical symmetry of the shell or in the illumination will grow in size and can cause the shell to break up before it has been fully compressed.

ICF reactors

In magnetic confinement fusion (MCF), the plasma has $n \sim 10^{20}$, so the confinement time τ_E required to satisfy the Lawson criterion for break-even is greater than $\frac{1}{2}$ second. For inertial confinement fusion (ICF), the initial density of the D–T fuel is $\sim$200 kg m^{-3}. As we have seen the implosion of the shell raises the pressure and density enormously. At the very centre the density is expected to be $\sim 10^5$ kg m^{-3}. This density corresponds to $n \sim 10^{31}$, so only $\tau_E > 5 \times 10^{-12}$ second is required for break-even.

In ICF τ_E is given by eqn (9.40) so the radius r_i of the very dense core of the plasma, where ignition occurs, must be greater than

$$r_i > \tau_E u_s.$$

The speed of sound is about $\sim$500 km s^{-1} (see Exercise. 9.11), so the initial burn would then occur within a sphere of radius of a few microns. This burn would heat up the remaining fuel to fusion.

EXAMPLE 9.5

Calculate how many D–T spheres would be required to fuse per second to provide energy for a 1 GW$_e$ power station. The radius of the D–T volume is 2 mm. Estimate the maximum Q.

The density of the fuel is $\sim$200 kg m^{-3} so the amount is 6.7 mg. A mole of D–T is 5 g so the total number N of D–T pairs of atoms is 8.0×10^{20}. When these fuse they release 17.6 MeV per D–T fusion, 2.25 GJ in total. A 1 GW$_e$ power station requires $\sim$3 GW thermal power, so a minimum of about 1.3 spheres per second would be required.

The energy stored in the compressed plasma at 10 keV is $6NkT \sim 7.68$ MJ as there are two electrons per D–T pair. The maximum Q is therefore $\sim$290. In practice the Q would be much lower due to losses.

Outlook for ICF

Laser-induced inertial confinement fusion is being actively pursued in several laboratories around the world. At the National Ignition Facility at Lawrence Livermore National Laboratory in California, there are 192 neodymium glass lasers under construction. These are designed to send 1.8×10^6 J pulses of energy at a power of 500×10^{12} W on to millimetre size spheres at the centre of a 10 m target chamber. The facility is due to be completed by mid-2009. A megajoule laser facility is also under construction in France.

However, current high-powered lasers have a low duty cycle. One of the most difficult tasks facing ICF is to design a device, called a driver, that can produce MJ pulses of energy of duration of about a nanosecond, at a repetition rate of 1–10 Hz. Good efficiency is also required. One alternative under consideration is to use particle accelerators.

Another difficulty is ensuring uniform illumination of the target. One technique under investigation is to enclose the D–T sphere in a gold cavity called a **hohlraum**. The laser light vapourizes the gold into a plasma which radiates X-rays strongly. These X-rays are then absorbed in the surface of D–T sphere and cause it to implode. This indirect method is predicted to give good uniformity.

Although there are still problems to solve in ICF, the potential benefit to mankind from harnessing fusion power argues for pursuing both MCF and ICF research.

SUMMARY

- Fusion has a very high energy density: 100 kg of deuterium and 150 kg of tritium would produce 1 GW for one year. Tritium is produced by neutron reactions in lithium.
- Resources of lithium and deuterium are sufficient to provide the world with 10 TWy of energy annually for $\sim$1000 years.
- Fusion requires plasma at temperature of $\sim 10^8\,^\circ$C, equivalent to $\sim$10 keV, and the plasma must be contained away from reactor walls.
- Tokamak magnetic field configuration cancels the effects of magnetic drift of charged particles in a toroidal field and provides good macroscopic stability and containment.
- Divertor tokamaks have a magnetic field configuration that diverts particles diffusing outwards on to plates away from the main central region. With sufficient plasma heating, improved containment is possible (H-mode).
- Plasma heating obtained initially by resistive heating and then by neutral current injection and rf heating.
- For a self-sustaining fusion plasma the particle density n, the temperature T, and the containment time τ_E must satisfy the Lawson criterion:

  $n\tau_E T \geq 3 \times 10^{21}\,\text{m}^{-3}\,\text{s keV}$ (T in keV).

  For $T \sim 10$ keV and $n \sim 10^{20}\,\text{m}^{-3}$, $\tau_E \geq \sim 3$s.
- The increase of $n\tau_E T$ over last 35 years has been four orders of magnitude, with the Joint European Torus (JET) obtaining a ratio of fusion to input power of about 0.6 and a τ_E of $\sim$1 s in 1997 with a D–T plasma.
- Inertial confinement fusion in which pellets of D–T are heated by intense laser pulses is also under development. In this method τ_E is much shorter than in magnetic confinement, but n is much larger. The inertia of the extremely compressed ions gives the confinement.

- The International Thermonuclear Reactor (ITER), which is a tokamak design, has now (2006) been approved and will be built in the South of France. Data from this facility will be used to design a demonstration reactor (DEMO). It is hoped that the first commercial reactor will be operational by 2050.
- The importance of controlled fusion as a means of providing an almost unlimited supply of carbon-free energy cannot be overstated: both for reducing the risk of catastrophic climate change and for improving the standard of living worldwide.

FURTHER READING

Boyd, T.J.M. and Sanderson, J.J (2003). *The physics of plasmas*, Cambridge University Press, Cambridge. A good textbook on plasma physics.

Chen, F.F. (2006). *Introduction to plasma physics*, Springer, Berlin. A good introduction to plasma physics.

WEB LINKS

www.jet.efda.org/documents/wesson/wesson.html Interesting description of the science of JET by J. Wesson.

farside.ph.utexas.edu/teaching/plasma/lectures/lectures.html Good lecture notes on plasma physics by R.Fitzpatrick.

www-fusion-magnetique.cea.fr/gb/ Useful quantitative discussion of fusion.

LIST OF MAIN SYMBOLS

τ_E	energy containment time	p	plasma pressure
P_{fusion}	fusion power	T	plasma temperature
P_{ext}	external power	ρ	radius of gyration
Q	quality factor	ω_c	cyclotron frequency
W	total plasma energy	ω_p	plasma frequency
n	number of particles per unit volume		

EXERCISES

9.1 Calculate the amount of tritium required to fuel a 5 GW_e power station for 3 years.

9.2 Using the conservation of momentum, explain why in the fusion of D + T the alpha-particles receive 3.5 MeV and the neutrons 14.1 MeV.

9.3 Calculate the number density of a plasma at a temperature of 10^8 °C and a pressure of 2 bar.

9.4 What is the Larmor radius of a triton in a plasma at a temperature of 15 keV where the magnetic field is 5 Tesla?

9.5 The energy U of a dipole in a magnetic field is given by $-\mu \cdot \boldsymbol{B}$. The force acting in the z-direction on the dipole is given by $F_z = -\partial U/\partial z$. Deduce eqn (9.27) for the force acting parallel to the field (note the direction of the magnetic dipole moment).

9.6 The kinetic energies of the charged particles in a magnetic mirror remain constant and their magnetic moments are conserved. Deduce the condition on $\sin^2 \alpha_m$ given by eqn (9.30).

9.7 Calculate the ion cyclotron resonance frequency for tritons in a plasma when the magnetic field is 4 tesla.

9.8 A current of 3 MA flows in a tokamak plasma that has $N = 2 \times 10^{20}$. Estimate the temperature of the plasma.

9.9 Estimate the distance travelled by a triton in a 15 keV plasma contained for 5 seconds.

9.10 Describe the main features and principles of operation of a tokamak.

9.11 Estimate the speed of sound in the dense plasma formed after the implosion of a spherical shell of D–T material. Assume a plasma density of 10^5 kg m^{-3} and that the pressure of the plasma is 2×10^5 Mbar.

9.12 Compare and contrast inertial and magnetic confinement fusion.

10 Generation and transmission of electricity, energy storage, and fuel cells

List of Topics

- Electromagnetic induction
- Generators
- Transformers
- High voltage transmission
- AC versus DC
- Rectifiers and thyristors
- Electricity grids
- Energy storage
- Pumped energy storage
- Compressed air storage
- Flywheels
- Superconducting magnetic energy storage
- Batteries
- Fuel cells

Introduction

In the earlier chapters we considered the ways in which various forms of primary energy are exploited to do work on a turbine. We now describe how the rotational energy of the turbine is converted into electrical energy and how electricity is transmitted to consumers. We also investigate various forms of energy storage, batteries, and fuel cells.

10.1 Generation of electricity

The basic principle of electricity generation is illustrated in Fig. 10.1. Consider a planar loop of conducting wire of area A, rotating at a constant angular velocity ω in a uniform magnetic field B. Suppose the loop consists of N turns and subtends an angle $\theta = \omega t$ to the magnetic field at some instant t. The magnetic flux intersecting the loop is given by

$$\varphi = NBA \cos\theta = NBA \cos\omega t. \quad (10.1)$$

By Faraday's law of electromagnetic induction, the electromotive force (i.e. the work done on unit charge) is equal to the rate of change of magnetic flux. Thus

$$V = -\frac{d\varphi}{dt} = NBA\omega \sin\omega t. \quad (10.2)$$

If the electrical circuit is completed by connecting the ends of the wire across a resistance R, an alternating electric current flows through the resistance, given by

$$I = \frac{V}{R} = I_0 \sin\omega t \quad (10.3)$$

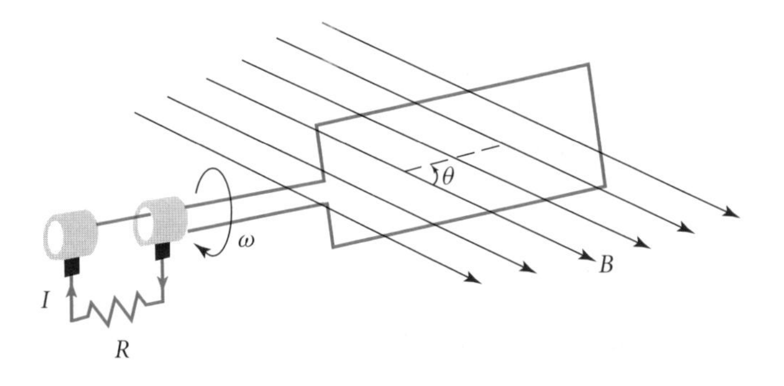

Fig. 10.1 Loop rotating in uniform magnetic field.

where

$$I_0 = \frac{NBA\omega}{R}. \tag{10.4}$$

In order to make electrical contact between the moving and stationary parts, slip rings and graphite brushes are inserted between the rotating conductor and the external circuit. It should be noted that the current induced in the wire is an effect that exists only while the conductor rotates in the magnetic field; thus $I \to 0$ as $\omega \to 0$.

EXAMPLE 10.1

A coil with 1000 turns of cross-sectional area 1 m^2 rotates at 50 Hz in a uniform magnetic field of 0.5 T. What is the maximum current flowing through a load of 1000 Ω?

From eqn (10.4), the maximum current is given by

$$I_0 = \frac{NBA\omega}{R} = \frac{10^3 \times 0.5 \times 1 \times 2 \times \pi \times 50}{10^3} \approx 157 \text{ A}.$$

In a typical large generator, the magnetic field is produced by passing a direct current through coils mounted on a central rotating shaft connected to the turbine, called the **rotor**. There is a small loss of energy due to resistance heating of the coils but it is more economic than using permanent magnets. The rotor is surrounded by a set of stationary coils wound around an iron core, called the **stator**. The rotating magnetic field due to the rotor intersects the stationary coils in the stator and induces a current. The frequency of generation is determined by the angular velocity of the rotor. It is chosen to be high enough to avoid flickering of electric lights—60 Hz in North America, South America, and parts of Japan and 50 Hz in Europe.

The configuration of the windings in the stator is very complicated. The general principle is to maximize the emf, but it is also necessary to minimize the flow of **eddy currents**: self-circulating currents that produce unwanted components of magnetic field and cause losses. Eddy currents are reduced by increasing the resistance of the paths through which eddy currents flow, i.e. by laminating the core using thin sections of steel alloys, thereby forcing the current to flow mainly through the laminations. The evolution of the design of rotors and stators has been something of a black art and details of particular machines tend to be commercially sensitive.

Electricity is usually generated as three-phase current rather than single-phase current; this is achieved by employing three independent sets of windings, 120° apart, around the generator (Fig. 10.2). The idea of using three phases was originated by Nikola Tesla, a pioneer in the early years of electricity generation and transmission. Unlike single-phase, three-phase power never drops to zero and the power delivered to a resistive load is constant in time [Exercise (10.5)]. Another advantage of three-phase current is that it needs only 75% of the material to conduct the same quantity of power as in single phase.

Heat is dissipated in the stator, which needs to be cooled in order to maintain it at a constant temperature; at higher temperatures the resistance of the wiring increases and the life of the insulation is shortened. For medium-sized machines, forced cooling using a rotor-mounted fan is sufficient, but for large machines stators are cooled using de-ionized water and hydrogen (which is a better heat conductor than air).

Mechanical stability is also an important consideration, particularly in large machines. A plant outage to repair a large generator is a time-consuming and costly business, and usually means that less efficient generating plant has to be used. It is therefore normal practice to monitor mechanical vibrations for early signs of metal fatigue and cracks before a major incident occurs.

Apart from electricity generation by conventional turbines (i.e. steam, gas, or water turbines) described earlier in the book, special-purpose generators are needed for exploiting certain kinds of alternative energy. The case of wind turbine generators is described in Box 10.1.

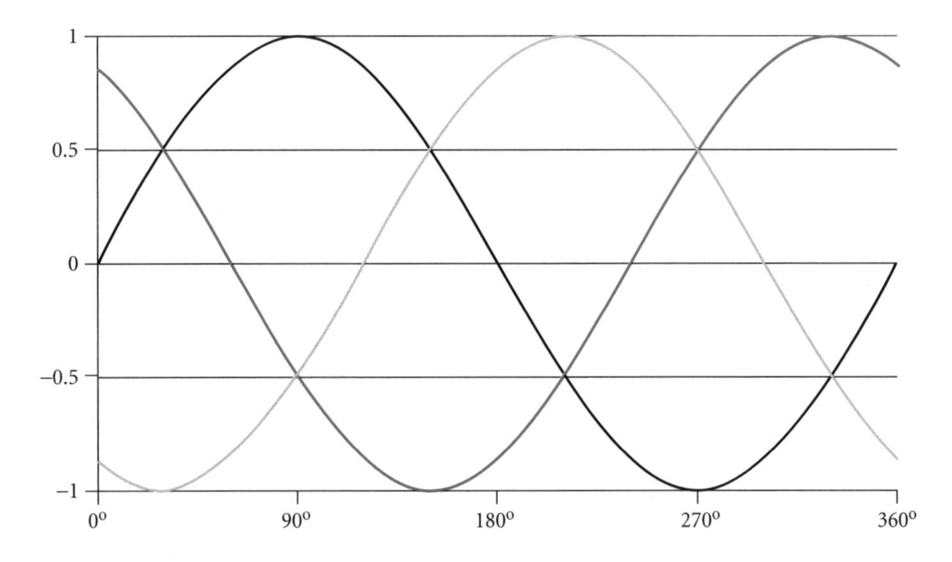

Fig. 10.2 Three-phase current.

Box 10.1 Wind turbine generators

Generators work by moving magnetic field lines through conductors and thereby inducing a current. The magnetic field can be provided by current flowing in coils, or by permanent magnets. The reduction in cost of rare earth magnets (in particular NdFeB magnets) has led to permanent magnet generators being built with capacities of several MW. NdFeB has a very high *BH* product where *B* and *H* are the magnetic flux density and the magnetic field strength, respectively, which means that only a small amount of magnetic material is required to produce a strong magnetic field. The magnets are attached to a cylindrical drum (the rotor) connected to the turbine shaft. Surrounding the rotor are conducting coils attached to the inside of a fixed cylinder (the stator). As the magnets rotate, an alternating current is induced in the coils.

In a variable speed turbine the alternating current is first rectified to direct current and then converted back to alternating current at a fixed frequency for connection to the grid. A torque T is required to turn the rotor, since the induced current produces a magnetic field that opposes the motion of the magnets (Lenz's law). The opposing tangential force per unit area (i.e. the shear stress), $\langle\tau\rangle$, depends on the surface current density K (amperes per metre) and on the strength of the magnetic field B. (N.B. For a single conductor of length L carrying a current I normal to a field B, the force is BIL.) The magnitude of the current density is limited by dissipation and the maximum B field is limited by magnetic saturation; as a result, $\langle\tau\rangle$ is typically between 25 and 100 kPa.

The size of generator required depends on the torque T produced by the wind turbine, which is given by the product of $\langle\tau\rangle$, the surface area of the rotor, and the radius of the rotor, i.e.

$$T = 2\pi rl\langle\tau\rangle r \quad \text{or} \quad T/V_R = 2\langle\tau\rangle$$

where V_R is the volume of the rotor. From Chapter 5, eqn (5.4), the power P absorbed from the wind is given by

$$P = \tfrac{1}{2}C_P\rho Au_0^3 = T\Omega$$

where C_P is the power coefficient, ρ is the air density, u_0 is the wind speed upstream of the turbine, Ω is the angular velocity, and A is the swept area of the turbine. Ω is related to the radius R of the turbine, the tip-speed ratio λ, and the wind speed u_0 by

$$\Omega = \lambda u_0/R.$$

Hence

$$V_R = C_P\rho Au_0^3/(4\Omega\langle\tau\rangle) = \pi R^3 C_P\rho u_0^2/\{4\lambda\langle\tau\rangle\}.$$

For a 2 MW turbine operating with $C_P = 0.45$ at a wind speed of $12\ \text{m s}^{-1}$, $R = 35$ m, and a tip-speed ratio $\lambda = 8$, the volume of the rotor with $\langle\tau\rangle = 80$ kPa is

$\sim$4.1 m^3 (ρ_{air} $\sim$1.2 kg m^{-3}), i.e. $\sim$2 m in diameter and 1.3 m long. We can see that the generators become very large for outputs of the order of MW but the simplicity of having no gearbox reduces maintenance costs (and noise) and improves reliability.

10.2 High voltage power transmission

To transmit large quantities of power the following issues need to be addressed.

- What is the optimum voltage for long-distance transmission?
- How can the voltage be increased and decreased?
- Is it better to transmit AC or DC?

To answer the first issue we consider the heating due to the resistance of a long-distance transmission line, known as **ohmic** (or **Joule**) **heating**. In principle, a superconducting cable would be the perfect solution, but no material has yet been found that is superconducting at ambient temperatures.

Consider a wire of resistivity ρ, cross-sectional area A, and length L. The total resistance of the wire is

$$R = \frac{\rho L}{A}. \tag{10.5}$$

Suppose the wire conducts a current I at a voltage V. The loss of power due to the resistance of the wire is

$$\Delta P = RI^2 = \rho \frac{I}{A} LI.$$

Putting $I = P/V$, we have

$$\frac{\Delta P}{P} = \rho \frac{I}{A} \frac{L}{V}. \tag{10.6}$$

In practice there is an upper limit to the current density I/A that can be conducted. Otherwise the wire would get too hot. Hence, the fractional loss of power $\Delta P/P$ is proportional to L/V, and it follows from eqn (10.6) that the operating voltage should be as high as possible to minimize the power loss. The total loss of power for a national grid due to long-distance high voltage transmission and local distribution is typically about 5–10%.

EXAMPLE 10.2

A power plant transmits 100 MW of power along a transmission line of length 50 km with a resistance of 0.01 Ω km^{-1}. Calculate the percentage loss of power if the line is at (a) 10 kV; (b) 400 kV.

The resistance of the complete line is $R = 0.01 \times 50 = 0.5\ \Omega$. For a line at 10 kV, the current needed to transmit a power of 100 MW is $I = \frac{P}{V} = \frac{10^8}{10^4} = 10^4$ A. The fractional power loss is $\frac{RI^2}{P} = \frac{0.5 \times 10^8}{10^8} = 50\%$. For a line at 400 000 kV, the current is $I = \frac{P}{V} = \frac{10^8}{4 \times 10^5} = 250$ A and the fractional power loss is $\frac{RI^2}{P} = \frac{0.5 \times 250^2}{10^8} \approx 0.03\%$.

High voltage overhead lines operate between about 110 and 1200 kV. In Europe the typical voltage for long-distance transmission is around 400 kV. The capital cost of overhead lines is typically about 10% that of underground lines for the same load capacity. Underground lines are used only where overhead lines are unacceptable, e.g. in built-up areas and underwater crossings. Increasing or decreasing the voltage is straightforward for AC transmission, using transformers (Section 10.3). However, in the case of DC transmission, electronic devices are required, which are more expensive (Section 10.4).

Power is normally transmitted as three-phase, each phase being conducted at a different height above the ground. The tops of the transmission towers are connected by a cable at earth potential, which helps to protect the transmission cables from lightning strikes. The height of the cables above the ground is determined such that the electric field at ground level is too low to endanger life beneath the transmission line.

The maximum voltage for transmission is determined by the electrical breakdown strength of air. The maximum electric field occurs at the surface of the cable and, if it becomes too large, the surrounding air becomes ionized, forming a **corona discharge** that conducts electric current to ground. The electrical breakdown strength of air is around 3×10^6 V m^{-1} for dry air, but is lower in damp conditions.

Overhead transmission cables consist of twisted strands of conducting wire (Fig. 10.3). The outer strands are usually made of aluminium (due to its low resistivity and low cost) and the inner strands are made of steel, for mechanical strength. The electric field at the outer surface varies inversely with the radius of the cable. For voltages over 110 kV, cables are usually configured in bundles, which reduces corona losses. Two-cable bundles are used for 220 kV lines and three- or four-cable bundles for 400 kV lines.

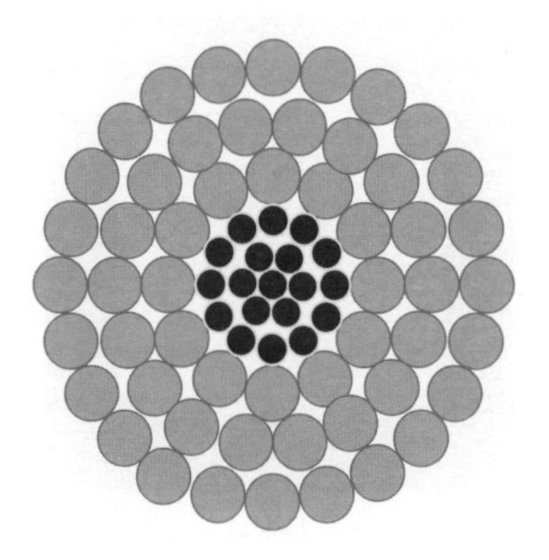

Fig. 10.3 Stranded conductor for high voltage power transmission.

There are important differences between AC and DC high voltage transmission systems. In AC transmission, the inductance and capacitance of the line present added complications that give rise to a reactive current, which produces extra losses. Capacitors and other components are required to control the reactive power flow and maintain the stability of the system voltage. For long distance power transmission, high voltage direct current (HVDC) transmission may be more economic than high voltage alternating current, due to lower ohmic and corona losses, and smaller construction costs, which offset the additional cost of converters at the ends of the line.

Public concerns have been expressed that overhead power lines are a health hazard to people living in close proximity. Workers in electric substations in the former USSR were paid an extra allowance because of perceived health risks, though these have not been substantiated elsewhere.

10.3 Transformers

The electricity in power stations is typically generated at around 18–20 kV. The power is fed into a substation, where a series of **step-up transformers** increase the voltage to that of the transmission line. Conversely, a **step-down transformer** is used to reduce the voltage from the transmission line to that of the local distribution network. The principle of a transformer is shown in Fig. 10.4.

A transformer basically consists of two coils of wire wrapped around a common iron core. In the case of a step-up transformer, the number of turns N_2 in the secondary coil is greater than the number N_1 in the primary coil, and vice versa for a step-down transformer. The iron core has a high magnetic permeability to ensure that the bulk of the magnetic flux is concentrated in the iron core.

A time-varying voltage applied across the primary coil produces a rate of change of magnetic flux, given by

$$V_1(t) = -N_1 \frac{d\varphi}{dt}. \tag{10.7}$$

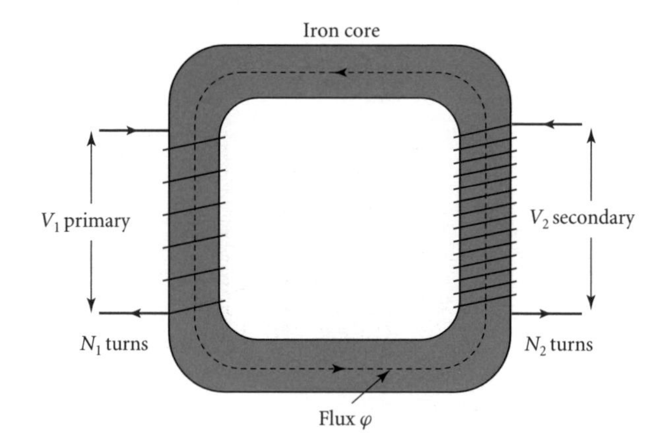

Fig. 10.4 Transformer.

Assuming there is no flux leakage, the voltage across the secondary coil is similarly given by

$$V_2(t) = N_2 \frac{d\varphi}{dt}. \quad (10.8)$$

Eliminating $d\varphi/dt$ between eqns (10.7) and (10.8) yields

$$V_2 = -\frac{N_2}{N_1} V_1. \quad (10.9)$$

Hence, for $V_2 \gg V_1$ we require $N_2/N_1 \gg 1$.

Large power transformers (over 50 MV A) are typically over 99% efficient. The main energy losses are due to the resistance of the windings; cooling systems are used to maintain transformers at a steady temperature.

EXAMPLE 10.3

Calculate the ratio of the number of turns in the secondary to the primary side of a transformer required to step-up the voltage from 20 kV to 400 kV.

From eqn (10.9), the required ratio is $\frac{N_2}{N_1} = \frac{V_2}{V_1} = \frac{400}{20} = 20$.

10.4 High voltage direct current transmission

Direct current is used for long overhead high voltage transmission lines and for underground or underwater cables, where reactive power losses for alternating current transmission would be high, such as the English Channel or the Baltic Sea between Sweden and Germany. Direct current is produced by rectifying alternating current. Until the mid-1970s, the main devices for rectification (AC to DC) and conversion (DC to AC) were gas discharge vessels called **mercury arc rectifiers**, but solid state devices called **thyristors** are now more common.

The principles of operation of a mercury arc rectifier are illustrated in Fig. 10.5. During the **on-phase**, electrons emitted from a mercury pool cathode are drawn through a steel

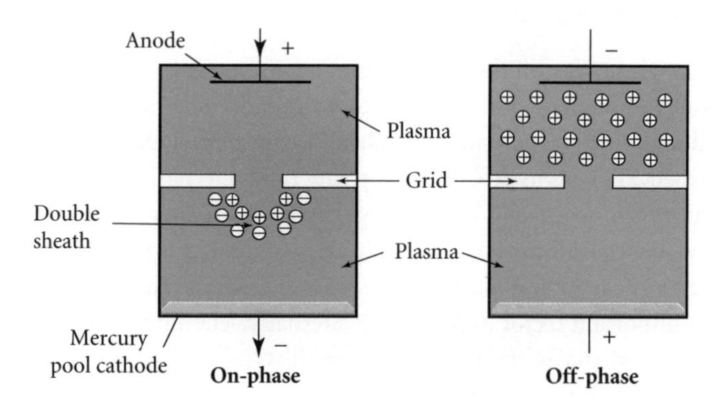

Fig. 10.5 Mercury arc rectifier.

grid towards a steel anode. The maximum current is controlled by the flow of electrons and positive ions moving in opposite directions across a charged layer covering the entrance to the grid, known as a **double sheath**. During the **off-phase**, the polarities of the anode and the cathode are reversed. Electrons are repelled from the space between the anode and the grid, leaving a residue of positive ions which accelerate towards the anode. The maximum voltage drop the device can withstand during the off-phase is determined by the need to avoid damage to the anode due to ion bombardment (sputtering). A thyristor also has two phases of operation. It consists of four semi-conducting layers of alternate n-type and p-type materials. The on-phase and off-phase are controlled by a gate that is biased positively or negatively.

10.5 Electricity grids

In most countries generating plants are interconnected by a 'grid' of high voltage transmission lines and substations. Electricity grids are complex networks that need to be managed by a central grid control unit, that decides how much power each plant can generate, taking account of the cost of the electricity generation, plant availability, the load distribution, and the need to minimize the energy losses along the transmission lines, subject to various constraints such as the maximum current rating for each line. It is also important to maintain stability of the grid, because unbalanced networks can produce excessive currents along certain lines and lead to blackouts. In order to prevent such events from occurring, transmission systems tend to be built with a degree of redundancy, with multiple pathways linking generating stations to large consumers. The loss of power along any line can then be overcome by diverting power through a different line. Nonetheless, when grid overloads do arise the results can be spectacular. In 2003, a massive power outage in parts of the United States and Canada affected about 50 million people and caused financial losses of around US$6 billion. A blackout in New York City in 1977 was followed 9 months later by a 35% increase in the birth rate.

10.6 Energy storage

Electricity grids need the flexibility of being able to store energy over time intervals varying from a few seconds to about a day. Energy storage reduces generation costs during periods of peak demand and enables the grid controllers to cope with sudden changes in electricity demand or unexpected losses in generation capacity until alternative generating units can be brought into action. Unfortunately, storing large quantities of energy in the form of electrical energy is not a viable option for large-scale electricity supply purposes, but there are many other ways that energy can be stored and converted into electrical energy when needed. Energy storage is also an important factor in the case of alternative energy sources of generation such as solar, wave, and wind, due to their inherent variability, i.e. it is important to be able to store the energy and to supply it to the grid when there is demand.

We now give a brief description of some of the different forms of energy storage.

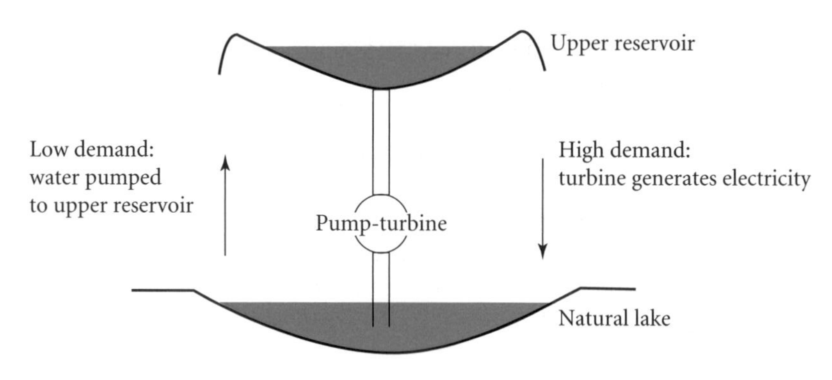

Fig. 10.6 Pumped storage.

10.7 Pumped storage

Nuclear reactors need to produce a steady output in order to maintain reactor stability and to minimize ageing effects. For countries with nuclear plants and suitable high level terrain, **pumped storage** can be an attractive option. At night-time, when electricity demand is lowest, water is pumped from a low level reservoir to an upper reservoir using (predominantly) nuclear electricity. During periods of peak demand, when electricity prices are at a maximum, water is discharged from the upper reservoir back to the lower reservoir. The same machine is used for pumping water to the upper reservoir and for generating electricity, i.e. it acts as a reversible pump-turbine. Pumped storage is essentially a form of hydropower, described earlier in Chapter 4. The operating principles are illustrated in Fig. 10.6.

A major advantage of pumped storage is that it can respond continuously to fluctuations in demand and to a sudden surge in demand, e.g. due to the loss of a generator elsewhere on the grid. The Dinorwig pumped storage plant in North Wales has a working volume of 6×10^6 m^3, a head of 600 m, and a total storage capacity of 7.8 GWh. There are six generating units, each of which can deliver 317 MW in 16 seconds from rest, or in 10 seconds if they are already spinning in air.

The environmental impact of pumped storage is confined to mountainous areas (since large heads are required) and is usually hidden from view. The capital cost of pumped storage tends to be high. In 2005, the global capacity of pumped storage was about 90 GW. In Japan, which has a large nuclear capacity, about 10% of electricity is generated using pumped storage. The potential for pumped storage is mainly in the developing world.

EXAMPLE 10.4

It is proposed to build a pumped storage plant with a head h of 500 m. How large a working volume is needed if the plant is required to generate 100 MW for 3 hours a day? Assume the efficiency of generation η is 90% and the density of water ρ is 10^3 kg m^{-3}.

From eqn (4.1) the volume throughput Q required to generate a power output P of 100 MW is given by

$$Q = \frac{P}{\eta \rho g h} \approx \frac{10^8}{0.9 \times 10^3 \times 10 \times 5 \times 10^2} \approx 22.2\ \text{m}^3\text{s}^{-1}.$$

The working volume V needed to deliver 100 MW for 3 hours is

$$V = Qt \approx 22.2 \times 3 \times 60 \times 60 \approx 240\,000\ \text{m}^3.$$

10.8 Compressed air energy storage

Another method of storing energy using cheap off-peak electricity is to pump air at high pressure into a large chamber until needed. Storing compressed air in large air-tight pressure vessels is uneconomic, but underground caverns or disused mines are feasible provided they are leak-tight. Aquifers (Chapter 2) and salt caverns are particularly suitable, salt being self-sealing under pressure.

The first large compressed air storage facility, at Huntdorf in Germany, uses an old salt cavern with a capacity of 300 000 m^3 and air is compressed to around 70 bar, and the plant can generate 300 MW for 2 hours. A compressed air energy storage in McIntosh, Alabama generates 110 MW, and a plant under construction near Cleveland, Ohio will generate 2700 MW.

Assuming that air acts like a perfect gas, the energy stored by compressing n moles of air at constant temperature from an initial volume V_i to a final volume V_f is given by

$$E = -\int_{V_\text{i}}^{V_\text{f}} p\text{d}V = -\int_{V_\text{i}}^{V_\text{f}} \frac{nRT}{V}\text{d}V = nRT\ln\left(\frac{V_\text{i}}{V_\text{f}}\right) = p_\text{f}V_\text{f}\ln\left(\frac{V_\text{i}}{V_\text{f}}\right). \tag{10.10}$$

The advantages of compressed air energy storage are quick start-up (typically 10 minutes) and long storage capability. The air is usually mixed with natural gas and burned in a combustor, after which it is expanded through a turbine and generates electricity. The option of expanding air alone (i.e. without gas) is not feasible because the temperature of the air at discharge would be very low, causing material degradation and seal leakage. A public concern over compressed air energy storage is the risk of an explosion.

EXAMPLE 10.5

A compressed air energy storage chamber has a capacity of 300 000 m^3 and compresses air from 1 bar to 70 bar. Assuming that the compression process is at constant temperature, estimate the energy stored in the chamber. Also estimate the electrical power output if the air is discharged over a period of 2 hours, assuming the overall efficiency of the plant is 24%.

From eqn (10.10), the energy stored is $E = 3 \times 10^5 \times 70 \times 10^5 \times \ln(70) \approx 8.9 \times 10^6$ MJ. The average power output over a 2 hour period is $P = \frac{0.24 \times 8.9 \times 10^6}{2 \times 60 \times 60} \approx 300$ MW.

10.9 Flywheels

Another means of storing mechanical energy is in the form of rotational kinetic energy. The idea is not new.

- Grid controllers use the 'spinning reserve' of rotors to make minor adjustments to power supply and frequency.
- Flywheel-powered buses were used in Switzerland in the 1950s.
- The flywheel in a car provides kinetic energy to keep the engine turning between piston strokes.

In recent years, flywheels for energy storage have been developed for niche markets, e.g. providing power for testing switchgear equipment, which would otherwise cause large disturbances to the local distribution network due to sudden drops in current.

Conventional flywheels for energy storage are metallic with mechanical bearings and rotate up to around 4000 rpm. By using materials that are both strong and light such as plastics, epoxies, and carbon composites, together with magnetic bearings in vacuum to minimize friction, up to 100 000 rpm can be achieved.

The kinetic energy of a flywheel with a moment of inertia I and angular velocity ω is given by

$$E = \tfrac{1}{2}I\omega^2. \tag{10.11}$$

The moment of inertia is of the form

$$I = kmr^2 \tag{10.12}$$

where m is the total mass of the flywheel situated and r is the outer radius. Hence

$$E = \tfrac{1}{2}kmr^2\omega^2. \tag{10.13}$$

$k = 1$ for a thin ring and $k = \frac{1}{2}$ for a uniform disc. Since the kinetic energy varies as the square of the radius but linearly with the mass, it is more effective to rotate flywheels faster than to make them heavier. For dynamic equilibrium, the centrifugal force is balanced by the tensile stress. The maximum angular velocity is determined by the maximum tensile stress σ_{max} that the material can withstand without breaking (see Exercise 10.14).

Typically, the energy storage capacity of flywheels made from epoxies or plastics is about 0.5 MJ/kg, i.e. about 10 times that of steel. For a 100 tonne flywheel, the storage capacity is about 15 MWh. The storage capacities of modern flywheels are comparable with those of batteries but flywheels can be energized and de-energized much more rapidly than batteries. The typical efficiency of a flywheel is about 80%. The major problems are safety and cost. Flywheel explosions due to material failure or bearing failure can be catastrophic, so strong containment vessels are essential, adding to the capital cost. The total capital cost of flywheels is typically double that of batteries or pumped storage hydropower.

EXAMPLE 10.6

A 1 tonne flywheel is a uniform circular disc of radius 1 m and rotates at 4000 rpm. Calculate the kinetic energy of the flywheel.

Putting $k = \frac{1}{2}$ in eqn (10.13), we have

$$E = \frac{1}{2} \times \frac{1}{2} \times 10^3 \times 1 \times \left(\frac{2\times3.14\times4\times10^3}{60}\right)^2 \approx 43.8 \text{ MJ}.$$

10.10 Superconducting magnetic energy storage

Superconducting magnetic energy storage (SMES) is the storage of energy in the magnetic field due to the flow of direct current in a superconducting material. The energy stored in the magnetic field is released by discharging the current in the coil. Since superconductors have no resistance, the current (and the associated magnetic field) does not decay with time once a direct current has been induced to flow in a superconducting coil.

Essentially, a SMES system consists of three components: a superconducting coil; a cooling system; and a power conditioning system (which converts AC to DC, and vice versa). The overall efficiency of an SMES is typically 95%, after allowing for losses in AC/DC conversion and cryogenic cooling of the superconducting material.

The advantages of SMES are that there are no moving parts, power is available almost immediately, and a very high power output can be delivered for short periods. The disadvantages are that superconductors operate only at low temperatures, the capital cost is high, and the energy content is fairly small. The magnetic energy stored per unit volume is

$$E = \frac{B^2}{2\mu_0} \approx 4 \times 10^5 B^2 \text{ J m}^{-3}.$$

Thus, a magnetic field of 4 teslas has an energy density of about 6.4 MJ m^{-3}.

The main application of SMES is for improving power quality for electricity supply utilities, e.g. smoothing fluctuations due to intermittent loads on transmission lines, and for manufacturing processes where ultra-smooth power supply is important.

10.11 Batteries

Batteries are electrochemical devices for storing energy in a form that can be readily converted into electrical energy. The chemicals are stored within the device from manufacture, unlike a fuel cell where the chemicals are renewed continuously. Batteries are either non-rechargeable (primary) or rechargeable (secondary), but only rechargeable batteries are of interest for large scale energy storage.

A battery is essentially a series of cells, each containing two electrodes immersed in an electrically conducting medium called the **electrolyte**. The lead–acid battery has two porous

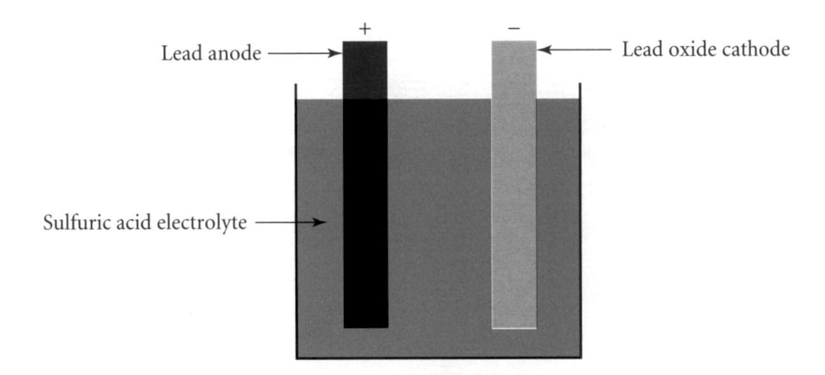

Fig. 10.7 Lead–acid battery.

electrodes; a lead anode and a lead oxide cathode, immersed in an electrolyte of sulfuric acid (Fig. 10.7). A lead–acid battery can be recycled several hundreds of times, depending on the discharge rate. They are used in the power supply industry for stand-by applications, e.g. for controlling switchgear and for providing emergency lighting. The typical energy density of a good lead–acid battery is 30 Wh kg^{-1}.

In the charging phase, the reactions are

$$Pb^{2+} + 2e^- \rightarrow Pb \qquad \text{(cathode)},$$
$$Pb^{2+} + 2H_2O \rightarrow PbO_2 + 4H^+ + 2e^- \qquad \text{(anode)},$$

and in the discharging phase the reactions are reversed, i.e.

$$PbO_2 + 4H^+ + 2e^- \rightarrow Pb^{2+} + 2H_2O \qquad \text{(cathode)},$$
$$Pb \rightarrow Pb^{2+} + 2e^- \qquad \text{(anode)}.$$

New types of batteries are being developed for large-scale applications. In Japan, **sodium sulfur batteries** are used by electricity supply utilities and by industries where non-interruptible electricity supplies are essential. The energy density of a Na–S battery is about 3–4 times that of a lead–acid battery, the materials used for construction are inexpensive, and the efficiency of conversion is around 95%. Another development is the **flow battery**, in which electrolyte is pumped around the cell. One example is the vanadium-redox battery, which can be left discharged and then recharged after long periods without damage. The energy-to-volume ratio is low but the capacity is virtually unlimited. Also, the device can respond very quickly to changing loads. The vanadium-redox battery is well-suited to large power storage applications, for dealing with surges in demand, and for smoothing fluctuations due to variable generation sources, such as wind or solar power.

10.12 Fuel cells

A **fuel cell** is an electrochemical device that can be used to generate electricity or store energy in the form of hydrogen. Unlike a battery, a fuel cell is fed continuously by chemicals. The

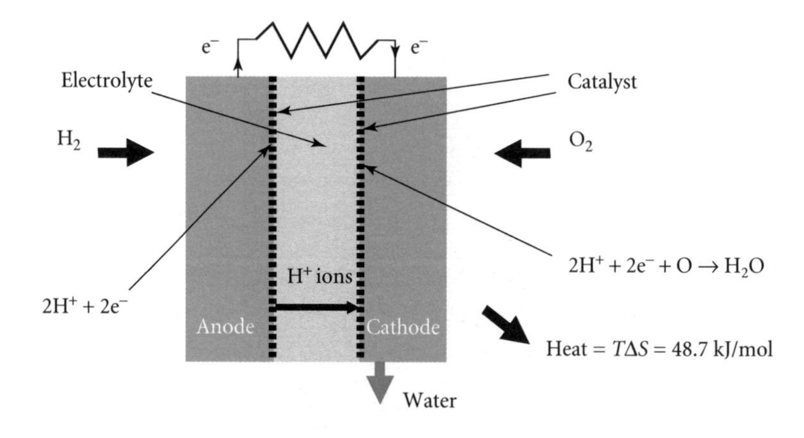

Fig. 10.8 PEM fuel cell (schematic). The electrolyte is a thin plastic membrane and the anode and cathode are porous.

chemical feed consists of hydrogen and oxygen and the fuel cell acts as a means of combining them to make water, i.e. the opposite of electrolysis. There are two electrodes (the anode and cathode), where the chemical reactions take place, and a catalyst speeds up the reactions at the electrodes. There is virtually no pollution and the only byproduct is water. Also, since a fuel cell is not a heat engine operating in a closed cycle, the Carnot limit to its efficiency does not apply.

The fuel cell was invented in 1839 by William Grove but interest in it soon fell away as cheap fossil fuels became available. It was not until the 1920s and 1930s that significant progress was made on its development. In the early 1960s, NASA decided to use fuel cells to provide electricity for the Gemini and Apollo space capsules, which led to further improvements in fuel cells. Because of their carbon-free energy and high efficiencies, interest in fuel cells is now very high. Fuel cell development has centred on the materials used for the electrodes, the catalysts, and the choice of electrolyte. Some devices use liquid electrolytes (alkalis, molten carbonate, and phosphoric acid), whilst others use solid electrolytes (proton exchange membrane (PEM) and solid oxide). We concentrate here on PEM cells, since they illustrate the physical principles involved (see Box 10.2). They are also well-developed and suitable for use in cars, though not yet cost-effective compared with gasoline or diesel fuelled engines. The main features of a PEM cell are shown in Fig. 10.8.

10.13 Storage and production of hydrogen

As yet, there is no way of storing hydrogen very compactly. It can be compressed up to 34 MPa ($\equiv$ 340 bar) in lightweight polycarbonate bottles, but even at that pressure there are only 31 g/litre of hydrogen. Liquefying H_2 is expensive and only increases the density to 71 g/litre. Metal hydrides such as $LaNi_5H_6$ and $NaAlH_4$ can hold 1.3 wt% and 3.7 wt% of hydrogen, respectively, which can be released at temperatures suitable for PEM fuel cells ($\sim$80°C).

Box 10.2 PEM fuel cell

Hydrogen is introduced on one side of the cell and flows through a porous anode, where it dissociates into hydrogen ions and electrons on contact with a catalyst. The porous electrodes are often made of a porous carbon impregnated cloth or paper 100–300 micron thick. The membrane acts as the electrolyte since it is permeable to hydrogen ions but is impermeable to electrons. One such membrane is perfluorosulfonic acid. This is a plastic made from polytetrafluoroethylene (PTFE), by reacting it with sulfonic acid, which adds ion clusters of $SO_3^-H^+$ to the polymer. Water is attracted to these clusters and, once the plastic has absorbed water, the hydrogen ions (but not the SO_3^- ions) become mobile. A hydrogen ion bonded to a water molecule can move from one SO_3^- ion to another. Consequently the hydrogen ions can migrate through the plastic. However, since the plastic is an insulator, with no free electrons, electron transport is blocked. The membrane is typically ~100 microns thick. The electrons flow through the external circuit, after which they combine on the catalyst with oxygen and the hydrogen ions, which have passed through the plastic electrolytic membrane, to form water. The overall reaction is

$$H_2 + \tfrac{1}{2}O_2 \rightarrow H_2O, \qquad \Delta H = -285.8\,\text{kJ/mol}$$

where ΔH is the change in enthalpy. The reaction occurs at constant pressure, so ΔH gives the energy released from when the hydrogen and oxygen combine and from the work done from when the volumes of the gases decrease (as explained in Chapter 2). The entropy of the gases decreases in this process, since the number of moles is reduced. In a reversible process the entropy of the entire system, reactants and surroundings, remains constant. As a result, an amount of heat equal to $T\Delta S$, where ΔS is the change in the entropy of the gases, is transferred to the surroundings. The amount of energy available as electrical energy is the change in the Gibbs free energy ΔG, where

$$\Delta G = \Delta H - T\Delta S = (-285.8 + 48.7) = -237.1\,\text{kJ/mole}.$$

In this reversible (ideal) process the efficiency of conversion is $\Delta G/\Delta H$ =83%. The voltage V generated is given by equating ΔG to the total charge ΔQ that flows multiplied by the voltage V, i.e.

$$\Delta G = \Delta Q \times V = nN_A eV$$

where n is the number of electrons released per mole, N_A is the number of molecules in a mole (Avogadro's number), and e is the charge on an electron. The product $N_A e \equiv F$ (known as the **Faraday constant**) is equal to the amount of charge in one mole of electrons and has the value 9.65×10^4 coulombs per mole. Since two electrons are released per molecule of hydrogen, $n = 2$, the voltage generated is

$$V = 237.1 \times 10^3/(2 \times 9.65 \times 10^4) = 1.23\text{ volts}.$$

This is the open circuit voltage of a PEM cell but, when a current flows through an external circuit, the voltage drops as shown in Fig. 10.9.

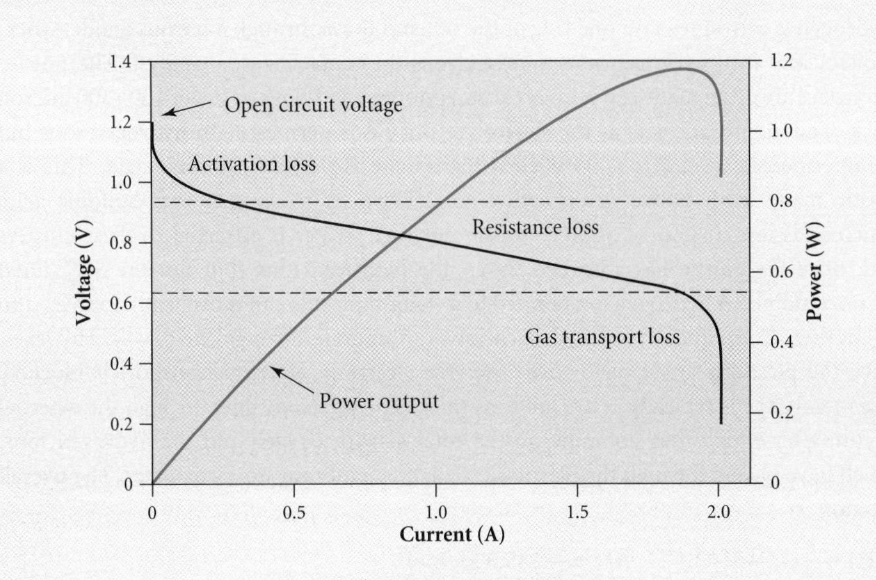

Fig. 10.9 Polarization curve of a PEM fuel cell.

The initial drop in voltage reduces the activation energy governing the catalysed electrochemical reactions. The rate of these reactions, and hence the current, increases as the voltage drops (activation loss). As the current rises further, the resistance to current flow within the cell causes a linear drop in voltage with increasing current (ohmic loss). Eventually, the flow of gas through the electrodes becomes the limiting factor (gas transport loss), since the formation of water occurs at such a rate as to block the reaction sites on the catalyst. The resulting curve is called a **polarization curve** and these losses are often called **activation**, **ohmic**, and **concentration polarization**. A catalyst layer about 10 microns thick with a loading of ~0.15 mg/cm^2 of Pt can produce ~ $\frac{1}{2}$ A/cm^2 at a voltage of 0.7 V.

A power of one watt is achieved when the current drawn is 1333 mA and the voltage is 0.75 V. Higher voltages are obtained by stacking a number of fuel cells in series. Oxygen for the PEM cell is obtained from the air, while hydrogen has to be produced and stored in a container alongside the fuel cell system. There are a number of ways of producing and storing hydrogen, which are described in section 10.13.

However, these densities correspond to only 13 and 37 g/kg of hydrogen. Some hydrides release hydrogen when water is added, an example being $NaBH_4$, which can hold 4 wt%.

The energy required to store and release the hydrogen needs to be considered—it takes energy to compress hydrogen to 340 bar—approximately 10% of the stored amount. There is

Table 10.1 Energy storage capacity of various fuels and stores

Fuel/Store	Wh/litre	Wh/kg	Fuel/Store	Wh/litre	Wh/kg
Diesel	9950	11 890	Liquid H_2	1400	1885
Petrol*	8990	12 070	H_2 (34.5 MPa)	700	2000
Dry Wood	2720†	5430	Hydride‡	1600	1880
Ethanol	5910	7490	NaNiCl§	160	100
Methanol	4430	5580	Lead–acid	70	30

* 'Gasoline' in American usage.
† Density 0.5 kg/l.
‡ 300°C.
§ known as a ZEBRA battery.

considerable research and development on new materials to store hydrogen, e.g. clathrates. The aim of the US Department of Energy is 6 wt% by 2010 and 9 wt% by 2015, corresponding to 2 and 3 kWh/kg, respectively. Currently around 1 kWh/litre and 1 kWh/kg have been achieved for hydrogen storage. Table 10.1 compares the energy densities of various fuels and storage devices. The energy density by volume is most relevant when size is more important than weight.

Hydrogen can be produced in a number of ways:

(a) electrolysis of water;
(b) reaction of a hydrocarbon with steam (a process called **reforming**);
(c) from biomass.

Hydrogen can be made in (a) using energy from carbon-free sources, such as hydropower, solar, wind, and nuclear power. The associated carbon dioxide emissions are low. In (b) there is no net production of CO_2 if the hydrocarbon is produced from biomass, e.g. bio-ethanol or bio-diesel. An example of the steam-reforming reaction is with methanol using copper-based catalysts

$$CH_3OH + H_2O \rightarrow CO_2 + 3H_2.$$

As described in Chapter 7, hydrogen can also be produced from biomass using micro-organisms.

Hydrogen can also be made within the fuel cell itself. This is the process used in the direct methanol fuel cell (DMFC). This cell contains a plastic electrolyte that is permeable to H^+ ions, as in a PEMFC, but methanol rather than hydrogen is injected into the anode. The temperature of operation is slightly higher than in a PEMFC at 120°C, at which temperature the methanol reacts with water in the presence of a platinum catalyst to produce carbon dioxide, hydrogen ions, and electrons. At the cathode the reactions are the same as in the PEMFC, i.e.

$$CH_3OH + H_2O \rightarrow CO_2 + 6H^+ + 6e^- \quad \text{at anode,}$$
$$\tfrac{3}{2}O_2 + 6H^+ + 6e^- \rightarrow 3H_2O \quad \text{at cathode,}$$
$$CH_3OH + \tfrac{3}{2}O_2 \rightarrow CO_2 + 2H_2O \quad \text{overall reaction.}$$

The DMFC is less developed than the PEMFC and has a lower operating voltage of ~0.4 V, but does not require hydrogen storage or separate cooling to remove heat.

10.14 Outlook for fuel cells

Fuel cells are receiving a lot of attention throughout the world. They provide carbon-free electricity with very low emissions and good efficiencies of ~50%. A fuel cell has no moving parts, so it is vibration-free, quiet, and reliable. There are many types of feedstock that can either provide hydrogen directly or indirectly as in the direct methanol fuel cell. Many of these supplies can be obtained using renewable sources such as hydro, wind, and solar power, or nuclear reactors. Hydrogen can be transported by pipeline or truck or can be reformed locally.

However, fuels cells are currently too expensive to compete without subsidies: the cost of a fuel cell power supply is ~$4500/kW compared with $1000/kW for a diesel generator. For use in vehicles the present weight and size of fuel cells are too large. Materials and construction need to be improved to reduce costs. In the USA, the DOE hopes to reduce the cost to $400/kW by 2010. As the cumulative global production of fuel cells increases, cost reductions are expected (see Chapter 11, Section 11.2.2). Considerable improvements have been made since the developments for the NASA space programmes—the amount of catalyst required per square centimetre is now over a factor of ten smaller and the platinum catalyst is no longer the main cost. Fuel cells are now being used in portable electronics, such as laptops, and in some cars and trucks, and many more uses are expected to be found by the end of this decade.

SUMMARY

- The emf generated by a coil of N turns and area A, rotating in a uniform magnetic field B with angular velocity ω, is given by
  $V = NBA\omega \sin \omega t$.

- Large-scale electrical generators consist of a rotor and a stator. The magnetic field is produced in the rotor by passing direct current through coils, and the rotating magnetic field due to the rotor induces a current in the stationary coils of the stator.

- Electricity is usually generated as three-phase current, which never drops to zero, delivers constant power through a resistive load, and uses less material to conduct a given quantity of power.

- The fractional loss of power along a transmission line is inversely proportional to the operating voltage.

- Transformers provide a convenient and efficient means of increasing or decreasing the voltage of a transmission line. To transform from a voltage V_1 to a voltage V_2, the ratio of the number of turns is given by
  $N_2/N_1 = V_2/V_1$.

- Most high voltage transmission is AC but DC is used for very long lines and for underground and underwater transmission.

- Mercury arc rectifiers and thyristors are used to rectify high voltage AC to DC and to convert back from DC to AC.
- Pumped storage uses cheap electricity to pump water from a low level reservoir to an upper reservoir. The water is discharged during periods of peak demand or to compensate for the sudden loss of a generating unit on the electricity grid.
- Compressed air energy storage facilities are useful for smoothing out power fluctuations, but finding suitable sites is a problem.
- Flywheels are efficient devices for smoothing power fluctuations but the capital cost is significant, so applications are restricted to niche markets. Advances with strong lightweight materials and magnetic bearings have enabled 100 000 rpm to be achieved.
- Superconducting magnetic energy storage (SMES) is an efficient system providing very high power for short periods and for manufacturing processes where ultra-smooth power supply is important. The disadvantages are that it only operates at low temperatures and the capital cost is high.
- Lead–acid batteries are used as a back-up supply. Newer battery systems, such as the Na–S battery and the vanadium-redox battery, are extending the range of large-scale applications.
- Fuel cells provide carbon-free electricity with very low emissions and efficiencies of around 50%. Fuels cells are currently too expensive for large-scale applications, but are expected to become competitive within a few years.

WEB LINKS

www.energy.gov US Department of Energy website; appraises current programmes of research on batteries and fuel cells.

www.fuelcellsuk.org UK fuel cell association.

www.wikipedia.org Online encyclopaedia, general information on topics in this chapter.

LIST OF MAIN SYMBOLS

A	area	t	time
B	magnetic field	T	torque
E	energy	V	voltage
G	Gibbs free energy	θ	angle
I	current, moment of inertia	φ	magnetic flux
N	no. of turns	μ_0	magnetic permeability of free space
P	power	ρ	resistivity
p	pressure	ω	angular velocity
R	resistance		

? EXERCISES

10.1 How many turns are needed for a coil of cross-sectional area 0.1 m^2 rotating at 50 Hz in a uniform magnetic field of 0.5 T to generate a current of 1000 A?

10.2 Discuss the feasibility of replacing the brushes in a generator by a liquid metal contact.

10.3 Prove that the sum of the currents in a 3-phase supply is zero.

10.4 Propose a circuit to generate 3-phase electricity.

10.5 Prove that the power dissipated through a resistance by a three-phase current is independent of time.

10.6 A power plant transmits 2000 MW of power along a transmission line of length 10 km with a resistance of 0.02 $\Omega\,km^{-1}$. Calculate the percentage loss of power if the line is at: (a) 110 kV; (b) 400 kV.

10.7 Why can HVDC transmission carry more power than HVAC transmission for a given cable?

10.8 Compare the advantages and disadvantages of using copper instead of aluminium for overhead high voltage lines.

10.9 It is desired to increase the reliability of part of a grid by installing two extra components. Assuming the components have a probability of failure of 0.01, what is the total probability of failure if the two components are connected: (a) in series; (b) in parallel?

10.10 It is desired to reduce the voltage of a transmission line from 400 kV to 12 kV. Calculate the ratio of the number of turns from the primary to the secondary side of a transformer.

10.11 A pumped storage plant has a head of 600 m and a working volume of 500 000 m^3. How much power can be generated if the plant is required to operate at maximum output for 2 hours a day? Assume the efficiency of generation is 85% and the density of water is $10^3\ kg\,m^{-3}$.

10.12 Estimate the energy stored in a compressed air energy storage chamber with a capacity of 100 000 m^3, if the pressure of the air in the chamber is 50 bar. Assuming the overall efficiency of the plant is 30%, estimate the electrical power output if the air is discharged over a period of 1 hour.

10.13 A uniform cylindrical flywheel of mass 10 kg and radius 10 cm rotates at 100 000 rpm. Calculate the kinetic energy of the flywheel.

10.14 Consider a ring of inner radius a and outer radius $a + t$ with a square cross-section ($t \ll a$) that is rotating at an angular velocity ω. The material has a maximum tensile stress of σ_m and a density ρ. Deduce that the maximum ω, $\omega_{\max}$, is given by $\omega_{\max}^2 = \sigma_m/(a^2\rho)$ and that the maximum kinetic energy per unit mass equals $\frac{1}{2}\sigma_m/\rho$.

10.15 Write an article of 1000 words for a popular science magazine on the prospects for fuel cells. Do not assume that the readers have any prior knowledge of the subject.

11 Energy and society

List of Topics

- ☐ Global warming
- ☐ Impact of more CO_2
- ☐ Discounted cash flow analysis
- ☐ Learning curve estimation
- ☐ Life cycle analysis
- ☐ Risk–cost–benefit analysis
- ☐ Designing safe systems
- ☐ Carbon abatement policies
- ☐ Stabilizing the amount of CO_2

Introduction

The amount of energy consumed per capita by any country is closely related to its standard of living. As the developing countries of the world become more industrialized and their population grows, their demand for energy increases. Most of our energy is currently produced from fossil fuels. If this does not change, the projected increase of carbon dioxide in the atmosphere is such that the world is at risk from dangerous climate changes due to increased global warming.

We begin by reviewing the environmental impact of energy production, in particular, that of global warming. We then discuss the economics of energy production, which is the key factor affecting the uptake of more carbon-free energy. We also show how the current and future costs of electricity can be estimated and look at how the risks and benefits of a particular strategy can be assessed.

The current known reserves of fossil fuels are in excess of 40 years for oil, 70 years for gas, and 250 years for coal. The alternatives are more expensive at present, so there is little economic incentive to reduce consumption of fossil fuels, although there may be reasons of energy security for a reduction. Priority therefore needs to be given to mitigate the world's dependence on fossil fuels and we outline some of the policies, such as carbon emissions trading, that are already being implemented. Finally, we discuss strategies to avoid the threat of damaging climate change that current projections indicate.

11.1 Environmental impact of energy production

11.1.1 Global warming

The concentration of carbon dioxide in the Earth's atmosphere has risen sharply over the last 50 years. As explained in Chapter 1, water vapour and carbon dioxide are the two main greenhouse gases that trap the infrared radiation emitted by the Earth and thereby raise the temperature of the Earth's surface. The current level of CO_2 is 375 ppm (2005), while in 1955 it was 312 ppm and, in the pre-industrial era (before ~1750), 280 ppm (Fig. 11.1(a)).

Analysis of ice cores shows that, over a much longer period, there have been fluctuations in the concentration of CO_2 in the atmosphere (Fig. 11.1(b)). These variations are related to slight periodic changes in the Earth's orbit about the Sun (and hence solar intensity), called **Milankovitch cycles**, after the Serbian mathematician who calculated their effect on climate in the 1920s. These cycles are linked to previous ice ages—the change in solar intensity triggering a sharp change in temperature (Fig. 11.1(b)).

The Milankovitch cycles illustrate a mechanism by which an increase in CO_2 concentration is caused by a rise in temperature. The increase in CO_2 will affect the temperature and the overall effect will also depend on taking into account all the sources and sinks of CO_2. The existence of a correlation between global temperature and atmospheric carbon dioxide concentration does not tell us which one caused the other or, indeed, whether both were the result of another change. In order to address these questions we need to look closely at what is affecting the Earth's climate.

The rise in temperature of about 0.5°C witnessed over the last 30 years is closely correlated with the anthropogenic (human-induced) increase in CO_2 during this period. While some of

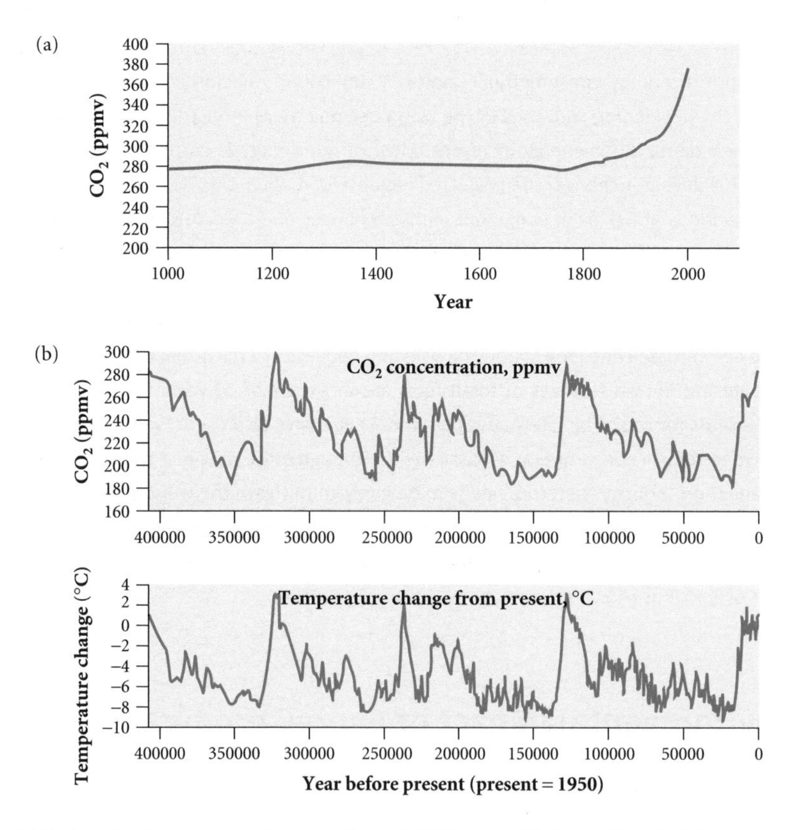

Fig. 11.1 (a) Carbon dioxide concentration over the last 1000 years, (b) Fluctuations in CO_2 concentrations and temperature over the last 400 000 years.

this temperature rise is caused by variations in solar intensity and natural phenomena such as volcanic activity, the rise over the last 30 years can only be explained if the anthropogenic change in CO_2 concentration is taken into account. As we discussed in Chapter 1 (see Fig. 1.9), the predicted temperature variation from natural causes over the last 30 years shows a slight fall. But the actual temperature has been rising during this period, with four of the warmest years on record occurring since 1998. It is this temperature rise over the last few decades that is referred to as **global warming**.

Besides the higher temperatures, there is a considerable body of other evidence for global warming in recent years. The ice cover in the Arctic regions has decreased significantly and glaciers worldwide are retreating—during the winter of 2005–06 the area of ice cover in the Arctic ocean decreased by 300 000 sq km and, for example, in Argentina, the Uppsala glacier, once the biggest in South America, is now retreating at the rate of 200 m a year. Snow cover has decreased globally by 10% since the 1960s.

The warming climate has affected coastal ice shelves in the Antarctic, but as these are afloat they do not affect the sea level when they melt. Inland in the Antarctic, nearly all of the ice is so cold and high up that it remains frozen all the time.

An analysis of ocean temperatures over the last 40 years has shown an increase of 0.5°C in surface temperatures and 0.15°C at greater depths, in close agreement with climate models that include the anthropogenic increase in CO_2. Ice cores in Greenland have shown a decrease in the salinity of Arctic waters from the freshwater from melting snow and ice. The Greenland ice sheet is disappearing at a rate of 45 cubic kilometres a year. Meanwhile, the salinity of waters near the equator has risen due to increased evaporation of the warm tropical and subtropical oceans.

Marine life has also been affected. The population of whales and walruses has dropped sharply off Alaska, affecting the livelihood of the Eskimos. Many species of plankton, an important part of the marine food chain, are moving north to cooler waters. Large amounts of coral worldwide are dying; there has been a large loss of coral reef in the Caribbean off Puerto Rico and the Virgin Islands. Hundred of thousands of birds have died because of the displacement of plankton in the Bering Sea by inedible plants that thrive in warmer waters and require a higher than normal CO_2 concentration.

The effects of large-scale global warming could be catastrophic for many parts of the world. It could accelerate the melting of glaciers in the Andes and western China, depriving millions of people of enough water. The rise in sea levels would flood low-lying regions, leading to massive movements of populations. Over the last century alone sea levels have risen 10–20 cm. The predicted rise over the period 2000–2100 is about 0.4 m for a temperature rise of 3.6°C; a rise of between 1.4°C and 5.8°C is predicted by the Intergovernmental Panel on Climate Change (IPCC). A warming in excess of 3°C could initiate the melting of the Greenland Ice Sheet and lead to its disappearance, raising sea levels by 7 m within a thousand years. A 1 m rise would submerge a substantial fraction of Bangladesh.

Further warming would increase the intensity and frequency of severe weather conditions, such as hurricanes, El Niño events, and heat waves. The change in the salinity of the oceans from ice melting could affect the ocean flow patterns. The Gulf Stream carries warm water up to the North Atlantic from the Tropics and the Gulf of Mexico, where it heats the atmosphere, which, in turn, helps to keep Europe warm. The evaporation and freezing of the surface water in the Gulf Stream in the North Atlantic increases its salinity, and hence its density, and the water sinks and then flows south at depth. This **thermohaline circulation** could be slowed by a decrease in the salinity of northern waters, and that could have significant climatic effects.

There would also be an increased threat to human health from higher temperatures, particularly in tropical and subtropical regions. Biodiversity is also be likely to be altered with the extinction of some endangered species.

11.1.2 Impact of more CO_2 in the atmosphere

The amount of carbon as CO_2 in the atmosphere is currently around 800 Gt. For every Gt emitted into the atmosphere, approximately half is absorbed by the Earth, mainly by the oceans. The burning of fossil fuels has emitted some 225 Gt of carbon since 1955, so the amount of carbon in the atmosphere has increased by about 110 Gt during the last 50 years. This increase corresponded to a 63 ppm rise in CO_2 and a rise in average global temperature of about 0.5°C. The amount of carbon currently emitted into the atmosphere each year (2005) is ~7 Gt, and this is predicted to rise to ~14 Gt by 2055 if there is no action to reduce carbon emissions. Assuming the same fraction is absorbed by the Earth, these emissions would add around 260 Gt of carbon to the atmosphere. Such an increase would cause about a 1.2°C rise by 2055, assuming the temperature was linearly related to the amount of CO_2, and that the absorption by the Earth of CO_2 remained the same.

However, there are effects that can amplify or diminish the interdependence. Particularly important amplifying effects are the increase in areas of open (dark) sea, due to ice melting, and the increase in water vapour, due to increased evaporation. Water vapour is the major greenhouse gas in the atmosphere and solar radiation is absorbed by open sea, rather than being reflected by ice, so both will cause further warming. A rise in temperature could also lead to the release of methane frozen under the Siberian permafrost. Methane could also be released from methane hydrate deposits in ocean sediments. As methane is more than 20 times as effective a greenhouse gas as CO_2, any release of methane would further increase the warming. These are examples of positive-feedback effects. Should one of these cause global warming to accelerate, then we would have what has been called a **tipping point**, which is an effect that triggers dangerous climate change.

Water vapour can form clouds and these reflect solar radiation. The conclusion of the IPCC is that this negative-feedback from reflection is outweighed by the positive feedback arising from the increased trapping of the infrared radiation from the Earth's surface. But the effects of clouds do cause a significant uncertainty in their predictions.

An important negative-feedback effect is that the oceans absorb heat from the atmosphere. However, the rate of absorption is expected to decrease as the surface temperature of the oceans increases, because mixing of the less dense warm surface water with the colder denser deep water decreases. The solubility of CO_2 also decreases as the temperature of the water rises, reducing the ability of the ocean to act as a carbon dioxide sink. The ocean is the largest sink of CO_2 and has absorbed about half of the around 300 Gt of carbon that has been emitted from the burning of fossil fuels over the last two centuries. The deep cold water has the pre-industrial concentration of CO_2 and full mixing of the ocean takes longer than a thousand years. Hence the ocean will continue to act as a vital sink for many centuries. It is currently sinking about 2 Gt of carbon a year.

Absorption of CO_2 by the land has depended on the relative scale of land clearance for agriculture and on the uptake by re-growth of vegetation, which may be aided by higher CO_2 concentrations. But growth could be adversely affected by drier and hotter climates. It has been estimated that the land will act as a net sink of ~1–2 Gt carbon per year, though this estimate is very uncertain.

Pollutants from burning forests, crop wastes, and fossil fuels can form aerosols that disperse worldwide. (Aerosols are tiny liquid and solid particles suspended in the air.) Many of these are damaging to health but the overall effect of these aerosols is to reduce the rise in temperature by shading the Earth. As it is expected that the levels of aerosols are likely to increase less quickly than those of CO_2, due to emissions controls, the effects of CO_2 on temperature will be exacerbated. One possible important negative-feedback effect suggested by Lindzen from MIT is that global warming might cause a drying out of the upper levels of the troposphere. The decrease in water vapour would reduce the greenhouse effect and hence reduce the temperature rise.

The prediction of the IPCC in 2001 was that the global temperature would rise over the period 1990–2100 by between 1.4°C and 5.8°C, the amount varying according to different model assumptions. There was a smaller variation between models over the shorter period to 2050, with an estimated temperature rise of 1.0–1.8°C and a CO_2 concentration of 475–575 ppm.

As a result of these forecasts there is now an overwhelming consensus in the scientific community that action must be taken to reduce carbon emissions if we are to reduce the risk of potentially catastrophic effects from climate change. We have seen in the earlier chapters that there are technologies that can provide low carbon energy. Why these are not more widespread is largely a question of cost and in the next section we summarize some of the main economic considerations that determine the cost of energy.

11.2 Economics of energy production

For all energy technologies we can break down the money involved into three parts. First, capital is required to build the plant. Second, money is needed to maintain and operate (M & O) the plant, and, finally, there is the revenue obtained from selling the energy. A simple measure of the competitiveness of a plant is the time required to pay back the cost of manufacture and installation, allowing for the annual cost of M & O. However, this payback time does not quantify the return on the investment. To calculate this we must take account of the fact that the value of revenue received in the future is worth less than if it were received today. For example, 100 UK pounds invested today at 5% interest would be worth £105 after 1 year; conversely, the value of £105 of revenue in a year from now would have a present value of £100.

11.2.1 Discounted cash flow analysis

The translation of future money to its present value is called **discounted cash flow analysis** and the interest rate used is called the **discount rate** R. For N equal annual amounts A, the present value V_P is given by

$$V_P = A[1 - (1 + R)^{-N}]/R. \quad (11.1)$$

Derivation 11.1 Discount rate and the present value

An amount of Q pounds invested today will be worth $Q(1 + R)^n$ pounds after n years if the interest (discount) rate is R. So if A pounds are received in n years time, its present value is $A/(1 + R)^n$. The present value V_P of a series of annual payments A made over N years is therefore

$$V_P = A/(1+R) + A/(1+R)^2 + A/(1+R)^3 + \cdots + A/(1+R)^N.$$

Put $x = 1/(1+R)$; then

$$V_P = Ax(1 + x + x^2 + \cdots + x^{N-1}).$$

Let $S = 1 + x + x^2 + \cdots + x^{N-1}$; then

$$xS = x + x^2 + \cdots + x^N = S - 1 + x^N.$$

So $S = (1 - x^N)/(1 - x)$ and

$$V_P = Ax(1 - x^N)/(1 - x)$$
$$= A[1 - (1 + R)^{-N}]/R.$$

Discounting is particularly important when revenue is expected over many years, as from a wind turbine scheme. In Example 11.1 we consider a simple case for a wind turbine where we ignore M & O costs and only take into account the capital cost of manufacture and installation and the annual revenue.

EXAMPLE 11.1

Calculate the present value of the revenue generated over a 30 year lifetime from a 3 MW turbine. The capital cost $C_{capital}$ of the turbine is 2000 k$ and the discount rate R is 6%. The cost of electricity charged by the wind turbine operator is 4 US cents per kWh. The capacity factor of the turbine is $\frac{1}{4}$.

At a capacity factor of $\frac{1}{4}$ the energy produced per year, E, will be

$$E = 24 \times 365 \times \tfrac{1}{4} \times 3 \times 10^3 = 6.57 \times 10^6 \text{ kWh}.$$

At 4 US cents per kWh the revenue per year A is $(0.04 \times 6.57 \times 10^6) = 263$ k$.
Substituting in eqn (11.1) with $N = 30$, gives the present value V_P of the total revenue as

$$V_P = 263[1 - (1.06)^{-30}]/0.06 = 3620 \text{ k\$}.$$

This amount is less than half of what we would have calculated had we neglected the effect of the interest rate.

We are now in a position to calculate the cost of producing energy by this wind turbine and the rate of return on the capital invested to build it. If we subtract the capital cost from the present value of the revenue we get the **net present value** V_{NP}

$$V_{NP} = V_P - C_{capital}. \tag{11.2}$$

In this example $C_{capital} = 2000$ k$ and $V_P = 3620$ k$ so $V_{NP} = 1620$ k$. As V_{NP} is positive the rate of return is greater than the discount rate, which is generally a requirement for investment. The **rate of return** R_{return} is given by finding the discount rate R_{return} that makes V_{NP} zero, i.e. the value when

$$C_{capital} = A[1 - (1 + R_{return})^{-N}]/R_{return}. \tag{11.3}$$

In the above example, with the revenue per annum $A = 263$ k$ and $N = 30$, this is when R_{return} is 12.8%.

We can calculate the **cost of energy** C_{cost} by finding what annual revenue A_{cost} at the given discount rate R would make V_{NP} equal to zero. This revenue A_{cost} is given by

$$C_{capital} = A_{cost}[1 - (1 + R)^{-N}]/R. \tag{11.4}$$

The cost of energy C_{energy} is then obtained by dividing A_{cost} by the energy produced per year E, i.e.

$$C_{energy} = A_{cost}/E. \tag{11.5}$$

In our example, the cost of energy is 2.2 cents/kWh.

Equations (11.4) and (11.5) show that the cost of energy is effectively the cost, per unit of energy generated, of tying up capital in the project for the time that the energy is generated. The difference between the revenue per unit and this cost is the profit per unit.

In Example 11.1 we have ignored the time taken to build the wind turbine. To compare costs and revenues that occur at different times we measure them all in terms of their value at a particular time. For the revenues we have chosen the time as when the turbine starts to generate. So, if it takes 2 years to manufacture and install a wind turbine, then we need to discount the capital to the time when the turbine starts to produce energy. This discounted amount will be larger than the sum of the annual capital payments, since these would accrue interest if invested elsewhere, e.g. $1000 spent 2 years ago with an interest rate of 10% would be worth $\$1000 \times (1.1)^2 = \$1210$, at the start of production. For equal annual capital payments of $C_{capital}/N$ over N years we need to increase the capital sum $C_{capital}$ in eqns (11.2)–(11.4) by the factor $F_{discount}$ given by

$$F_{discount} = (1 + R)[(1 + R)^N - 1]/NR. \tag{11.6}$$

A long construction time costs money and increases the cost of energy. From eqn (11.6) the effect of increasing the construction time by 1 year is to increase $C_{capital}$ by approximately $(1 + R/2)$. We will illustrate the importance of construction time by considering its effect on the cost of energy from a nuclear power plant.

EXAMPLE 11.2

A 1 GW_e nuclear power plant cost 2500 million US dollars to build. The plant has a capacity factor of 85% and a lifetime of 30 years. Assume a discount rate of 10% and that the annual cost of fuel and M & O is \$5/MWh and \$10/MWh, respectively.
Calculate the cost of electricity when the construction period is: (a) 5 years; (b) 4 years. By how much are these costs reduced when: (c) the construction cost is reduced by 25%; (d) the discount rate is reduced to 8%?

At a capacity factor of 85% the energy produced per year E will be

$$E = 24 \times 365 \times 0.85 \times 10^9 = 7.45 \times 10^6 \text{ MWh}.$$

(a) From eqn (11.6) for a construction period of 5 years, $F_{\text{discount}} = 1.343$, so C_{capital} becomes 3.358×10^6 k\$.
Substituting in eqn (11.4) with $N = 30$, gives the annual revenue required to repay the capital, $A_{\text{costcapital}}$, as

$$A_{\text{costcapital}} = 3.358 \times 10^9 \times 0.1/[1 - (1.1)^{-30}] = 356 \times 10^6\$.$$

There is an additional cost of \$15/MWh for fuel and M & O so the total annual cost is

$$A_{\text{cost}} = (112 + 356) \times 10^6\$.$$

The cost of electricity $C_{\text{electricity}}$ is therefore given by

$$C_{\text{electricity}} = A_{\text{cost}}/E = 6.28 \text{ cents/kWh}.$$

(b) From eqn (11.6) for a construction period of 4 years, $F_{\text{discount}} = 1.276$. So C_{capital} becomes 3.191×10^6 k\$. $A_{\text{costcapital}}$ is therefore 338×10^6\$. The annual cost is

$$A_{\text{cost}} = (112 + 338) \times 10^6\$.$$

The cost of electricity is therefore $C_{\text{electricity}} = A_{\text{cost}}/E = 6.04$ cents/kWh.

(c) Cutting construction costs by 25% reduces $A_{\text{costcapital}}$ by 25% and the cost of electricity becomes 5.09 cents/kWh for a 5 year construction period and 4.91 cents/kWh for a 4 year one.

(d) Repeating the calculations for (a) and (b) with a discount rate reduced to 8% gives the cost of electricity as 5.28 and 5.13 cents/kWh, respectively.

These simple examples illustrate some of the key points, but an actual economic analysis is much more complicated. In practice, a particular uncertainty concerns future revenues, and projections would be made using different assumptions on revenue. The actual discount rate is difficult to predict—one estimate would be the long-term interest rate (e.g. 30 year rate on bonds) minus the expected rate of inflation over the period of the project. But, as the example above shows, the discount rate directly affects the return.

The relative cost of alternative ways of producing power can also affect the profitability; for instance, a significant drop in the price of gas could have a large effect. Renewable energy has an additional value in saving a utility power company from using other fuels. These direct savings are called **avoided costs**. Its use will also give some reduction in the requirement for conventional generating capacity. The desire to reduce global warming can also result in requirements on utilities to use a certain percentage of renewable energy such as that from wind power.

One of the challenges to a government or to a regulator that wants a particular source of energy to be used (e.g. nuclear or wind) is to devise a pricing regime that guarantees a certain rate of return but still maintains competition among utilities. Estimates of future costs are also very important in planning an energy strategy. Some of the technologies that we have discussed are relatively young and prices are dropping. In order to predict what the costs will be in the future, a technique called **learning curve estimation** is useful.

11.2.2 Learning curve estimation

It is common experience from a broad range of technologies that costs fall as production increases. In particular, costs decrease roughly linearly with cumulative production when plotted on a log/log scale. The learning rate for a technology is the percentage reduction in costs for a doubling in cumulative production; typically the learning rate is between 10 and 30% for industrial products and tends to be larger in the developing stages. It is the global cumulative production that matters, not the production in a particular country. Although it appears rather simplistic, learning curves have been shown to provide good cost estimates—the technique was first noted in the construction of aircraft in the 1930s. The tendency for the slope of the learning curves to decrease with time as a technology becomes more mature needs to be allowed for when extrapolating into the future. To work out the time when the cost will have fallen to a particular value requires assumptions about the rate of production.

As an example, the learning curve for global onshore wind is shown in Fig. 11.2; the learning rate between 1990 and 2002 was 15%. This is a slight fall from that in the EU over the period

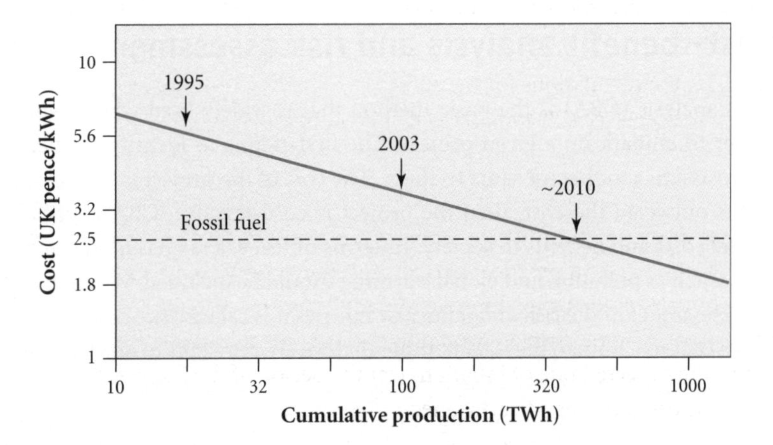

Fig. 11.2 Learning curve for global on-shore wind.

1980–95 when it stood at 18%. The average price for wind-generated electricity was 3–4 UK pence/kWh in 2004 when the global cumulative production was ~100 TWh. If a learning rate of 15% were maintained, wind would become competitive with CCGT at a price of about 2.5 pence/kWh when the cumulative production reached around 400 TWh. This amount of production corresponds to approximately 150 GW of installed capacity and would occur in about 2010, if the present annual increase in capacity of approximately 25% were sustained.

11.2.3 Life cycle analysis (LCA) and external costs

Besides the costs involved in manufacture, installation, and operation there are related costs that can affect the cost of electricity. If there were a tax on the emission of carbon, that would increase the cost of electricity from fossil fuel power stations. There is a cost in reducing other emissions that are damaging, such as SO_2 from coal, which gives rise to acid rain. Renewable energy has substantial environmental benefits and these can be given a monetary value from the amount of CO_2 and other emissions saved. The external cost (mainly environmental) of coal-fired generation is estimated to be 4 euro cents/kWh compared to 0.2 euro cents/kWh for wind energy. If this external cost were included, the price of electricity in Europe from coal would roughly double.

In the combustion of municipal solid waste (MSW), although the fuel is mainly organic, the combustion is not carbon-neutral because some of the material is derived from fossil fuels (typically 20–40%). The analysis of all the emissions involved in producing energy from a particular source is called a **life cycle analysis** (LCA). It calculates the amount of CO_2 (and other gaseous emissions) per kWh of energy produced. While the burning of agricultural wastes give less than 30 g per kWh, that of municipal solid waste, MSW, also called energy from waste, EfW, gives ~360 g per kWh compared with ~970 g per kWh from coal and ~450 g per kWh from a natural gas CCGT power plant.

There are costs and benefits as well as risks associated with energy production and we now look at how these can be analysed. The perception of risk and how to evaluate the risks associated with a technology are also very important and these are also considered.

11.3 Cost–benefit analysis and risk assessment

Cost–benefit analysis (CBA) is the basic method that is widely used when trying to decide upon whether to embark on a given project. The first step is to identify the benefits of the project and to assign a monetary value to them. The cost of the project is then calculated and, if the benefits outweigh the cost, then the project is cost-effective. CBA is usually used for calculating the costs and benefits to society. In terms of these, a tax has no net effect, whereas external costs such as pollution and global warming should be included. When the project also involves an assessment of the risk of fatalities or injuries it is called risk–cost–benefit analysis.

Assigning a cost to a human life is something that we are reluctant to do. A more acceptable approach is to assign a cost to avoiding a human fatality or injury. We can estimate this cost by looking at situations where people accept higher risks for more pay, e.g. on oil rigs. We can then determine the value people put on accepting risks and hence the cost they place on reducing risks. These costs may well be controversial and difficult to estimate, but it is important to recognize and try to evaluate them.

A simple example of CBA is the building of a motorway. The main cost is buying the land and building and maintaining the road. The benefits include the economic improvements and the journey-time savings that the motorway brings. These can be given a value by comparison with previous similar projects. The road may also reduce the number of car accidents on journeys between places linked by the new motorway.

The benefits and risks of a project may well affect different groups of people. In the siting of a nuclear power station, the local population is at most risk from an accident at the site, whilst a much larger group benefits from the power produced. Who should decide on whether the nuclear plant should go ahead? Should it just be the local population, a panel of experts, scientists from other disciplines, or an even wider involvement? Experts may have vested interests in the technology they are assessing.

There needs to be a balance between a widespread consultation and a consequently slow decision process, and a small review panel that could produce a much quicker decision. The latter helps government planning on, for example, national energy security, as well as on the international issue of tackling global warming, while the former is more likely to gain public confidence.

The public perception of risk is often different from that of experts who focus on the probability of an accident. Risks are perceived to be greater when they are involuntary, uncontrollable, and potentially catastrophic. These concerns contribute to what has been called the **dread factor**. Conversely, risks are more acceptable if they are voluntary, controllable, and limited. Figure 11.3 shows a plot of perceived fatalities associated with various activities against actual fatalities. Nuclear power is an example of an industry with a high **dread factor**. The actual number of deaths per year due to nuclear power is low compared with much more familiar incidents such as fatal car crashes.

There is also a tendency to think that if something can occur then it will, which increases the perceived danger of risks with a very low probability. Conversely, there is a reduction in the perceived danger from risks with a higher probability. Awareness of these differences is important when considering any project.

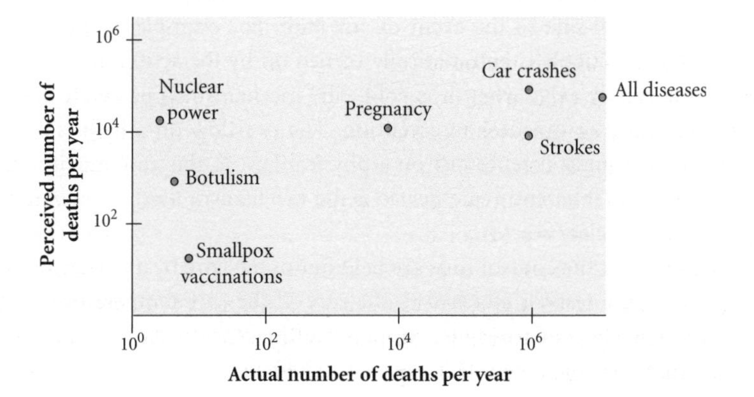

Fig. 11.3 Perceived risks versus actual risks.

Public involvement and trust in the utility companies and government agencies involved are key factors that can help speed the approval of a project. It is important that the public are aware of the benefits that the project will bring, both to themselves and to others. It is also important to not assume that the general public cannot understand technological arguments—if you believe a project to be safe then you should be able to explain why you think so. This raises the question of how do you make a system reliable and safe? We now consider methods used in designing safe systems.

11.4 Designing safe systems

One of the ways that a system can be made safer is by adding **redundancy**. If we have a key component, e.g. a relay, that has a probability of failure of q, then, by adding another relay in parallel, so that both would need to fail for the system to fail, the reliability (i.e. the probability of not failing) is improved from $(1 - q)$ to $(1 - q^2)$, see Exercise 10.9. For example, if $q = 0.1$ the reliability improves from 90% to 99%. However, it is important that both components are independent. The reliability would be less, for example, if both components were from the same batch using the same type of bad resistor, or if the main power supply and the backup cables ran through the same duct (because they would both fail if there were a fire in the duct). These are examples of **common-mode** or **common-cause failures**.

The introduction of new technology can initially result in unexpected failures as the experience of the new technology is lacking. This can lead to a very conservative approach. For example, the most important requirement for an aircraft engine is reliability and an engine designed 40 years ago may still be in operation today since it has a proven track record (an 'if it works then don't fix it' philosophy).

In designing a large complex system it is important that safety is incorporated in the design right from the start—it is often much easier to allow for the failure of a component at the design stage than afterwards. One method that is used to assess the risk of failure is **probabilistic risk assessment**. In this technique faults are identified and characterized by the probability of failure and the severity of their consequences. The possible causes of any fault are then identified (**fault-tree analysis**). As a further check on the identification of all significant faults, the consequences of the failure of any component are analysed (**event-tree analysis**).

An ideal system is **fail-safe** in the event of any fault. For example, if the pilot light goes out in a boiler, the gas supply is automatically turned off by the action of a bi-metallic strip that closes off the supply valve when it is cold. The mechanism is **passively activated**, as it requires no mechanical or operator intervention. The overflow on a bath is an example of an **inherently safe** design as it relies just on a physical law, in this case gravity, to work. An illustration of a passive inherently safe device is the mechanism used to operate the control rods in a pebble-bed nuclear reactor.

In this type of reactor the control rods are held out of the core by an electromagnet whose current flows through a resistor attached to the core. If the core temperature rises, then the current flow through the electromagnet drops since the resistance of the resistor increases with temperature. This causes the rods to be released when the temperature rises too far and they drop under gravity and shut down the reactor.

A further degree of protection in a system is to incorporate a number of units that will carry out the same function should one fail, a **fault tolerant** system. These can either be in parallel, when they give redundancy, or in series, when another unit or units (**defence in depth**) come into operation should the first unit fail. An example of the latter is an uninterruptible power supply (UPS). This is a power distribution unit connected to the main power supply that includes a large battery that provides power if the main supply is cut. Another example is the outer containment building in a nuclear reactor plant.

It is important to look at the system as a whole and not just its components. Incorrect interactions between subsystems can lead to accidents, e.g. if the software controlling a unit detects a fault, the programmed response should take into account the state of other connected units. Software must also do exactly what is expected of it—called **safety critical software**. For some operations controlled by computer, it may be required that the operation can also be carried out manually, providing defence in depth.

One of the difficulties with probabilistic risk assessment is taking into account feedback between subsystems and, in particular, the effect of the management's attitude on safety and the safety culture of the workforce. The overall safety culture of an industry has been identified as a very important factor in reducing accidents. In France, a safety culture was encouraged 150 years ago in the explosives industry by a law requiring the manufacturer and his family to live on the premises!

The desire to save time and be more productive and efficient can lead to procedures being adjusted and safety being compromised. In the Bhopal chemical factory disaster, cost-cutting, workforce reductions, and other actions had reduced the safety margins over a period of time and led to the worst industrial accident on record. During the Three Mile Island nuclear reactor incident, the operators had incorrect and insufficient information on the state of the nuclear reactor core—a signal was sent from one valve to say that it had received power to close it, but not that it had closed. The lamp indicating the true position had been eliminated during construction to save time.

It is important that a 'best practice culture' is established in power plants, where comments on safety are encouraged. As those of you who have watched Homer Simpson in his control room will know, repetitive operations need to be made interesting to reduce boredom. It is also important that both scientists and engineers are on the staff to increase the chances that all aspects are considered in unexpected situations. A consideration of human behaviour under stress also needs to be made ('**human factors**').

One way of dealing with possible conflicts between safety and profitability is to appoint a **regulator**: an independent authority that can enforce safety requirements, maintenance schedules, and inspections. Preventive maintenance is a well established way of reducing the chance of an accident—replacement of components in a car after fixed distances is an example of this. All plant modifications should be justified beforehand by a formal **safety case**: a thorough explanation of physical processes should be given by experts in the field and challenged by independent experts appointed by the regulator. They then judge whether or not the safety case is convincing.

Independent regulation enables the approval of a plant design to be separated from planning consent. This can help in speeding up the planning process. Public confidence would be helped by getting internationally agreed safety criteria for the plant, as would allowing the public to be involved in the decision making process.

Even if a power plant is accepted as safe and desirable on environmental grounds, the fact remains that most of the alternatives to fossil fuels are currently uneconomic. As many countries realize the importance of reducing CO_2 emissions, what policies have been proposed to abate carbon emissions?

11.5 Carbon abatement policies

At the first UN international Earth Summit held in Rio de Janeiro, Brazil, in 1992, the United Nations Framework Convention on Climate Change (UNFCCC) was adopted. Its provisions on tackling global warming were strengthened in 1997 in Kyoto, Japan, where an amendment was negotiated, called the **Kyoto Protocol**. Under this protocol, industrialized countries agreed that they would reduce their emissions of greenhouse gases by 5.2% compared with those of 1990. If a given country's emissions were not falling in line with this reduction, then that country would be allowed to engage in emissions trading. The greenhouse gases specified are carbon dioxide, methane, sulfur hexafluoride, HFCs, and CFCs.

The protocol came into force in February 2005. Over 160 countries have ratified the agreement with the notable exception of the United States, which is concerned that the reduction in emissions would be damaging to its economy. The American government also felt that the protocol should have included targets for developing countries as well as the developed countries. In 2003 China emitted an estimated 3.5 Gt of CO_2 compared with 5.8 Gt by the United States. However, on a per capita basis, China is emitting less than a tenth of what the United States is emitting. The protocol also acknowledged that the bulk of the increased levels of CO_2 in the atmosphere had been produced by the developed nations. As a result India, China, and other developing countries are not required to reduce emissions.

11.5.1 Emissions trading

Each country that is part of the Kyoto Protocol has set limits on the amount of greenhouse gases that it can emit, but many countries have limits above their current production. The surplus amount can be purchased by other countries, which allows them to not reduce their emissions if they so wish. Countries also receive credit for any CO_2 sinks that they develop, such as creating a forest, or for work on carbon abatement in developing countries.

Setting up the scheme has proved difficult since it requires records and monitoring of emissions. A smaller scheme for emissions trading has been set up by the European Union.

11.5.2 Carbon tax

A simpler scheme to set up than emissions trading is to impose a tax (collected by the government) on all fuels that emit CO_2. The tax is based on the amount of carbon emitted, so coal would have a higher tax per kWh than gas. This scheme, which was introduced in Sweden, Finland, The Netherlands, and Norway in the 1990s, provides an inducement for everyone to reduce emissions and applies to transport, domestic consumers, and industrial consumers. It also demonstrates the importance attached to reducing global warming by the government concerned. It is more costly for countries with less efficient energy usage, which was why the United States was in favour of emissions trading rather than a carbon tax.

11.5.3 Renewables obligation

A renewables obligation is a mechanism set up in the UK to promote the growth of renewable energy. It requires electricity suppliers to obtain a specified fraction of their energy from generators using renewable energy sources. This fraction will increase to 10% by 2010. The obligation sets up a market in tradeable green certificates (renewable obligation certificates, ROCs). Suppliers have to buy a certain number of ROCs to show that they have obtained the specified fraction of renewable energy. This they can do by buying ROCs directly from the renewable generator or on the open market. Alternatively, they can pay a buyout price of 3 pence/kWh to make up any shortfall.

Renewable generators of electricity will earn revenue from selling both electricity and ROCs, which will give them extra income. The idea is that, as the price of ROCs rises, developers will be encouraged to build more renewable energy generators. Market forces will generate competition and favour the most economic renewable sources. Furthermore, the increase in production will help reduce prices as the technology improves (see Section 11.2.2). The renewable sources included are solar, small hydro, wave power, tidal energy, geothermal energy, biofuels, and on- and off-shore wind power. The buyout price limits the amount of financial support for renewables to 3 pence/kWh, compared with a wholesale price of currently ~1.8 pence/kWh.

How well this policy will work will depend not only on the supply and demand, but also on the ease of obtaining planning permission, the integration of renewables into the electricity grid, and on not penalizing the intermittent nature of renewables in favour of the predictability of fossil-fuel generated electricity. The renewables obligation does not discriminate between different renewable sources, so it is not encouraging the development of emerging technologies such as wave power and tidal energy. Investment in a wide range of new technologies is important for stimulating innovation, building up expertise, and for not pre-judging what is the best technology.

The need for a reduction on CO_2 emissions is now well established and widely recognized. But the amount of reduction required is very large, and we conclude by outlining a strategy, called stabilization wedges, put forward by Pacala and Socolov in 2004, that would hold the overall increase of CO_2 to around 500 ppm, a level thought to reduce the threat of dangerous climate change significantly.

11.6 Stabilization wedges for limiting CO_2 emissions

Stabilization at about 500 ppm requires holding the current emissions of 7 Gt C y^{-1} constant for the next 50 years, rather than letting them rise to around 14 Gt C y^{-1}, which is roughly the rate expected if we allow the situation to follow the current trend. Assuming the same constant fraction of close to a half is absorbed by the Earth during this period, as has been the case over the last 50 years, then we would estimate the level of CO_2 to rise to around 475 ppm. After 2055, Pacala and Socolov assume that the amount of CO_2 would fall linearly until there is no net gain per year after 2105. In our simple model the effect of this would be that the level of CO_2 would stabilize at around 525 ppm.

A fall in the amount emitted each year would be possible if by 2055 we had developed carbon capture and sequestration, expanded the use of nuclear power, and were using a much greater fraction of renewable energy sources, in particular, solar and wind. By 2055 it is just feasible that the first commercial fusion reactors could have been built, which could be used to supply both power and hydrogen.

Pacala and Socolov propose stabilizing emissions at 7 Gt C y^{-1} by means of seven different methods, each of which is already established. Each method is capable of reducing emissions by 1 Gt C y^{-1} by 2055 and they are illustrated in Fig. 11.4. Each method grows linearly with time, producing a wedge. Pacala and Socolov identified reductions through energy efficiency and conservation, the reduction and capture of carbon emissions, the use of more carbon-free and carbon-neutral energy, and the enhancement of biological storage.

11.6.1 Energy efficiency and conservation

In 2002 transport produced 22% of all CO_2 emissions, with industry and buildings producing 36% and power generation 42%. A wedge would be provided by improving fuel efficiency from an expected 30 miles per gallon in 2055, from its present value of just over 20 to 60 miles per gallon, or by reducing the average annual mileage from 10 000 to 5000 miles. Another efficiency wedge could be provided by building more energy-efficient buildings. For

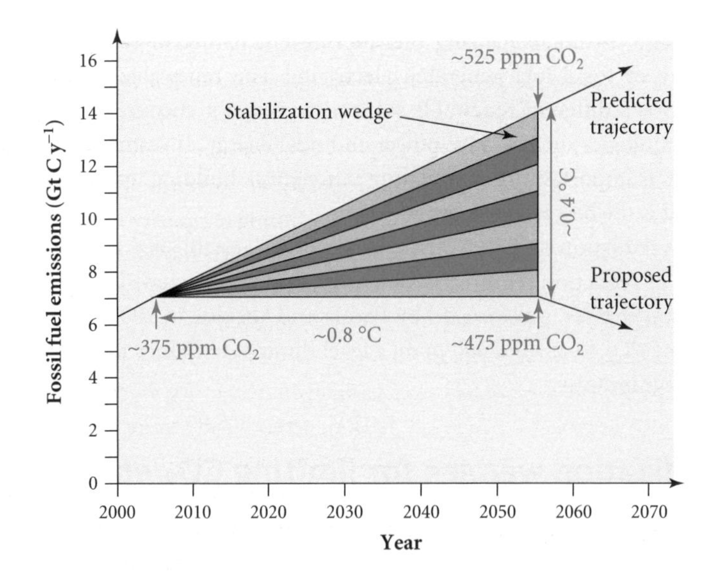

Fig. 11.4 Predicted and proposed carbon emissions using stabilization wedges. Future CO_2 levels and temperature rises based on simple linear model (see text).

example, installing compact fluorescent instead of filament bulbs could provide a quarter of a wedge.

11.6.2 Reduction and capture of carbon emissions

Switching about 1400 GW of power from coal-fired to natural-gas-fired plants would also provide a wedge, since the amount of carbon from gas per kWh is about half that from coal. Of more long-term importance is the identification of a wedge from increased carbon capture and storage (CCS). One technique first produces hydrogen and carbon dioxide from the fossil fuel. The hydrogen is then burnt to provide power, or could be used in fuel cells, and the carbon dioxide is stored underground. Another technique separates the CO_2 from the flue gas and then compresses it before pumping it underground. Currently, CO_2 is injected into oil reservoirs to enhance oil recovery (EOR). An increase of about 100 in the amount of EOR would produce another wedge.

Besides storage in old oil and gas reservoirs, storage in underground aquifers is also possible. In the North Sea off Norway, about one million tons of CO_2 a year is being stored under the sea in an underground saline aquifer lying above the Sleipner oil field. There is a large number of such formations in the world and the Sleipner project is being used to evaluate the stability and characteristics of storage of CO_2 in such aquifers. The Utsira aquifer at Sleipner is a massive sandstone aquifer 200–250 m thick capable of storing an estimated 600 Gt of CO_2, equivalent to 160 Gt of carbon. Carbon dioxide needs to be stored for 1000–10 000 years, by which time the anthropogenic emission of CO_2 is expected to have essentially ceased. In oil and gas fields the gas has been stored for millions of years and it seems unlikely that leakage of CO_2 stored underground will be a problem.

11.6.3 Increase in carbon-free and carbon-neutral energy

Introducing 700 GW_e of nuclear power by 2055 would yield another wedge of CO_2 reduction. This is about double the total nuclear capacity deployed today. While the public is becoming less opposed to nuclear power, serious concerns remain about safety, waste disposal, and proliferation. The wind turbine generating capacity is currently about 40 GWp, which corresponds to about 14 GW continuous power, assuming a capacity factor of $\frac{1}{3}$. This would need to increase by 50 times over the next 50 years to produce another wedge. There has been public opposition to siting wind farms in some areas due to their effect on the landscape and on the bird population. There is a strong case to be made that we should accept a relatively short-term disadvantage (wind turbines could be easily fully dismantled at the end of their use) because of the serious need to take action that will help everyone now.

Solar power could also make a wedge but there is currently only about 3 GWp of photovoltaic (PV) power installed. In very sunny locations a capacity factor of close to a third can be realized, though generally it is closer to 10–20%. PV has considerable potential, particularly in rural areas, as well as being integrated in new buildings. Geothermal power would need to expand by about a factor of 100 to provide a further wedge. There are considerable reserves of geothermal energy and it can be used both on a large scale and on a domestic scale using heat pumps to provide more efficient heating and cooling of buildings.

Currently, nearly all the world's transport uses fossil-based fuels. Biomass can be used to produce ethanol or biodiesel, but this requires huge land areas to produce a significant amount. Pacola and Socolow estimate that 250 million hectares would be required to produce one wedge. This is three-quarters of the size of India! An alternative would be to use hydrogen, which could be produced using nuclear power or fossil-fuel based power with carbon capture and storage (CCS). The hydrogen would then be used in fuel-cells in cars and trucks. Reducing the CO_2 emissions from vehicles and planes is of major importance. It would be easier to produce sufficient biofuel for planes, as they use less than vehicles. Furthermore, there is no simple alternative fuel.

11.6.4 Enhancement of natural carbon sinks

Finally, CO_2 reduction can be obtained by enhancing natural sinks on land. Ending deforestation by 2055 and replanting some 250 million hectares of tropical forest, or a slightly larger area of plantations, would produce one wedge. When land is tilled for replanting, about half of the stored carbon in the soil is lost, mainly through aeration. Extending conservation tillage, in which seeds are planted without re-ploughing the land, to all the world's cropland, coupled with using ground cover crops that help prevent soil erosion, would provide a further wedge.

11.7 Conclusions

Pacola and Socolow have shown that there are several established ways by which we can hold CO_2 emissions at a constant level. Wind, solar, nuclear, carbon capture and storage, and fuel savings in transport are all methods that could make a significant difference soon. However, these all require a determination by governments and a realization by the public to curb our CO_2 emissions immediately. The Kyoto Protocol is a step in the right direction but needs all nations to be involved, and the rate of CO_2 abatement needs to be speeded up.

If the world carries on as it has been there is the real threat of dangerous climate changes within 100 years. It would be unwise to rely on the lower predictions being correct for global temperatures rises. We might just be fortunate but the consequences if we are not are too serious to risk. People need to understand and support a significant shift in their energy supply and use. Limits on CO_2 emissions are already agreed among nations and quotas are being introduced on a smaller scale. Carbon taxes would bring home the seriousness of the situation to individuals. Unless there is a significant improvement we may even need to consider rationing carbon-emitting energy such as gasoline.

Hopefully, voluntary changes in use will avoid the necessity for such extreme measures. It will require a diverse approach and strong leadership. Fusion and breeder reactors could enable us to reduce our emissions after 2055 and it is important that research in these as well as in more established areas is strongly supported. We have seen in this book that there are a number of technologies that can significantly reduce our carbon emissions. By employing them we will be able to limit carbon emissions, which is vital for the well-being of future generations.

SUMMARY

- There is strong evidence that anthropogenic emissions of CO_2 have caused a significant global temperature rise of about 0.5°C over the last 50 years. The IPPC have predicted a rise of 1.4–5.8°C by 2100, with a risk of catastrophic climate change.
- Measures to limit the anthropogenic contribution to global warming have been endorsed by the vast majority of nations (but not, by 2005, the United States) in the Kyoto Protocol.
- Economics is the key factor in limiting the uptake of more carbon-free energy. To address this, carbon emissions trading, carbon taxes, and renewable obligations have been introduced.
- Cost–benefit analysis and probabilistic risk assessment of projects together with public involvement are important in developing energy strategies.
- To reduce the risk of dangerous climate change will require limiting the CO_2 level in the atmosphere to about 500 ppm. This will require holding carbon emissions constant for 50 years and then significantly reducing them.
- Many different carbon mitigation strategies are required in parallel to hold the level of CO_2 to 500 ppm. Wind, solar, nuclear, carbon capture and storage, and fuel savings in transport are all methods that could make a significant difference soon.

FURTHER READING

Cassedy, E.S. and Grossman, P.Z. (1998). *Introduction to energy*. Cambridge University Press, Cambridge. Informative section on risk and decision taking.

Botkin, D.B. and Keller, E.A. (2003). *Environmental science*. Wiley, New York. Interesting chapter on climate and global warming.

Twidell, J. and Weir, T. (2006). *Renewable energy resources*. Taylor and Francis, London. Good discussion of institutional and economic factors.

Boyle, G. (ed.) (2004). *Renewable energy*. Oxford University Press, Oxford. Interesting comments on renewable resources.

Ramage, J. (1997). *Energy, a guidebook*. Oxford University Press, Oxford. Good survey of energy sources and options.

Sorensen, B. (2004). *Renewable energy*, 3rd edn. Academic, New York. Good advanced reference book.

Borowitz, S. (1999). *Farewell fossil fuels*. Plenum, New York. Interesting book on the need to reduce our use of fossil fuels.

Shepherd, W. and Shepherd, D.W. (2003). *Energy studies*. Imperial College Press, London. Useful overview of the different energy sources.

Pacala, S. and Socolow, R. (2004). *Stabilization wedges*. *Science* **305** 968. Original article on stabilization wedges.

Jaccard, M. (2005). *Sustainable fossil fuels*. Cambridge University Press, Cambridge. Interesting book on using fossil fuels cleanly with carbon capture.

WEB LINKS

www.newscientist.com/channel/earth/climate-change Good articles on climate change.

EXERCISES

11.1 Discuss the evidence for an enhanced greenhouse effect occurring over the last hundred years.

11.2 Explain three possible global catastrophes if nothing is done to stop global warming.

11.3 Discuss what can be deduced from the observation in the climate record that the levels of CO_2 are correlated with the global temperature over a period of time.

11.4 Give two examples of effects that could amplify the rise in global temperatures from an increased level of CO_2 in the atmosphere.

11.5 Show from eqn (11.6) that the cost of electricity rises approximately by $(1 + R/2)$ per year of construction time, where R is the discount rate.

11.6 Consider a wind farm consisting of 20 2-MW turbines with a capacity factor of 0.3. The price per kWh is 4 US cents and the cost of each turbine is 1500 k$. Neglect other costs and take the discount rate to be 8%. Calculate for a 25 year lifetime of the wind farm: (a) the present value of the revenue V_P; (b) the net present value V_{NP}; (c) the rate of return; (d) the cost of energy.

11.7 A 4 GW_e nuclear power plant cost 8000 million dollars to build. The capacity factor of the plant is 90% and its lifetime is 40 years. Assume a discount rate of 7% and that the annual cost of M & O and fuel is 12 \$/MWh. Calculate the cost of electricity for a construction period of (a) 4 years; (b) 5 years.

11.8* Decommissioning the plant described in Exercise 11.7 will cost 1000 million dollars. Work out what annual payment would be required to have a value of 1000 million dollars after 40 years. Hence calculate the cost of electricity allowing for the decommissioning costs.

11.9 Explain why the effect of inflation at I% is to reduce the discount rate from R% to $(R - I)$%.

11.10* (a) Show that the learning curve can be expressed as

$$C = aP^b,$$

where C is the cost per kWh, P is the cumulative production in TWh, and a and b are constants. (b) Find a and b for the data in Fig. 11.2 and predict at what P the cost C will be 2 UK pence/kWh. (c) Relate b to the learning rate.

11.11 A person is willing to accept \$500/y less in salary for a reduction in the risk of a fatality from 10 in 100 000 to 5 in 100 000 per year. What is the implied value of a life (called the **value of a statistical life**)?

11.12 Is it necessary for a government to adopt a *BANANA* (Build Absolutely Nothing Anywhere Near Anyone) policy to counter the *NIMBY* (Not In My Back Yard) attitude that a project can engender?

11.13 What type of safety device is: (a) an electrical fuse box; (b) an airbag in a car; (c) a written procedure; (d) a parachute?

11.14 Discuss the relative merits of carbon emissions trading, carbon taxes, and renewable obligations.

11.15 Show that a reduction of about 1 Gt/y of carbon could be achieved by: (a) improving fuel efficiency from 30 to 60 miles per gallon for an annual mileage of 10 000 miles and assuming there are 2×10^9 cars globally; (b) substituting about 1400 GW_e of coal-based generating capacity with gas-based power (assume a plant efficiency of 50% and that a plant of 33% efficiency produces 1 kg of CO_2 per kWh from coal and 0.5 kg of CO_2 per kWh from gas); (c) placing wind turbines over about 15 million hectares; (d) planting about 400 million hectares of land with cellulose-based crops.

11.16 Comment *critically* on the following statements.

(a) The risks associated with burning fossil fuels are much greater than those associated with nuclear power.

(b) The 'hydrogen economy' does not help with the problems of global warming.

(c) Carbon capture technology is sufficiently developed that the world can carry on burning fossil fuels without any concern about triggering a dangerous climate change.

(d) Money should be spent on mitigating the effects of global warming, e.g. by building dykes or relocating people away from low-lying lands, rather than on preventing global warming.

Numerical answers to exercises

Chapter 1

1.2 80 m

1.4 2×10^4 N

1.10 (a) 948 Btu (b) 1.08kW

Chapter 2

2.1 11 W

2.2 5×10^3 s

2.4 298 K

2.5 $Re \approx 21000$

2.6 14.9 W

2.8 15 kJ

2.9 861 kJ

2.11 (a) 0.56; (b) 778 kJ kg^{-1}; (c) 342 kJ kg^{-1}; (d) 1015 kJ kg^{-1}

2.12 (a) 18 kJ kg^{-1}; (b) 3704 kJ kg^{-1}; (c) 1882 kJ kg^{-1}; (d) 0.50

2.13 854 K, 47%

2.14 0.75 km; 4 bar

Chapter 3

3.3 0.44 m s^{-1}

3.6 10^4

3.10 6.67:1

3.14 1.07×10^6 N

3.15 62.8 MW

Chapter 4

4.2 980 MW

4.3 $a = \frac{1}{2}; b = \frac{3}{2}$

4.10 (a) 44% increase; (b) 44% decrease

4.13 $a = \frac{1}{2}; b = \frac{1}{2}$

4.14 31 m s^{-1}

4.17 10 m s^{-1}

Chapter 5

5.1 9.85 MW

5.4 43 tonne-weight

5.8 3.45 m, 9.9°; 1.77 m, 2.6°; 1.19 m, 0.1°; 0.89 m, −1.2°

5.11 $N = 3.6 \times 10^8$

5.12 97–131 Pa

5.13 (a) 0.045 MW ha^{-1}; (b) 0.051 MW ha^{-1}

5.14 (b) ~9; (c) ~266 kW; (d) ~0.042 MW ha^{-1}

Chapter 6

6.1 42%

6.2 (a) 300 W m^{-2}; (b) 520 W m^{-2}

6.4 (a) 5777 K

6.8 (a) 84.6 mW, 3.61 Ω; (b) 83.0 mW

6.9 ~34 m^2

6.10 13.63 V; 14.15 V; 14.45 V; 14.66 V; 14.83 V

6.11 ~300 ha

6.12 (a) 1.25 eV and 0.69 eV; (b) 38.2 mW; (c) 38.2%

6.14 61.9%

6.16 26.7% versus 18.0%

6.18 0.28

6.19 3500 euros

6.20 (a) 23.8 eurocent per kWh (b) 29.3 eurocent per kWh

6.21 0.43 MW

6.22 941 m

Chapter 7

7.1 ~1.2×10^5 ha

7.2 ~600 litres

7.3 260 Mt

7.4 40.3 MJ kg^{-1}

7.5 ~22 Mt

7.8 ~110 Mt

7.9 ~2×10^4 sq km

7.11 (a) 5.3 Mt; (b) 3.2 Mt

7.12 (a) 22 Mt, 6 Mt; (b) 8.3×10^5 sq km

7.13 0.74 kg

7.14 (a) 62 mpg; (b) 38 mpg

Chapter 8

8.1 ~209 MeV

8.2 0.25 t

8.3 $\sim$80 kt

8.4 16 kt

8.10 (b) 1.76 m

8.11 $\sim$32 h

8.12 (a) 0.015 mSv; (b) 1414 m

8.13 1.5 ms

8.14 67 s

8.15 1.6%

8.16 1.7×10^4 TBq

8.17 (a) 6.8×10^7 m^{-2} s^{-1}; (b) 400 μ Sv h^{-1}

8.18 19 y; 27.4 y

8.19 (a) $\sim$12 m^3; (b) $\sim$320 m^3

Chapter 9

9.1 2400 kg

9.3 1.45×10^{20}

9.4 6.1 mm

9.7 128×10^6 rad s^{-1}

9.8 8×10^7 K

9.9 2500 km

9.11 $\sim$450 km s^{-1}

Chapter 10

10.1 64

10.6 (a) 13% (b) 0.25%

10.9 (a) 0.0199 (b) 0.0001

10.10 33:1

10.11 347 MW

10.12 19.56×10^{11} J; 163 MW

10.13 2.74 MJ

Chapter 11

11.6 (a) 44.9 M$; (b) 14.9 M$; (c) 13.7%; (d) 2.7 cents per kWh

11.7 (a) 3.46 cents per kWh; (b) 3.54 cents per kWh

11.8 (a) 3.48 cents per kWh; (b) 3.56 cents per kWh

11.10 (b) $a = 12.0$; $b = -0.264$; $P = 885$ TWh

11.11 10 M$

Index

A

B

E

F

G

H

Q

R

S

T

U

V

W